T0321958

Energy Systems Design for Low–Power Computing

Rathishchandra Ramachandra Gatti
Sahyadri College of Engineering and Management, India

Chandra Singh
Sahyadri College of Engineering and Management, India

Srividya P.
RV College of Engineering, India

Sandeep Bhat
Sahyadri College of Engineering and Management, India

A volume in the Advances
in Computer and Electrical
Engineering (ACEE) Book Series

Published in the United States of America by
IGI Global
Engineering Science Reference (an imprint of IGI Global)
701 E. Chocolate Avenue
Hershey PA, USA 17033
Tel: 717-533-8845
Fax: 717-533-8661
E-mail: cust@igi-global.com
Web site: http://www.igi-global.com

Library of Congress Cataloging-in-Publication Data

Names: Gatti, Rathishchandra, 1980- editor.
Title: Energy systems design for low power computing / edited by
   Rathishchandra Gatti, Chandra Singh, Srividya P, Sandeep Bhat.
Description: Hershey, PA : Engineering Science Reference, [2022] | Includes
   bibliographical references and index. | Summary: "This book is intended
   to be reference book for both industries and academia. It is meant for
   all those researchers and technologists who strive to reduce the
   computing power by all the possible use of trending low-powered
   technologies. This book strives to be useful reference for data center
   designers to reduce the PUEs by use of renewables and lowering power of
   non-computing data center infrastructure, VLSI designers to build energy
   efficient ULP chips, IoT designers to build ULP IoT device networks.
   This book will also be useful for researchers and students who want to
   venture in low-powered computing which is booming field in computing
   hardware"-- Provided by publisher.
Identifiers: LCCN 2022043695 (print) | LCCN 2022043696 (ebook) | ISBN
   9781668449745 (h/c) | ISBN 9781668449752 (s/c) | ISBN 9781668449769
   (eISBN)
Subjects: LCSH: Low voltage integrated circuits. | Electronic digital
   computers--Circuits. | Computer systems--Energy conservation. |
   Electronic apparatus and appliances--Power supply.
Classification: LCC TK7874.66 .E54 2022 (print) | LCC TK7874.66 (ebook) |
   DDC 621.38150285/53--dc23/eng/20221128
LC record available at https://lccn.loc.gov/2022043695
LC ebook record available at https://lccn.loc.gov/2022043696

This book is published in the IGI Global book series Advances in Computer and Electrical
Engineering (ACEE) (ISSN: 2327-039X; eISSN: 2327-0403)

British Cataloguing in Publication Data
A Cataloguing in Publication record for this book is available from the British Library.

For electronic access to this publication, please contact: eresources@igi-global.com.

Advances in Computer and Electrical Engineering (ACEE) Book Series

ISSN:2327-039X
EISSN:2327-0403

Editor-in-Chief: Srikanta Patnaik SOA University, India

MISSION

The fields of computer engineering and electrical engineering encompass a broad range of interdisciplinary topics allowing for expansive research developments across multiple fields. Research in these areas continues to develop and become increasingly important as computer and electrical systems have become an integral part of everyday life.

The **Advances in Computer and Electrical Engineering (ACEE) Book Series** aims to publish research on diverse topics pertaining to computer engineering and electrical engineering. **ACEE** encourages scholarly discourse on the latest applications, tools, and methodologies being implemented in the field for the design and development of computer and electrical systems.

COVERAGE

- Circuit Analysis
- Applied Electromagnetics
- VLSI Fabrication
- Computer Science
- Sensor Technologies
- Computer Architecture
- Microprocessor Design
- VLSI Design
- Digital Electronics
- Computer Hardware

IGI Global is currently accepting manuscripts for publication within this series. To submit a proposal for a volume in this series, please contact our Acquisition Editors at Acquisitions@igi-global.com or visit: http://www.igi-global.com/publish/.

Titles in this Series

For a list of additional titles in this series, please visit:
www.igi-global.com/book-series/advances-computer-electrical-engineering/73675

Artificial Intelligence Applications in Battery Management Systems and Routing Problems in Electric Vehicles
S. Angalaeswari (Vellore Institute of Technology, India) T. Deepa (Vellore Institute of Technology, India) and L. Ashok Kumar (PSG College of Technology, India)
Engineering Science Reference • © 2023 • 342pp • H/C (ISBN: 9781668466315) • US $260.00

Handbook of Research on AI Methods and Applications in Computer Engineering
Sanaa Kaddoura (Zayed University, UAE)
Engineering Science Reference • © 2023 • 632pp • H/C (ISBN: 9781668469378) • US $335.00

Structural and Functional Aspects of Biocomputing Systems for Data Processing
U. Vignesh (Information Technology Department, Vel Tech Rangarajan Dr. Sagunthala R&D Institute of Science and Technology, Chennai, India) R. Parvathi (Vellore Institute of Technology, India) and Ricardo Goncalves (Department of Electrical and Computer Engineering (DEEC), NOVA School of Science and Technology, NOVA University Lisbon, Portugal)
Engineering Science Reference • © 2023 • 246pp • H/C (ISBN: 9781668465233) • US $270.00

5G Internet of Things and Changing Standards for Computing and Electronic Systems
Augustine O. Nwajana (University of Greenwich, UK)
Engineering Science Reference • © 2022 • 308pp • H/C (ISBN: 9781668438558) • US $250.00

Theory and Applications of NeutroAlgebras as Generalizations of Classical Algebras
Florentin Smarandache (University of New Mexico, USA) and Madeline Al-Tahan (Lebanese International University, Lebanon)
Engineering Science Reference • © 2022 • 333pp • H/C (ISBN: 9781668434956) • US $270.00

701 East Chocolate Avenue, Hershey, PA 17033, USA
Tel: 717-533-8845 x100 • Fax: 717-533-8661
E-Mail: cust@igi-global.com • www.igi-global.com

Table of Contents

Section 3
Low-Power Technologies in Energy Infrastructure

Detailed Table of Contents

Section 1
Low-Power Computing in VLSI

Chapter 1

Preethi, Presidency University, India
Sapna R., Presidency University, India
Mohammed Mujeer Ulla, Presidency University, India

Due to the fact that low-power gadgets are currently dominating the electronics sectors, researchers are studying their design. Power management is a crucial parameter for designing VLSI circuits since it is essential for estimating the performance of devices, especially those utilized in biomedical and IoT applications. To achieve greater performance, designing a low-power system on a IC is becoming increasingly challenging due to the reduction in size of chip, increases in chip density, and rise in device complexity. Furthermore, for the less than 90 nm node, due to its increasingly complicated design, the total power factor on a chip is turning into a significant difficulty. Leakage current also has a significant effect on how low-power VLSI devices manage their power. Leakage and dynamic power reduction are increasingly being prioritized in VLSI circuit design in order to improve the battery life of electronic portable devices. The many methodologies, tactics, and power management schemes that can be employed for the design of low-power circuit systems are discussed in this chapter.

Chapter 2

Srividya P., RV College of Engineering, India

At present, the transistor size is reduced to a few tens of nanometers, as larger transistors demand a large die area and power. Power is an important design parameter in multi-gigahertz communication and ASIC/SOC designs. To deliver higher performance with lower power, various technologies are adopted in semiconductor industry. SOI is one such technology that helps in achieving higher performance. It offers a platform to integrate digital and RF circuit onto a single chip. Adopting SOI technology, faster chips with lesser power can be designed. This extends the battery life of handheld devices. The SOI structure is comparable to MOSFET except for an added buried oxide (Box) layer beneath the device region. The Box layer isolates the top and the base silicon layers and reduces the junction capacitances. This reduction accelerates the speed, lowers the power consumption, allows higher transistor stacking, and improves the device performance. These capabilities have led SOI usage in RF circuits. This chapter discusses the SOI technology in building energy- and power-efficient designs.

Chapter 3

A register is basically known as a storage device for units in circuits. In data processing systems, they are used to immediately transfer data by using CPU. In digital electronics, shift registers are known as the sequential logic circuits that are used to store data temporally and transfer the data to its output for each and every clock pulse. Shift registers are found as digital memory unit storage in such devices as calculators, computers, etc. Based on shifting data, shift registers are classified in two types: universal shift register and bidirectional shift register. This chapter dealt with design and implementation of 8-bit universal shift register with CG scheme for minimizing power. Circuit operation is performed by Xilinx-14.7 software tool and simulated with I-SIM simulator tool using VHDL language. XPE tool is used to optimize power in the circuit. Results improved the power consumption in circuit by 40.65%. Also 4.76% of area was increased due to adding external circuitry and delay was reduced by 12.93% in the proposed design.

Chapter 4

The demand for portable device applications has grown immensely. For such applications, low voltage and low power operation is an essential prerequisite to prevent overheating and ensure reliable functioning. Low voltage operation curtails the total number, weight, and dimensions of batteries, and low power consumption extends battery life. The shrinking size of MOS transistors in CMOS processes necessitates the use of lower supply voltages. Since the threshold voltage of MOS transistor is not diminished at the same rate as the power supply voltage, analog designers face problems due to shrinking voltage headroom. One of the findings that can overcome the issues introduced by comparably high threshold voltages is based on the enactment of body bias approach. In such a solution, a relatively small potential is applied at body terminal of a MOS transistor to adjust its threshold voltage. This chapter discussed that body bias approach is an attractive opportunity for utilizing the body effect positively to improve the performance of low voltage-integrated circuits.

Chapter 5

S. Darwin, Dr. Sivanthi Aditanar College of Engineering, India
E. Fantin Irudaya Raj, Dr. Sivanthi Aditanar College of Engineering, India
M. Appadurai, Dr. Sivanthi Aditanar College of Engineering, India
M. Chithambara Thanu, Dr. Sivanthi Aditanar College of Engineering, India

The rapid advancement of integrated circuit (IC) technology in the recent decades paved the path for miniaturization of electronic devices. Nowadays all the handheld devices are battery operated, which moves the researchers to develop the devices with low power utilization, high-speed operating capability, and low cost. The advancement in technology scaling is crucial for enhancing the effectiveness of IC in the areas of latency, power dissipation, and signal processing. The chapter provides an outline of the history of nano electronic device development and emphasizes the potential of the single electron transistor (SET) as a new nano device that will eventually replace more traditional ones.

Chapter 6

Shankara Murthy H. M., Sahyadri College of Engineering and
Management, India
Niranjana Rai, Canara Engineering College, India
Ramakrishna N. Hegde, Srinivas Institute of Technology, India

Through the ongoing downsizing and fast growth of heat flow of electronic components, cooling concerns are confronting severe tasks. This chapter examines the recent advancements and modernization in the cooling of electronics. The most popular electronic cooling technologies, which are classed as direct and indirect cooling, are examined and described in depth. The best prevalent methods of indirect cooling by employing heat pipes, microchannels, PCM are discussed. The efficiency of cooling strategies for various levels of electronic cooling requirements, as well as approaches to increase heat transfer capabilities, are also discussed in depth. Meanwhile, by considering the intrinsic thermal characteristics, optimization approaches, and pertinent uses, the advantages and disadvantages of various thermal management systems are examined. Furthermore, the present issues of electronic cooling and thermal management technologies are discussed as well as the prospects for future advancements.

Chapter 7

Jothimani K., Graphic Era University, India
Bhagya Jyothi K. L., KVG College of Engineering, India

Convolutional neural network (CNN) systems have an increasing number of applications in healthcare and biomedical edge applications due to the advent of deep learning accelerators and neuromorphic workstations. AIoT and sense of care (SOC) medical technology development may benefit from this. In this chapter, the authors show how to develop deep learning accelerators to address healthcare analytics, pattern classification, and signal processing problems using emerging restrictive gadgets, field programmable gate arrays (FPGAs), and metal oxide semiconductors (CMOS). Neuromorphic processors are compared with DL counterparts when it comes to processing biological signals. In this study, the authors focus on a range of hardware systems that incorporate data from electromyography (EMG) and computer vision. Inferences are compared using neuromorphic processors as well as integrated AI accelerators. In the discussion, the authors examined the issues and benefits, downsides, difficulties, and possibilities that various acceleration and neuromorphic processors bring to medicine and biomedicine.

Section 2
Low-Power Technologies in IoT, WSNs, and Embedded Systems

Chapter 8

Wilma Pavitra Puthran, Microsoft, India
Sahana Prasad, T-Systems International GmbH, Germany
Rathishchandra Ramachandra Gatti, Sahyadri College of Engineering
and Management, India

Energy harvesting has been the empowering innovation in the internet of things to power the wireless sensors envisioned to be deployed ubiquitously. In recent decades, there has been an increasing drift towards remote sensor systems from wired networks in commercial and industrial applications due to expensive cabling and their non-feasibility in remote locations. The challenge is to convert these remote sensor systems into self-powered wireless sensor networks using energy harvesters. A brief review of current trends in the applications of energy harvesting in remote sensor systems is discussed in this chapter. A generic architecture of the energy harvesters and their transduction mechanisms and the design methodology of energy harvesters is introduced. The existing business products and the potential prototypes of the energy harvesters with their application domains are reported.

Chapter 9

C. Padmavathy, Sri Ramakrishna Engineering College, India
Dankan Gowda V., B.M.S. Institute of Technology and Management, India
Vaishali Narendra Agme, Bharati Vidyapeeth College of Engineering, India
Algubelly Yashwanth Reddy, Sree Dattha Group of Institutions, India
D. Palanikkumar, Dr. N.G.P. Institute of Technology, India

Nearly all application fields are paying increased attention to the internet of things (IoT). Nearly 20 billion devices are now linked to the internet. With several applications ranging from smart buildings and smart cities to smart devices, IoT has progressed over the last few decades. As a result, the quantity of sensors, sensor nodes, and gateways has increased, making these battery-powered devices power-hungry. It will be a laborious operation to change the battery in remote monitoring applications for these smart sensors or nodes. By gathering RF energy from the environment and converting it to DC power, RF energy harvesting is a cost-effective method of extending the lifespan of wireless sensor networks (WSNs). A brand-new, IoT-based smart universal charger is suggested in this chapter for charging multichemistry batteries. The suggested charger has an advantage over traditional chargers since it can charge both already installed batteries and any future batteries.

Chapter 10

Anil Sharma, College of Computing Sciences and IT, Teerthanker
Mahaveer University, India
Dankan Gowda V., B.M.S. Institute of Technology and Management, India
A. Yasmine Begum, Sree Vidyanikethan Engineering College, Mohan
Babu University, India
D. Nageswari, Nehru Institute of Engineering and Technology, India
S. Lokesh, PSG Institute of Technology and Applied Research, India

The objective of deploying a wireless sensor network is to collect data about the environment in which it will be utilised and then to transmit that data to a distant sink where it will be used to estimate or reconstruct the environment or event. In order for the centralised sink to be able to accurately reconstruct or estimate the event and take the necessary actions on time, the wireless sensor and actuator network must be able to guarantee delivery of a sufficient amount of the information gathered by the deployed sensor nodes in a time-bound and coherent manner. In addition to the above-mentioned fundamental problem, reliability also refers to the network's capacity to tolerate defects up to a certain point without compromising performance. This chapter introduces a brand-new, dependable data acquisition technique that makes use of wireless sensor and actuator networks.

Chapter 11

Manjula Gururaj Rao, N.M.A.M. Institute of Technology (Deemed), India
Sumathi Pawar, N.M.A.M. Institute of Technology (Deemed), India
Priyanka H., People's Education Society University, India
Hemant kumar Reddy, VIT-AP University, India

Any internet of things (IOT) deployment must have connectivity; this is accomplished by WSNs (wireless sensor networks). A few factors need to be taken into account when choosing a wireless technology for an IoT device: the maximum throughput, the distance range, the availability in the deployment zone, as well as the power consumption. The aim of this research is to maximize the lifetime of the nodes of WSN and to reduce the energy consumption. The system is also focused on managing WSN nodes with huge residual energy, small routing distance, and with maximum number of neighbors. This system makes use of LEACH protocol, and this protocol is hierarchical clustering protocol and also energy efficient. The cluster is constructed in such a way that average dissipation of energy in each node is minimized, and speed of the network is increased. To connect sensor networks with gateways and transfer data from these sensor networks, other technologies such Sub-1 GHz are Zigbee can also be employed.

Chapter 12

Sandeep Kumar Hegde, N.M.A.M. Institute of Technology (Deemed), India
Rajalaxmi Hegde, N.M.A.M. Institute of Technology (Deemed), India

The primary goal of the chapter is to modernize the security requirements for smart homes while building a complete home security system based on the internet of things. This system provides better security and dependability at a lower cost when compared to other security systems. Using the registered data to unlock the door increases security by preventing unauthorized unlocking. Security is ensured using two different techniques: first, the user can utilize facial recognition technology; second, they can provide access using a control app. The system provides the user with safer and more secure locking and unlocking technologies than the traditional technique.

Chapter 13

> N. M. G. Kumar, Sree Vidyanikethan Engineering College, Mohan Babu
>   University, India
> Ayaz Ahmad, National Institute of Technology, Mahendru, India
> Dankan Gowda V., B.M.S. Institute of Technology and Management, India
> S. Lokesh, PSG Institute of Technology and Applied Research, India
> Kirti Rahul Rahul Kadam, Institute of Management Kolhapur, Bharati
>   Vidyapeeth (Deemed), India

Many modern items that are in widespread use have embedded systems. Due of embedded processing's ability to provide complex functions and a rich user experience, it has grown commonplace in many types of electronic products during the last 20 years. Power consumption in embedded systems is regarded as a crucial design criterion among other factors like area, testability, and safety. Low power consumption has therefore become a crucial consideration in the design of embedded microprocessors. The proposed new method takes into consideration both the spatial and temporal locality of the accessed data. In the chapter, the new cache replacement is combined with an efficient cache partitioning method to improve the cache hit rate. In this work, a new modification is proposed for the instruction set design to be used in custom made processors.

Section 3
Low-Power Technologies in Energy Infrastructure

Chapter 14

> Hasan Huseyin Coban, Ardahan University, Turkey

Electromobility is considered the technology of the future due to its ecological and environmental advantages. Modern society relies on the movement of goods and people, but current transport systems have adverse effects on human health and the environment. There are also studies showing that the manufacture of batteries used in electric vehicles can also have a significant environmental impact. How large this environmental impact is affected by which batteries are used and their capacity, among other effects. At the same time, rapid development is taking place in the region, and information on environmental impact risks are rapidly becoming out of date. It is essential for lawmakers to provide up-to-date data on the environmental impact of the manufacture and charging of batteries and how infrastructure design affects the system. For the case study, an electric pickup truck belonging to a chain market in Turkey was used, and emissions from battery production and energy consumption were presented.

Performance benchmarking of electricity distribution markets is essential for improving industry performance parameters. In a benchmarking study, the most important problem is that regulators often do not have accurate, specific, and sufficient information to determine current input use to achieve expected amount of output. The study combines statistical symmetric error structure with stochastic chance constrained DEA models and compares deterministic data envelopment analysis (DEA) models with stochastic chance-constrained DEA models within random input and output variables. The proposed models were applied on Turkey's electricity distribution units for assessment of energy efficiency. Study revealed that the results obtained with random data softened efficiency frontier. This study contains symmetric error structure and random inputs and outputs for performance benchmarking of electricity distribution markets by stochastic data envelopment analysis within a symmetric error structure in Turkey.

Preface

In tandem with the development of computing technologies, the demand for energy has increased. Data centers and computing devices account for approximately 3% of total energy consumption. This proportion will increase as more internet of things (IoT) devices become web connected. The processing of these data requires immense computing power. Hence, there is a dire need to reduce the energy consumption at circuit level and device new energy methodologies such as energy harvesting and advanced batteries. This book focuses on the design of energy systems for low power computing. This book is designed to disseminate the most recent research and cutting-edge technologies, topologies, standards, and techniques for deploying energy intelligence in edge computing, distributed computing, and centralized computing infrastructure.

Overall, this book delves into the latest methodologies and strategies for designing VLSI circuits with low power consumption. The chapters cover a range of topics including SOI technology, clock gating techniques, dynamic body bias, single electron transistors, electronic cooling, AIoT and deep neural network-based accelerators for healthcare and biomedical applications, energy harvesting for autonomous wireless sensor nodes, wireless sensor and actuation networks, and energy efficient keyless approaches to home security using the Internet of Things. The book also includes a review of applications of energy harvesting for autonomous wireless sensor nodes, an exhaustive analysis of energy harvesting absorbers and battery charging systems for the Internet of Things, minimizing energy consumption for communication protocols in IoT and an enhanced method for running embedded applications in a power-efficient manner. Additionally, it explores the production and use of electric vehicle batteries and the use of Stochastic Data Envelopment Analysis in measuring the efficiency of electricity distribution companies.

The book serves as a valuable resource for researchers, engineers, and students working in the field of low power computing. It is useful to trigger new research in the areas of designing low power and ULP VLSI circuits, ULP/LP sensor nodes, energy harvesters and other energy efficiency improvement strategies.

This book is divided into three sections—"Low-Power Computing in VLSI" (Section 1), "Low-Power Technologies in IoT and WSNs and Embedded Systems" (Section 2), and "Low-Power Technologies in Energy Infrastructure" (Section 3)—covering the three scales of VLSI level ,device level and plant level design topologies in energy efficiency.

Researchers are looking at the design of low power devices because they are currently dominating the electronics markets. When developing VLSI circuits, power management is an important consideration since it is used to predict the performance of devices, such as those used in biomedical and Internet of Things (IoT) applications. As chip size decreases, chip density increases, and device complexity rises, developing a low power system on an IC to achieve higher performance becomes more difficult. In addition, the total power factor of a chip is becoming increasingly challenging at nodes smaller than 90 nm because of the increased complexity of their designs. To a lesser extent, leakage current also affects the efficiency with which low-power VLSI devices handle electricity. For the sake of extending the runtime of portable electronics, designers of very large-scale integrated circuits are placing a greater emphasis on minimizing leakage and dynamic power consumption. In the first chapter titled "Low-Power Methodologies and Strategies in VLSI Circuits", some of the best practices, strategies, and power management tools for designing low-power circuit systems are discussed.

Larger transistors require more energy and more space on the chip's die, therefore today's transistors are just a few tens of nanometers in size. Multi-gigahertz communications and ASIC/SOC designs place a premium on minimizing power consumption. The semiconductor industry has implemented several different technologies in order to provide better performance while using less energy. One such technology that aids in gaining greater performance is SOI. It provides the infrastructure necessary to combine digital and radio frequency (RF) circuitry on a single silicon chip. SOI technology allows for the development of speedier circuits that consume less power. The runtime of portable electronics is therefore improved. The only difference between the SOI structure and the MOSFET structure is the presence of a buried oxide (Box) layer directly below the active device region. The junction capacitances are lowered by the Box layer because it separates the top and bottom silicon layers. This decrease enhances device performance in terms of speed, power consumption, transistor stacking, and device density. Because of these qualities, RF circuits have increasingly used SOI. The second chapter titled "SOI Technology in Designing Low-Power VLSI Circuits" provides an overview of how SOI technology can be used to create more efficient energy and power systems.

The third chapter titled "Low-Power High-Speed Eight-Bit Universal Shift Register Design Using Clock Gating Technique" discusses the design and implementation efforts of an energy efficient 8-bit Universal Shift Register.

Shift Registers are sequential logic circuits used in digital electronics to temporarily store data and send that data to its output at regular intervals in sync with a clock signal. Calculators, computers, and other electronic devices that use digital memory units store their data in shift registers. There are two main kinds of shift registers, the Universal Shift Register and the Bidirectional Shift Register, both of which are defined by the direction of the data they shift. This chapter focuses on building an energy efficient 8-bit Universal Shift Register by using Xilinx-14.7 for coding the circuit and I-SIM for VHDL simulation. The proposed low-power register decreased its power consumption by 40.65 percent with only a marginal space increase of 4.76% due to minor additional circuitry.

There has been a dramatic increase in interest in software designed for use on mobile devices. In order to avoid overheating and maintain dependable performance, low voltage and low power operation are needed for such uses. Low voltage operations reduce the size, weight, and quantity of batteries required, while low power consumption lengthens the life of the batteries. Because of the miniaturization of MOS transistors in CMOS technologies, lower supply voltages are now the norm. The voltage headroom available to analogue designers is decreasing, yet MOS transistor threshold voltages don't drop as quickly as power supply voltages. The use of a body bias technique is the basis for one of the discoveries that can solve the problems caused by relatively high threshold voltages. Specifically, a MOS transistor's threshold voltage can be adjusted by applying a modest potential to the transistor's body terminal. In the fourth chapter titled "Dynamic Body Bias: A Transistor Level Technique for Design of Low Voltage CMOS Analog Circuits," we will look at how the body bias technique may be used to enhance the functionality of low-voltage ICs by taking advantage of the body effect in a constructive manner.

Miniaturization of electrical devices has been made possible by the rapid development of Integrated Circuit (IC) technology in recent decades. All portable devices are now battery-operated, prompting scientists to design new models with improved efficiency, faster processing speeds, and lower prices. Improvements in latency, power dissipation, and signal processing are all possible thanks to the scaling up of existing technologies. The fifth chapter titled "A Detailed Study on Single Electron Transistor in Nano Device Technologies" gives a brief background on the evolution of nanoelectronics devices and highlights the promise of the Single Electron Transistor (SET) as a promising new nano gadget that may one day replace older, bulkier ones.

As electronic components continue to shrink and heat outputs to rapidly increase, cooling issues are presented with formidable challenges. The sixth chapter titled "Electronic Cooling" delves into the innovations and updates in electrical cooling in the recent past. Direct and indirect cooling, the two most common approaches to cooling electronic components, are dissected and detailed at length. Heat pipes, microchannels, and phasor-coupled-cooling (PCM) were discussed as some of the most effective current indirect cooling approaches. Strategies to improve heat transfer are also explored at length, along with the effectiveness of cooling methods for varying degrees of electronic cooling requirements. The benefits and drawbacks of different thermal management systems are analyzed by thinking about their inherent thermal properties, optimization methodologies, and relevant applications. The current problems and potential solutions in electronic cooling and thermal management technologies are also examined.

The seventh chapter titled "AIoT and Deep Neural Network-Based Accelerators for Healthcare and Biomedical Applications" demonstrates how to design low powered deep learning accelerators for use with new, resource-constrained devices like Field Programmable Gate Arrays (FPGAs) and Metal Oxide Semiconductor (MOS) logic to solve difficulties in healthcare analytics, pattern categorization, and signal processing (CMOS). When it comes to handling biological signals, neuromorphic processors are compared to their deep learning (DL) equivalents. This chapter focuses on a variety of hardware setups that combine electromyography (EMG) and AI-based computer vision. Neuromorphic processors and built-in AI accelerators are contrasted to evaluate the inferences .The pros, cons, challenges, and opportunities that various acceleration and neuromorphic processors present for the medical and biomedical fields are discussed in detail.

In order to power the wireless sensors that are expected to be widely deployed as part of the internet of things, energy harvesting has emerged as a game-changing invention. The high cost of cabling and the impracticality of laying wires in inaccessible areas have contributed to a shift in commercial and industrial applications away from wired networks and toward wireless remote sensor systems in the past couple of decades. The difficulty lies in transforming these wireless sensor networks into a self-sustaining wireless sensor system using energy harvesters for remote sensing. The eighth chapter titled "Review of Applications of Energy Harvesting for Autonomous Wireless Sensor Nodes" provides a brief overview of recent developments in the use of energy harvesting in remote sensor systems. We provide an approach for designing energy harvesters and present a generic architecture for energy harvesters and their transduction processes. Information about current commercial offerings and future energy harvester prototypes, along with examples of prospective uses, is provided.

Nearly every industry is starting to take the Internet of Things seriously (IoT). In the present day, about 20 billion gadgets are connected to the web. IoT has developed over the past few decades with numerous uses including smart buildings, smart cities, and smart devices. Consequently, there is a rise in the number of sensors, sensor nodes, and gateways, all of which are quite demanding on their respective batteries. However, in remote monitoring applications where these smart sensors or nodes are used, replacing the battery will be a time-consuming process. Cost-effectively prolonging the lifetime of wireless sensor networks, RF energy harvesting collects RF energy from the environment and converts it to DC power (WSNs). The ninth chapter titled "An Exhaustive Analysis of Energy Harvesting Absorbers and Battery Charging Systems for the Internet of Things" proposes a novel approach to charging multi-chemistry batteries: a smart universal charger built on the Internet of Things. The proposed charger is superior to current alternatives since it can be used to power both existing and future batteries.

The purpose of setting up a wireless sensor network is to gather information about the area where the network will be put to use, send that information to a central location (the "sink"), and utilize that information to make educated guesses about or reconstructions of the surrounding area or event. Delivery of a significant amount of the information gathered by the deployed sensor nodes in a time-bound and coherent manner is critical for the centralized sink to accurately reconstruct or estimate the event and perform the necessary actions in a timely fashion. Reliability, in addition to referring to the underlying issue described above, also describes the extent to which a network can accept flaws without suffering a significant drop in performance. In the tenth chapter titled "Wireless Sensor and Actuator Networks Based Reliable Data Acquisition Mechanism", a novel and trustworthy method of data gathering employing wireless sensor and actuator networks is presented.

With a WSN, you may link your Internet of Things (IOT) devices (Wireless Sensor Networks). When deciding on a wireless technology for our IoT device, we need to take into account a number of criteria, including the maximum throughput, the distance range, the availability in the deployment zone, and the power consumption. The eleventh chapter titled "Energy Consumption Minimization Approach for Communication Protocol in IoT" focuses on finding ways to cut down on power usage while extending the life of WSN nodes. Managing WSN nodes with a large amount of remaining power, a short routing distance, and a large number of neighbours is another key emphasis of the system. The LEACH protocol, a hierarchical clustering mechanism that also boasts excellent efficiency, is used in this system. By carefully crafting its architecture, the cluster reduces the average rate at which energy is lost at each node and boosts the network's throughput. Sub-1 GHz and Zigbee are two alternative technologies that can be used to link sensor networks to gateways and convey data from these networks.

The twelfth chapter titled "An Energy-Efficient Keyless Approach to Home Security Using Internet of Things" aims to update the security standards for smart houses while developing a comprehensive IoT-based home security system. When compared to competing security systems, this one offers more protection and durability at a more reasonable price. Increased safety is achieved since unauthorized entry is blocked when the door is unlocked with the registered information. The user has the option of either relying on facial recognition technology to grant access, or using a control app to do so. Locking and unlocking devices provided by the system are more secure and safer than the conventional method.

Embedded systems can be found in many everyday modern products. In the last 20 years, embedded processing has become standard in many different kinds of electronic products due to its capacity to perform sophisticated operations and a rich user experience. Alongside aspects like size, testability, and safety, power consumption is considered a significant design criterion for embedded systems. Consequently, low power consumption has emerged as a primary concern throughout the development of embedded microprocessors. In the thirteenth chapter titled "An Enhanced Method for Running Embedded Applications in a Power-Efficient Manner", a novel method is proposed that accounts for both the physical and temporal proximity of the data being retrieved. In the approach proposed, the cache hit rate is enhanced by combining the novel cache replacement with an effective cache partitioning strategy. In this chapter, a different approach to the design of the instruction set for bespoke processors is offered.

Because of its environmental and ecological benefits, electromobility is often viewed as a cutting-edge technology of the future. Transportation is essential in today's world, yet the methods now in use endanger both people and the planet. Moreover, research has shown that the production of batteries for use in EVs might have a major effect on the natural environment as well. The size of this influence on the environment is affected by a number of factors, including the type and capacity of batteries utilised. Simultaneously, growth in the region is progressing quickly, which means that data on potential negative environmental impacts is quickly becoming outdated. Legislators must offer the most recent information on the ecological effects of battery production and charging, as well as the system's sensitivity to changes in infrastructure design. In the fourteenth chapter titled "Production and Use of Electric Vehicle Batteries", emissions from battery production and energy usage were described, and a Turkish chain market's electric pickup truck was employed for the case study.

Electricity distribution performance must be benchmarked in order to raise industry standards. The most significant issue in a benchmarking study is that regulators frequently lack accurate, detailed, and sufficient information to assess current input utilisation to attain predicted quantity of output. The fifteenth chapter titled "Stochastic Data Envelopment Analysis in Measuring the Efficiency of Electricity Distribution Companies" contrasts deterministic Data Envelopment Analysis (DEA) models with stochastic chance constrained DEA models inside random input and output variables, and blends statistical symmetric error structure with stochastic chance constraints. The presented models were used to evaluate the energy efficacy of Turkey's power distribution units. The efficiency barrier was shown to be loosened when researchers used random data. In order to compare the efficacy of different power distribution markets in Turkey, this research used a stochastic data envelopment technique with a symmetric error structure and random inputs and outputs.

It is our hope that the present book will be helpful for researchers and industry personnel in gaining insights into the design of electronic computing systems in energy efficient way. It instils the reader that the focus is not only on reduction in energy consumption but also energy generation and efficient transmission that will converge to cater to the ever-growing energy needs of ubiquitous computing.

We would like to thank all the authors for their contributions, the reviewers for their selfless service in reviewing the chapters and the trust and support of the publishing team of IGI Global.

Rathishchandra Ramachandra Gatti
Sahyadri College of Engineering and Management, India

Chandra Singh
Sahyadri College of Engineering and Management, India

Srividya P.
RV College of Engineering, India

Sandeep Bhat
Sahyadri College of Engineering and Management, India

Section 1
Low–Power Computing in VLSI

Chapter 1
Low-Power Methodologies and Strategies in VLSI Circuits

Preethi
Presidency University, India

Sapna R.
Presidency University, India

Mohammed Mujeer Ulla
Presidency University, India

ABSTRACT

Due to the fact that low-power gadgets are currently dominating the electronics sectors, researchers are studying their design. Power management is a crucial parameter for designing VLSI circuits since it is essential for estimating the performance of devices, especially those utilized in biomedical and IoT applications. To achieve greater performance, designing a low-power system on a IC is becoming increasingly challenging due to the reduction in size of chip, increases in chip density, and rise in device complexity. Furthermore, for the less than 90 nm node, due to its increasingly complicated design, the total power factor on a chip is turning into a significant difficulty. Leakage current also has a significant effect on how low-power VLSI devices manage their power. Leakage and dynamic power reduction are increasingly being prioritized in VLSI circuit design in order to improve the battery life of electronic portable devices. The many methodologies, tactics, and power management schemes that can be employed for the design of low-power circuit systems are discussed in this chapter.

DOI: 10.4018/978-1-6684-4974-5.ch001

INTRODUCTION

Since the transistor era, which laid the groundwork for low-power consuming gadgets, the microelectronics industry has experienced a significant boom. The electronic components (IC) reduced size while also enhancing the circuits' performance. This causes the area of the power in the components to increase. There is a great demand for low power consuming gadgets is driven by a striking increase in battery oriented complicated life rescuing devices. Nowadays more concentration is to develop low power devices and design methodologies. On the other hand, with high power consumption devices, the silicon failure rate doubles for every 10-degree increase in temperature. As deep submicron nodes and nanoscale technologies advanced, the unprecedented development of power reduction gained significance.

Very-Large-Scale Integration is a technique that builds an integrated circuit (IC) by fusing thousands of transistors with other devices onto a single chip. Today, designing low power consumption is the top priorities for intricate very large-scale integration (VLSI) circuits. Table 1.1 lists various VLSI technology and its representation. Evolution of IC Technology précised in Table 1.2.

Rapid and creative advancements in low-power design have occurred recently rising the popularity of mobile devices. Hence, there is a necessity to reduce power consumption in very-high density chips. The motivations for these advances include portable applications such laptops, portable devices, and personal digital assistants that require low power consumption and high throughput. In these situations, it is necessary to achieve both the challenging objectives of high chip density and throughput in addition to the low power consumption requirements. As a result, the field of CMOS design for low-power digital integrated circuits has become quite active and is expanding very quickly.

Low-power consumption in digital systems, achieved using a variety of approaches, ranging from device to algorithm level. Reduced power consumption is mostly a result of device parameters (such as threshold voltage), device geometries, and connection qualities. To reduce power dissipation at the transistor level, circuit-level techniques can be utilized. Circuit-level includes the selection of circuit design styles appropriately, lowering the voltage change, and clocking methods. Smart management of power technique of different blocks, the use of pipelining, parallelism, and the bus structure are examples of architecture-level controls. Finally, system power consumption can be reduced by carefully choosing the data processing algorithms and limiting the amount of switching events for a given activity.

Performance of a processor measured as million instructions per second (MIPS), has previously been equated with processing power or circuit speed. When designing ICs, power consumption was only a minor consideration. However, power has emerged as the most crucial issue in nanoscale technology due to:

Table 1. VLSI Technology

Name of Technology	Representation
Micron	1μm, 2μm, 3μm, etc
Sub-Micron	0.8μm, 0.6μm, 0.35μm 0.25μm etc
Deep Sub-Micron	0.18μm, 0.13μm
Nano	90nm, 65nm etc

- Increasing the number of transistors count
- Faster speed of operation
- Higher device leakage currents

Heat is the primary form in which power is lost. The heat is transferred to the environment by cooling methods like air conditioning. Initiatives like the EPA's Energy Star initiative, which resulted in a power management for desktops, smart phones and laptops, have emerged to lessen the negative impact on the environment.

The circuit complexity and speed of evolving VLSI technology implies an increase in power consumption (Sadhu et al., 2022). Among all, the main elements that affects the dynamic power dissipation in CMOS VLSI circuits is the charging and discharging of internal node capacitances owing to switching activity. Optimization is necessary at all stages of the design process in order to decrease power, increase area, and improve speed. To achieve the desired low power design principles, a variety of design techniques are explored here.

Table 2. Evolution of IC Technology

Year	Technology	Components count	Name of Product
1947	Transistor Invention	1	-
1950–1960	Discrete components	1	Junction diodes and transistors
1961-1965	Small-scale integration	10-100	Planner devices, logic gates and Flip flops
1966-1970	Medium-scale integration	100-1000	Counters, MUXs, decoders, adders
1971–1979	Large-scale integration	1000–20,000	8-bit μp, RAM, ROM
1980-1984	Very-large-scale integration	20,000-50,000	DSPs, RISC processors
1985– till today	Ultra-large-scale integration	> 50,000	64-bit μp, dual-core μP

Applications have different reasons for wanting to cut back on power consumption. The objective is to maintain an appropriate battery lifetime, weight, and packaging cost for the class of micro-powered battery-operated portable applications, such as cell phones. The objective is to cut the power dissipation of the electronics component of the system to a level that is around half of the overall power dissipation for high performance portable computers like laptops. The general purpose of power minimization for high performance non-battery-operated systems, like workstations, is to lower the system cost while ensuring long-term device reliability. Process technology has pushed power to the forefront of all elements in such designs for such high-performance systems (Kaur & Noor, 2011).

Sources of Power Dissipation

The three basic concepts of any hardware accelerators are power, energy and area. The power usage of devices is attributed for two subcomponents: static and dynamic power (Bonamy et al., 2011). The contribution of static power is the power leakage. The impact of FPGA-based accelerators on the design of dynamic power consumption for a given system is another crucial consideration.

The power consumed by circuits during idle state is known as static power. Generally, static power leads to leakage current. Static power is a generally consistent property of a certain device does not depend on input conditions or capacitance. The leakage current can be measured as current flow through the transistors in idle state. This can be observed by considering transistor attributes.

Dynamic power (Preethi et al., 2021) is the amount of energy used by the transistors, passed through circuit when there is a change in states. Transistors in digital CMOS circuits use the greatest power during this transition period because the maximum current is flowing through them. Increased clock frequencies often equate to higher energy utilization in synchronous digital circuits since the majority of energy is used during clock transitions. Clocking the logic as infrequently as feasible is one way to reduce power usage.

By parallelizing portions of the application's duties so that more work is done per clock cycle and decreasing the clock rate, one might potentially reduce the amount of dynamic power that a design uses. Dynamic power is heavily dependent on the clock frequency. Only when the power saved by lowering the clock rate is more than the power absorbed by the additional logic needed to parallelize the task is the total power lowered.

Hardware accelerators (Preethi et al., 2021) can be useful in a various parameter other than area to improve performance. Hardware accelerators are an additional power-saving device. If adding hardware accelerators to your system improves performance while maintaining the same clock frequency, adding hardware accelerators also preserves the initial performance while reducing the clock frequency. A higher performance can be exchanged for power savings because of the tight correlation between clock frequency and power usage.

A single hardware accelerator can frequently result in significant power savings over a processor alone, but sometimes adding numerous accelerators can further save power. Because adding a lot of inefficient hardware can actually decrease the system's power efficiency rather than increase it, it is even more crucial to consider the architecture of the complete system to allow numerous accelerators. For instance, if numerous accelerators are arbitrating for the same system resource continuously, the efficiency benefit of having multiple accelerators may be offset by the shared resource's excessive use.

More heat is produced by higher power usage, and electronics are harmed by heat. An electronic circuit may malfunction or perhaps get permanently damaged if the temperature is too high (Teubner & Woods, 2013). The simple equation (2.1) that follows shows how much power the CPU consumes (PCPU):

$$P_{CPU} = \underbrace{\frac{a \times C \times V_{dd}^2 \times f_{clk}}{}}_{dynamic\ power} + \underbrace{\frac{V_{dd} \times I_{leak}}{}}_{static\ power} \qquad (1)$$

Where C - capacitance at the load (F)

Vdd – operating voltage (V)

fclk - clock frequency (Hz)

Ileak- current leakage (A)

Ignore other components of power dissipation which are very negligible and are due to current from short circuits and glitches. The important two elements that are most crucial: dynamic power and static power. The energy used when transistors are switching, or when they change states, is referred to as dynamic power. Parameter quantifies the switching activity. Another important power parameter is static power. The amount of power dissipated even when the transistors are not in use since they contribute for small magnitude of current leakage (Ileak).

The static power experiences current leakage from transistor when they are within a static configuration, is a feature of CMOS-based devices that is generally constant (standby). The majority of the power consumed by transistors comes from the dynamic power, which results from the current that passes through them when they change states (Lin et al., 2017). The equation (2.2) provides (Shang et al., 2002):

$$Dynamic\ Power = C_t V_s^2 f_s A_f$$

Where Ct - capacitance of the transistor (F)

Vs- supply voltage (V)

fs - switching frequency (Hz)

Af - activity/switching factor

Power calculations must take both static and dynamic power into account. Although FPGA manufacturers are dedicated to offering devices with minimal power consumption, as process technologies advance from 130nm to 90nm to 65nm and beyond, transistors feature size reduces become leakier by nature, increasing static power utilization. Furthermore, the dynamic power utilization of systems embedded with FPGAs increases as a function of frequency and switching node count due to the exceptionally high-performance requirements of these systems.

Since parasitic capacitance is constantly being charged and discharged, heat caused by dynamic power adds for the majority of loss in power align with recent FPGAs (Shang et al., 2002). The power consumption is given by equation (2.3).

$$P_{avg} = \frac{1}{2} \sum_{n \in nets} C_n \cdot f_n \cdot V^2$$

where Pavg - average power consumption (W)

Cn- n net capacitance (F)

V - supply voltage (V)

fn - average toggle rate of net n (Hz)

Each capacitance and net's switching activity are required as two factors for the equation used to estimate power.

Common measures used to quantify power dissipation include the following:

1. Peak power: The maximum amount of power that a given gadget can use at any one time is known as peak power. The high peak power value is typically correlated with issues such connector melting and power-line kinks.
2. Average power: The mean of a device's power usage over a given time period is known as average power consumption. High average power values cause issues with VLSI chip cooling and packaging.

Architecture Level Low-Power Design

Reducing the voltage supplied is the most efficient way to lower the power consumption. Since the most dominating component is quadratic dependent on the supply voltage while other components have a linear relationship. Unfortunately, performance suffers as a result of this decrease in power dissipation. To implement

Figure 1. Low Power Design Flow

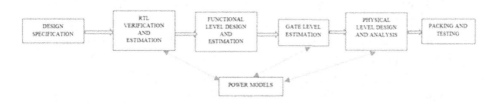

low power, high-performance circuits (Varadharajan & Nallasamy, 2017), it is imperative to design acceptable mechanisms to stop this performance loss caused by supply voltage scaling. Using appropriate methodologies at the various stages of the design hierarchy, namely the physical level, logic level, architectural level, and system level, the performance loss can be made up for. scaling of device feature sizes, pipelining and parallelism, architectural-level changes, dynamic voltage, and frequency scaling, among other methods.

Figure 1 shows a top-down technique (Varadharajan & Nallasamy, 2017) to low-power design flow including processes for power analysis and power estimation is made. It enumerates the series of actions required for a system to move from system to architecture-level specifications. Naturally, Low power design flow dictates how many steps an engineer must complete for a given application. Every level in design flow offers the chance to use approaches for power optimization. Also, each level can make a major contribution to further power decrease in VLSI circuits. However, combining low power design strategies, as opposed to just one, may produce good outcomes, which are highly dependent on the application needs.

a.   Design Specification: in recent days, there is a great demand for the portable electronic devices and rapid change in electronic technology. This results in Power dissipation, which is a important factor in low power VLSI circuit designs. The complex circuitry and evolving change in VLSI technology results in increasing power consumption.

b.   RTL Verification and Estimation: The most crucial component of circuit design is the development and verification of the Register Transfer Level (RTL). Verification's goal is to determine whether a design complies with all system requirements and specifications. Although RTL power estimation is quicker and does not require a netlist, the estimations' accuracy is reduced. Effective design-power compromises can be made early stage in the design cycle. This results in reliable power estimations during RTL phase.

c.   Functional Level Design and Estimation: At several levels of design abstraction, simulation software is used in VLSI design implementation. Simulation at a lower level of design abstraction delivers better accuracy at the cost of more intensive use of the computer. Full chip analysis can typically be handled via logic simulation, although the precision is not as excellent, and the execution speed is occasionally too slow. Simulation at the behavioral or functional level allows for quick analysis but at the expense of accuracy.

d.   Gate Level Estimation: A gate level simulation is carried out to improve the initial estimate once the gate-level design is available. If the initial estimate proves to be incorrect and the design doesn't meet the specifications, it is changed and verified once more. Up until the gate level estimate meets the specification, iteration continues.

e.   Physical Level Design and Analysis: The process of turning a netlist into a layout is called physical design. The floorplan, placement of all logical cells, clock tree synthesis, and routing are the primary physical design phases. Constraints on timing, power, design, and technology must be satisfied during this physical design phase.

f.   Packing and Testing: The wafer can proceed to the final processing stage—testing and packaging—once all the layers have been deposited and the connecting routes have been etched. Wafer evaluation: On each wafer, many hundred different ICs would have been created over the entire process.

g.   Power Models: Static and dynamic power dissipation are the two basic categories into which the power consumed in a VLSI circuit may be divided. Static power is the amount of energy used while a circuit is in a quiescent state, meaning there is no activity. To decrease the power consumption of a particular block, numerous power elements and their effects must be recognized. Maximum power dissipation, one of two types of power dissipation is related to peak instantaneous current, while average power dissipation is the other type. Peak current impacts on supply voltage noise because of the resistance of the power lines. This results in the gadget heating up, which lowers performance. This average power dissipation is more crucial when considering battery life.

LOW POWER STRATEGIES

Table 3.1 shows the different strategies available for low power at different level in VLSI design process (Kaur & Noor, 2011). At each design level there are strategies to be incorporated to achieve the low power.

Table 3. Low power design strategies

Design Level	Strategies
Technology Level	Reduction of threshold on multi-threshold devices
Circuit Level	Change in logic styles, transistor size reduction and energy recovery
Architecture Level	Pipelining, Redundancy and data encoding
Software Level	Locality, Regularity, concurrency
Operating System Level	Power down, Portioning

Low-Power Design Space

The VLSI design space has three degrees of freedom: voltage, physical capacitance, and data activity. In order to optimize for greater power, one or more of these elements must be reduced.

a. Voltage: Voltage reduction gives the better way to reduce power usage because to its quadratic relationship to power. The fact is two reductions in supply voltage results in a factor of four reductions in power consumption without the need for any specialized circuits and technology. Sadly, there is a speed cost for reducing the supply voltage, and delays dramatically rise as Vdd reaches the device's threshold voltage Vt. Modifying the devices' threshold voltages is a method for lowering the supply voltage without sacrificing throughput. Scaling down the supply voltage without experiencing speed loss is possible by lowering the Vt. The requirement to create suitable noise margins and restrict the increase in the subthreshold current (Alidina et al., 1994; Inukai et al., 2000; Yeo et al., 2000) sets the upper limit of Vt.

b. Capacitance: Linearly correlated with the physical capacitance being swapped is dynamic power usage. In other words, limiting capacitances is another method for reducing power consumption in addition to running at low voltages. By employing fewer and shorter cables, smaller devices, and less logic, the capacitances can be maintained to a minimum (Alidina et al., 1994; Inukai et al., 2000; Yeo et al., 2000).

c. Data activity: The average periodicity of data arrivals is determined by Fclk, and the number of transitions that each arrival will produce is determined by E(sw) (Graphics, 2009). These two factors together make up switching activity. Switch capacitance Csw=C is produced by combining the physical capacitance C and the data activity E(sw). E(sw), which represents the average capacitance charge throughout each data period, and Fclk, which establishes how much power the CMOS circuit uses, are both expressed in terms of sw.

STRUCTURED LOW POWER TECHNIQUES

Low power requirements can be met at the various level, listed as follows (Preethi et al., 2021):

i.   Circuit level
ii.   Gate level
iii. RTL level
iv.   System level.

This is a broad topic. Adopting the circuit level adjustments in a semi-custom design is challenging. Improvements at the level of logic gates are similarly challenging because semi-custom designs rely on foundry standard cell libraries that are already accessible. After several years of research, the industry has chosen low power solutions that are well-developed, easily adaptable to semi-custom design flows, and don't require any changes to the design or architecture. Clock gating, Power gating, and the use of Multi VDD and Multi Vth (Threshold Voltage) libraries are some examples of these techniques. Clock gating and multi-threshold are the only structured low power strategies used in this study on sorters. The structured low power techniques are described as follows:

1.   Clock gating: simply turning off the clock signal while the circuit is not in use, lowers dynamic power dissipation.
2.   Power gating: technique used for designing integrated circuits that save power consumption by cutting off the current to inactive circuit blocks.
3.   Multi-VDD: It is used to conserve the design's dynamic and static power and uses many voltage domains in the same system without compromising speed.
4.   Multi-Vth: This method uses Low Vth/High Vth cells while minimizing leakage power. Basic operation of this technique is reduce the critical path in threshold gates.

Clock Gating:

In most architectures, data is moved into registers only occasionally, whereas the clock signal is constantly provided. At each defined cycle, the clock signal will toggle constantly. The primary source of power dissipation during switching activity is

Figure 2. Schematic representation of clock gating

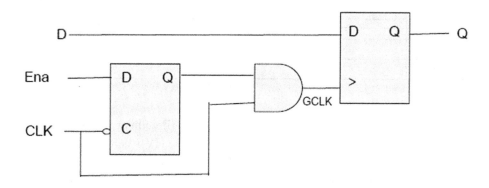

clock signal. The enormous load is driven by the clock signal. The clock is reduced by using a set of flip-flops (FF) in the design. This FF are enabled by the control unit. Power is not dissipated when the register is turned off using the gating function (Preethi et al., 2021).

There are various techniques to reduce Dynamic power consumption like reducing clock speed, reduced voltages or cut down the design activity. Clock gating network structure saves the dynamic power consumption required by circuit.

Clock gating can be achieved by stopping the clock signals to the register files that is unutilized module by using an extra AND gate (Preethi et al., 2021). This AND gate are controlled by an enable signal represented as Ena in Figure 2. There are two ways of representing extra AND gate with enabled control signal, one way is RTL and another way is using gate level netlist scripts. Automation (Rabaey & Rabaey, 2009) can be used to implement clock gating without disrupting the circuit operation. Figure 2 shows the clock gating representation with Data signal, enable signal, clocking signal and gated clock using AND logical gate.

The register must only accept a new value on the rising edge of the clock and the gate is high for a clock to be correctly gated. If the clock is high, simply ANDing the clock and the enable together will result in a false clock on the rising edge of the gate.

Multi-Threshold Libraries

Each standard cell in standard cell libraries has a separate MOS device threshold voltage. Different power and speed parameters are supported by the majority of common cell libraries. The Vth (Threshold voltage) of the logic gate, together with other factors like transistor size, determines the power and speed of the cell. The classes of cells known as High Vth (HVth), Standard Vth (SVth), and Low Vth

Table 4. Features of Multi-Vth cells

Vth	Leakage Power	Dynamic Power	Speed
LVth	Highest	Highest	Highest
SVth	Normal	Normal	Normal
HVth	Lowest	Lowest	Lowest

(LVth) are often supported in libraries. The table 4.1 (Preethi et al., 2022) lists the characteristics of the three various multi-Vth cells. The HVth cells have less speed, less dynamic power, and reduced leakage. The most leakage, dynamic power, and smallest delays will be found in LVth cells. SVth cells will operate at normal speeds and with typical power requirements.

The multi-threshold voltage approach is classified as follows:

i.   Standard Vth
ii.   Low Vth
iii.   High Vth

In this method, the critical path uses low threshold gates whereas the non-critical path uses high threshold gates. With this technology, operation speed is increased without an increase in power usage. Low and high Vth cells have the same area as standard Vth cells. Therefore, Standard Vth cells can be replaced with Low Vth or High Vth cells depending on the requirement in the critical paths or in any other paths of the design. With this method, performance is enhanced without compromising area or increasing power usage.

Multiple methods can be used to implement multi-Vth. It can be carried out by an engineering change order during layout, RTL synthesis, or at the gate level. The process of converting RTL Verilog code to functionally comparable gate-level netlist is known as logic synthesis.

To achieve multiple Vth cells for various timing paths in the design, the standard cell libraries with multiple Vth can be employed in synthesis. The best location to implement multi-Vths is during synthesis (Preethi et al., 2022).

Low-Power System Strategies

Electronic devices of today are made up of hardware platforms and a number of software layers. The interplay between the hardware and software layers is the basis for a number of system functions. Software often doesn't use energy while it is being executed, but when it is stored in hardware that uses memory, it will consume energy.

All activities are carried out via software execution, which also uses memory access for write and read operations. So, while employing software to execute programs, power is used for data storage, connection between devices, and mathematical calculations. Computer programs are also stored in semiconductor memory like dynamic RAM and SRAM, which require electricity for updating and regular operation. Program storage uses extremely little power and is well-predicted during the design process. As a result, shortening the program or reducing the number of lines reduces power consumption during software execution. Using compression techniques, the duration of the software is further reduced.

The low-level machine code of a program and the specifications of the hardware architecture determine how much power is used during execution. Software compilation transforms the source code into low level machine code. Normally, the software compilation process has an impact on machine code.

The number of clock cycles and energy needed to complete each clock cycle are used to indicate the number of software instructions. The processor's busy/idle status has a small impact on how much power is dissipated during each instruction (Tiwari et al., 1994). The data access from the memory will also cause a large rise in power consumption during memory-based instructions. The execution time of the machine code produced by the compiler is increased, and memory loss is reduced, by reducing the length of the program code.

Generally speaking, different styles and levels can be employed to produce software source code at the expense of power usage. Software that uses less power must have energy-efficient source code, which requires specific coding techniques or automated source code modification. With the use of a power-aware task scheduler, an energy-efficient operating system can be developed. In power-conscious operating systems, a dynamic power management system enables improved energy savings (Mowry, 1994).

Data and instructions are frequently prefetch. For instance, prefetching of instructions is carried out to boost the instruction-level parallelism necessary for the overlapping execution of instructions in pipelined processors. All modern high-performance processors make use of this. But software prefetching is less common than hardware prefetching. A method of putting prefetch instructions into memory codes are likely to cause a cache miss is known as software prefetching (Mowry, 1994; Tiwari et al., 1994). Either the programmer or a compiler accomplishes this.

Prefetch instructions are placed into programs at runtime. The system brings data into the cache memory ahead of time, results in memory overlapping with CPU calculation. Software prefetching improves efficiency by removing cache misses. However, because more prefetch instructions must be run, power consumption goes up. This causes a trade-off in power and decreases the performance of software prefetching-using programs.

CONCLUSION

The goal of electronic circuit design is to strike a balance between performance in terms of speed and power efficiency. VLSI circuit design for applications consuming low power is a complex issue since circuit designers must adhere to a number of degrees of freedom to achieve an acceptable power reduction. To reduce power consumption, a low power design flow must address power consumption concerns at every stage of the design process and at all abstraction levels. This chapter presents numerous approaches and strategies for reducing leakage and dynamic power. For the designers of low power VLSI circuits used in portable and real time applications, the ideas and methodologies outlined in this work are highly helpful.

ACKNOWLEDGMENT

This research received no specific grant from any funding agency in the public, commercial, or not-for-profit sectors.

REFERENCES

Alidina, M., Monteiro, J., Devadas, S., Ghosh, A., & Papaefthymiou, M. (1994). Precomputation-based sequential logic optimization for low power. *IEEE Transactions on Very Large-Scale Integration (VLSI) Systems*, *2*(4), 426–436.

Bonamy, R., Chillet, D., Sentieys, O., & Bilavarn, S. (2011). Parallelism level impact on energy consumption in reconfigurable devices. *ACM SIGARCH Computer Architecture News*, *39*(4), 104–105. doi:10.1145/2082156.2082186

Graphics, M. (2009). Low power physical design with Olympus SOC. *Place and Route White Paper*.

Inukai, T., Takamiya, M., Nose, K., Kawaguchi, H., Hiramoto, T., & Sakurai, T. (2000, May). Boosted Gate MOS (BGMOS): Device/circuit cooperation scheme to achieve leakage-free giga-scale integration. In *Proceedings of the IEEE 2000 Custom Integrated Circuits Conference (Cat. No. 00CH37044)* (pp. 409-412). IEEE. 10.1109/CICC.2000.852696

Kaur, K., & Noor, A. (2011). Strategies & methodologies for low power VLSI designs: A review. *International Journal of Advances in Engineering and Technology*, *1*(2), 159.

Klein, M. (2009). Power Consumption at 40 and 45 nm. *White Paper, 298*, 1-21.

Lin, J., Yu, W., Zhang, N., Yang, X., Zhang, H., & Zhao, W. (2017). A survey on internet of things: Architecture, enabling technologies, security and privacy, and applications. *IEEE Internet of Things Journal, 4*(5), 1125–1142. doi:10.1109/JIOT.2017.2683200

Mowry, T. C. (1994). *Tolerating latency through software-controlled data prefetching.* Stanford University.

Preethi, P., Mohan, K. G., Kumar, K. S., & Mahapatra, K. K. (2021, December). Low Power Sorters Using Clock Gating. In *2021 IEEE International Symposium on Smart Electronic Systems (iSES)(Formerly iNiS)* (pp. 6-11). IEEE. 10.1109/iSES52644.2021.00015

Preethi, P., Mohan, K. G., Kumar, K. S., & Mahapatra, K. K. (2022). Sorter Design with Structured Low Power Techniques. *SN Computer Science, 4*(2), 129. doi:10.100742979-022-01546-7

Rabaey, J., & Rabaey, J. (2009). Optimizing Power@ Design Time–Circuit-Level Techniques. *Low Power Design Essentials*, 77-111.

Sadhu, A., Das, K., De, D., & Kanjilal, M. R. (2022). Low power design methodology in quantum-dot cellular automata. *Computers & Electrical Engineering, 97*, 107638. doi:10.1016/j.compeleceng.2021.107638

Shang, L., Kaviani, A. S., & Bathala, K. (2002, February). Dynamic power consumption in Virtex™-II FPGA family. In *Proceedings of the 2002 ACM/SIGDA tenth international symposium on Field-programmable gate arrays* (pp. 157-164). 10.1145/503048.503072

Teubner, J., & Woods, L. (2013). Data processing on FPGAs. *Synthesis Lectures on Data Management, 5*(2), 1–118. doi:10.1007/978-3-031-01849-7

Tiwari, V., Malik, S., & Wolfe, A. (1994, October). Compilation techniques for low energy: An overview. In *Proceedings of 1994 IEEE Symposium on Low Power Electronics* (pp. 38-39). IEEE. 10.1109/LPE.1994.573195

Varadharajan, S. K., & Nallasamy, V. (2017, March). Low power VLSI circuits design strategies and methodologies: A literature review. In *2017 Conference on Emerging Devices and Smart Systems (ICEDSS)* (pp. 245-251). IEEE. 10.1109/ICEDSS.2017.8073688

Yeo, Y. C., Lu, Q., Lee, W. C., King, T. J., Hu, C., Wang, X., ... Ma, T. P. (2000). Direct tunneling gate leakage current in transistors with ultrathin silicon nitride gate dielectric. *IEEE Electron Device Letters*, *21*(11), 540–542. doi:10.1109/55.877204

KEY TERMS AND DEFINITIONS

Current Leakage: An electric current in an unwanted conductive path under normal operating conditions.

Synthesis: The process of converting the code (program) into a circuit.

Verilog: A hardware description language (HDL) used to model electronic systems. It is most used in the design and verification of digital circuits at the register-transfer level of abstraction.

Chapter 2
SOI Technology in Designing Low–Power VLSI Circuits

Srividya P.
https://orcid.org/0000-0002-3059-0078
RV College of Engineering, India

ABSTRACT

At present, the transistor size is reduced to a few tens of nanometers, as larger transistors demand a large die area and power. Power is an important design parameter in multi-gigahertz communication and ASIC/SOC designs. To deliver higher performance with lower power, various technologies are adopted in semiconductor industry. SOI is one such technology that helps in achieving higher performance. It offers a platform to integrate digital and RF circuit onto a single chip. Adopting SOI technology, faster chips with lesser power can be designed. This extends the battery life of handheld devices. The SOI structure is comparable to MOSFET except for an added buried oxide (Box) layer beneath the device region. The Box layer isolates the top and the base silicon layers and reduces the junction capacitances. This reduction accelerates the speed, lowers the power consumption, allows higher transistor stacking, and improves the device performance. These capabilities have led SOI usage in RF circuits. This chapter discusses the SOI technology in building energy- and power-efficient designs.

INTRODUCTION

The increase in the demand for compact portable electronic devices has given room for inventing number of techniques to reduce the power consumption while

DOI: 10.4018/978-1-6684-4974-5.ch002

delivering the same performance. This has paved way in investing lot of time and effort in developing low power techniques for computation intensive applications.

In recent days, SOI is grabbing a lot of interest in high-performance circuits design. It offers higher speed, low power, reliability and hardness well beyond the traditional technologies. The superior performance of SOI can be attributed to overall reduction in the capacitance and to lower device leakage. This paves way for using SOI in sophisticated IC designs working under Low Power-Low voltage conditions.

Devices built over Silicon on Insulator (SOI) wafer offers many potential benefits in microelectronics compared to the devices built on bulk silicon which has more volume of semiconductor material underneath the device and demands more charge to turn it on and off. The relative advantages of SOI wafers will be higher when operating at lower voltages (Kononchuk & Nguyen, 2014). SOI technology also minimizes the parasitic bipolar latch up effects. Reduction in latch up has reduced the size of the transistor and has increased the packaging density. For mixed signal and radio frequency applications, SOI technology offers lower noise and higher quality passives. Further cross talk can be reduced using ground plane silicon on insulator that is formed by incorporating silicide layers. These advantages of SOI technology is obtained owing to two key features:

a.   Reduction of junction parasitic capacitance.
b.   Total dielectric separation of the transistor elements.

SOI CONSTRUCTION, PRINCIPLES AND FEATURES

As SOI paves way to reduce the power requirements and accelerates the operational speed, it is now being used in hand held battery operated devices. SOI CMOS has 25% higher switching speed and approximately three times lesser power consumption than similar circuits built on bulk silicon. These improvements obtained using SOI CMOS is comparable to benefits obtained after one or two generations of transistors scaling built using bulk silicon.

SOI structure has a base layer which is a standard silicon wafer. It only provides the mechanical support to the structure. An insulation layer called buried oxide (BOX) layer and a very thin top layer of silicon film is then deposited on the base layer. All the functional structures are formed only in the top layer of thin silicon that is placed above the insulator. Each MOS device that is formed in the top layer is separated by surrounding BOX as shown in Figure 1. Hence neither a separate well nor field oxide deposition is required to electrically isolate the devices as in conventional CMOS structures. The silicon substrate is isolated from the top thin

Figure 1. SOI wafer structure

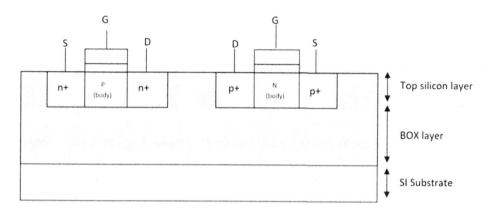

layer of silicon film by comparatively thick BOX. This is the main difference from traditional bulk process where every well has a separate contact.

The base silicon layer varies from few hundred μm to around 1mm in thickness. The Buried oxide (BOX) layer is produced either by oxidation of silicon or by implantation of oxygen into silicon. This layer is around 10 nm to 1 μm thick. But an ultra-thin Box layer will be around 10 to 25 nm thick. Finally, the top Silicon layer called Silicon on Insulator (SOI) layer varies between a few nm to 50 μm in thickness. A complete ultra-thin SOI wafer is usually in the range of 10 to 12 nm in its overall thickness. The substrate resistivity typically ranges between 10 Ωcm^{-1} to 10KΩcm^{-1} based on the application.

For logic devices, the top silicon layer typically ranges between 5 to 6 nm and the BOX layer between 20 to 25 nm. This layer thickness plays a significant role to provide good channel control with decreased Source Channel Effect (SCE). The SOI thickness is about 100nm and BOX layer thickness is about 145nm for RF applications. This provides good decoupling between the active thin film layer and the substrate.

Thickness of silicon layer and buried oxide layers for SOI wafers varies based on the application. SOI wafers with silicon layer thickness of greater than 1μm are used in high speed bipolar circuits, high voltage applications, varieties of power switching devices and in micro electro mechanical systems. If the silicon layer thickness varies between 50nm to 1μm it is classified as thin SOI wafer and CMOS devices are built on this. If silicon layer varies between 10nm to 50nm, it is classified as thin body device. If the silicon size is lesser than 10nm, it is referred to as nano SOI.

Various insulators like silicon-dioxide, silicon nitride, diamond or sapphire are used in building SOI wafers. The choice is based on the application type. For example short channel effects are lower when silicon-dioxide is used, thermal dissipation is provided by the use of diamond and sapphire can be used for high performance radio frequency applications.

SOI MANUFACTURING TECHNIQUES

SOI devices provided higher resistance to ionization due to solar wind radiations and voltage separation of the chips. This attracted their usage in early satellites and humans in space exploration systems in early 1960s. For the first time SOI structure was then used in radiation hard and high frequency applications. The underdeveloped manufacturing processes forced the usage of expensive materials to build active silicon on the top of the insulating layer. SOI devices made with silicon on sapphire were used initially.

The prime challenge in making a SOI structure was to manufacture a supreme quality, device grade single crystalline film of silicon above silicon oxide layer. To meet the requirement several manufacturing techniques were proposed. The early technique used was Zone Melt Re-crystallization (ZMR). In this technique, the thermally oxidized silicon wafer formed the substrate. The oxide was then patterned to create openings that exposed the underneath silicon surface. A thin layer of polycrystalline silicon was then deposited on the surface and was melted using a heat source. The layer was later allowed to solidify to form a single crystal on the silicon surface. This results in an extremely pure SOI wafers with lower defect densities. But this method could not meet the yield and reliability as demanded by the large-scale production of wafers.

Although quite a few methods are available for SOI manufacturing, at present only the following two methods are matured enough to meet the demands of the industry:

a. Bonded and etch back (BESOI)
b. Separation by implantation of oxygen (SIMOX)

A dielectrically separated thin silicon film above a silicon wafer can be made with both above-mentioned technologies. Apart from these methods, other methods of manufacturing techniques include smart cut and epitaxial layer transfer processes (ELTRAN).

Bonded and Etch Back: Direct bonding generally refers to joining two materials without any intermediate layer or external force. This process requires bonding together of two wafer surfaces that are flat, smooth and clean. The forces used for bonding can be capillary forces or electrostatic forces. Annealing of wafer pairs is then carried out at high temperature to convert physical forces to chemical bonds.

Steps involved in the process are as follows.

Figure 2. BESOI

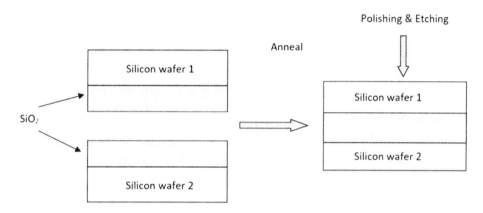

a. Initial preparation of the wafer
b. Formation of oxide layer
c. Surface smoothing and cleaning
d. Bonding of the two silicon wafers using an oxide layer in between them for isolation
e. Tapering down one of the wafers to attain SOI structure

The process of BESOI is illustrated in figure 2.

All these processes start with bonding together of two silicon wafers, but they differ in tapering of the structure till the top wafer meets the target requirement. The basic approach is to back grind and polish one of the wafers.

SIMOX: BESOI technique usually requires a silicon layer of minimum 1 micron thickness. This can be overcome by using SIMOX in which oxygen is directly implanted into the silicon wafer. This enables in making a very thin silicon device layers of few nanometers thick.

In SIMOX technique, BOX layer is produced by ion implantation of heavy dosage of oxygen into the silicon wafer. This creates a layer of stoichiometric silicon oxide. The developed structure is based on the dosage and the implanted ion energy. Ion implantation is carried out at temperature of around 400– 600 °C in order to prevent the changeover of silicon from crystalline form to amorphous form. This is followed by annealing at around 1300°C in order to dissolve the precipitate of oxygen and to acquire a planar structure in between buried oxide and silicon layers. Initially during the developmental stage, the defect density will be higher (more than 10^9 cm^{-2}). Futher, by combining multiple low doses of implantation, annealing and internal oxidation process, a material with thinner buried oxide layer and shallow

Figure 3. SIMOX

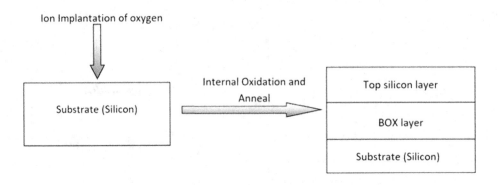

silicon layers could be achieved as shown in figure 3. This increases the throughput and reduces the defect density in the top layer of silicon.

Smart Cut Technique: In smart cut technique, implantation of hydrogen and wafer bonding is initially carried out. It is then followed by splitting along the hydrogen implantation line. This technique can be used in manufacturing large scale thin film SOI wafers. This technique is used to produce SOI wafer of uniform thickness that can be used in low power, low voltage and high-speed CMOS applications.

Major steps involved in Smart cut process are as follows:

a. Thermal oxidation of the first Silicon wafer.
b. Implantation of ions like H, He into the oxidized wafer in order to stimulate a weak buried zone.
c. Cleaning of the first implanted wafer and bonding it to the second wafer.
d. Thin layer is transferred into the second wafer from the first implanted wafer by inducing splitting in the weakened zone.
e. Removal of surface roughness by final treatment resulting in the final SOI structure.

ELTRAN: It is a bonding and etch back method that makes use of a very high etch rate of porous silicon. The process utilizes both structural and mechanical properties of Si layer and uses both direct wafer bonding and epitaxy techniques. It offers low threading defect densities because of the epitaxial layers. This is a major advantage of ELTRAN process.

The ELTRAN SOI layer thickness is greatly controllable by tuning the growing conditions. It ranges from a few nanometers (for MOSFET designs) to some micrometers (for MEMS applications). The thickness of the BOX layer is also controllable over a broad range, irrespective of the thickness of the SOI layer.

Major steps involved in ELTRAN process are as follows:

a.   On the surface of the initial donor wafer, porous silicon layers are formed.
b.   Epitaxial layers are then grown above the porous silicon layers.
c.   Partial oxidation of epitaxial layer by thermal oxidation is carried out to form the BOX layer in future. This creates a high-quality SOI- BOX interface.
d.   Bonding of donor wafer to Si wafer.
e.   Using mechanical stress, separation is induced in the porous silicon layer.
f.   Epitaxial SOI wafer is formed by removing the remaining porous silicon layers and by smoothing the surface by hydrogen annealing.

TYPES OF SOI

SOI devices are classified into two types:

a.   Fully Depleted SOI (FDSOI)
b.   Partially Depleted SOI (PDSOI)

Fully Depleted SOI: Over the last few decades, feature size of the transistor has decreased continuously resulting in superior performance and reduced power consumption. This has enabled manufacturing of electronic devices that are faster and more efficient. At present, as the feature size of the transistor has abridged below 10nm, it has progressively become much harder to meet the raising challenges of the forthcoming generation technology. FDSOI offers a potential solution to these challenges.

FDSOI, is a planar process technology that builds individual transistor component and in turn connects them together. This reduces silicon geometries and simplifies the manufacturing process. Building FDSOI is a two-step process: First, an ultra-thin BOX insulating layer ranging between 10nm to 25nm is placed above the base silicon. BOX layer reduces leakage and threshold voltage variations in chips as compared to conventional CMOS technology. It acts like a second gate allowing the chip makers to easily adjust the threshold voltage of the transistors. The transistor channel is then formed by depositing a thin silicon film. The deposited silicon film structure avoids channel doping, making the transistor fully depleted as shown in figure 4. In order to reduce gate leakage and power requirements, transistors with 5 to 6 nm Si/SiGe channel and a High-K metal gate can be built on the top silicon layer. These transistors have exceptional and attractive characteristics like enhanced switching speed compared to CMOS devices, even as the transistor size decreases.

Figure 4. Fully depleted SOI structure

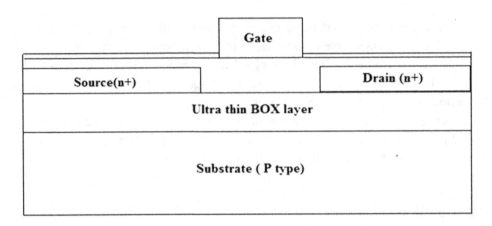

However, FDSOI demands specialized processes to create various layers. This makes them more expensive compared to bulk CMOS wafers.

FDSOI technology offers key advantages for upcoming technology nodes (Orouji & Abbasi, 2012). Few of the advantages are listed below:

a. Thin silicon film provides better electrostatic control of the gate that is present on the transistor channel, when compared to the conventional techniques like FinFETs and CMOS. This control is backed up by competent body biasing that offers superior switching speeds.
b. At circuit level, it provides better compromise between power consumption and performance.
c. It also offers scaling logic circuits required for mobile applications.
d. Switching from conventional technologies is easier using FDSOI as it is a planar technology, and its manufacturability is simpler compared to the FinFET.
e. The buried insulation layer in FDSOI provides the ability to dynamically modulate the device threshold voltage.

Partially Depleted SOI: As compared to FDSOI, in PDSOI the threshold voltage is less dependent on the thickness of the silicon film. However, in a PDSOI MOSFET, floating body effect is an inbuilt issue because the portion of silicon below the channel region is electrically floating as shown in figure 5. This is caused due to BOX layer that isolates the channel from the substrate and due to the accumulation of holes that are generated due to impact ionization in the vicinity of the drain region (Subrt, n.d.). This effect results in the reduction of the drain breakdown voltage, causes current instability during switching operation, kink effect and so on. Floating body effect can be mitigated by two different approaches:

- By providing body contacts - In this approach, the efficiency to absorb holes decreases swiftly as the channel width increases.
- By using techniques that are not dependent on the channel width.

Figure 5. Partially depleted SOI structure

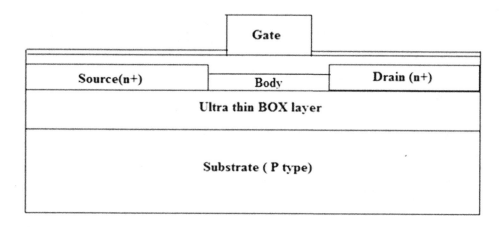

BENEFITS AND APPLICATIONS OF SOI

The benefits of SOI are summarized as follows:

a. Higher speed – It is mainly due to the elimination of sidewall parasitic capacitances and vertical stacking due to the thick BOX between the top thin silicon layer and the substrate.
b. Higher integration density – It is due to avoidance of latch up protection rings, substrate taps and the area required to isolate the internal devices.
c. Higher hardness and reliability – Latch up elimination provides higher reliability and when the device is exposed to radiations, most of the electron hole pairs are created in the thick silicon and the BOX layer provides the natural shield.
d. Low power consumption – It is due to lower parasitic capacitance.

SOI wafers find its applications in micro electromechanical systems (MEMS) and in photonics. SOI wafers have superior mechanical properties compared to polycrystalline films. This makes them to be a better option in fabricating MEMS components. The SOI wafers have higher refractive index variation between silicon and silicon-dioxide layers. This provides better photon confinement in smaller waveguides with sharper bends. This property has paved way for the use of SOI wafers in photonic applications. The potential benefits of SOI have also made it suitable for various products that finds applications in various fields as shown in Table 1 (Sadana & Current, 2006)

Table 1. SOI products and its applications

Product	Application
Digital SOI	Internet of things Cloud computing Mobile devices
RF SOI	High quality Front-End Modules LTE-Advanced front-end module ICs.
Power SOI	Switch mode power supplies Brushless motor drivers Class D amplifiers CAN/LIN Transceivers
Photonics SOI	High-speed optical transceivers in data centers
Imager SOI	Front-side imagers for near-infrared

DOWNSIDES OF SOI

SOI offers few disadvantages when advanced circuits or higher technology rules are considered. Few of them are as follows:

a. Kink effects are observed in PDSOI- When the voltage drop across the source and body is very high, the threshold voltage of the transistor decreases resulting in the current drop called kink in the drain characteristics. This causes disadvantage in precision analog design and in low power digital design. This effect can be overcome by creating body contact. This effect does not exist in FDSOI.

b. Self-heating is higher due to poor thermal properties of SOI substrate. This effect does not occur when the slew rate is greater than 20V/μS or when operated in pulse mode.

c. When the device is operated at the border of weak and strong inversion regions, hysteresis and latch effects are observed in SOI. Hysteresis also sometimes supports memory effect that does not turn off the transistor even when the gate voltage becomes zero. This is the result of the parasitic bipolar structure of SOI. This can be avoided using some advanced technologies.

d.   One main drawback of SOI technology compared to conventional semiconductor industry is its increased manufacturing cost. Hence are currently used by very few companies to manufacture high performance processors.

SUITABILITY OF SOI FOR LOW POWER AND LOW VOLTAGE (LP-LV) SYSTEMS

SOI is well suited for the operation in LP-LV devices because of its device characteristics. LP-LV circuits demands the operating point to be near (Sadana & Current, 2006) sub-threshold region. While operating in sub-threshold region, many of the drawbacks of SOI get vanished. Kink effect and heating effect disappears due to the reduction in the drain voltage and due to the operation near the threshold of the device. Latch or memory effect occurs only at high drain voltage. But at sub-threshold region they get cancelled.

SOI exhibits a lower sub-threshold slope due to its capacitance arrangement. This offers many advantages like better gate control and drain current in the sub-threshold region (Sun & Reano, 2010), lower device threshold voltage when compared to other bulk counterparts. Lower threshold voltage is most preferred for LP-LV applications as it enables the circuit to work at lower supply voltage (Vdd) which further helps in reducing the power dissipation.

The device isolation technique in SOI helps the device to have lower leakage. This is also supported by the fact that substrate junctions and well are also not present in SOI. These merits of SOI make it a potential candidate for the use in LP-LV systems.

CONCLUSION

Owing to the improved electrostatic control, both FINFET and SOI have shown classic improvements in performance over the traditional planar transistors. FINFETS are the best choice while making large chips with lot of wire capacitance. Whereas for smaller chips, gate capacitance becomes the major issue and FINFET offers a bit of disadvantage. FINFET technology involves much more complex manufacturing process, and it increases for RF and analog designs. The market demands for low cost, high performing and energy efficient technologies. SOI is one such technology that fulfills the market demands. SOI transistors not only exhibit exceptional electrostatic behavior but also have low mismatch properties. These are the two prerequisites for high performance at low voltage. SOI transistors offer power/performance that is optimized for low voltage. SOI technology is superior to bulk technologies in terms of performance. It has 50% faster operation and 18% lower power consumption compared with bulk technologies. Although bulk CMOS technology has the lowest cost, FINFET technology has the highest performance, SOI technology has the best power, performance and cost tradeoff.

REFERENCES

Kononchuk, O., & Nguyen, B. Y. (2014). *Silicon-on-insulator (soi) technology: Manufacture and applications*. Elsevier.

Orouji, A. A., & Abbasi, A. (2012). Novel partially depleted SOI MOSFET for suppression floating-body effect: An embedded JFET structure. *Superlattices and Microstructures, 52*(3), 552–559. doi:10.1016/j.spmi.2012.06.006

Sadana, D. K., & Current, M. I. (2006). Fabrication of Silicon-on-insulator (SOI) and strain-Silicon-oninsulator (sSOI) wafers using ion implantation. Ion Implantation: Science and Technology.

Subrt, O. (n.d.). *Silicon-On-Insulator-A perspective on low-power, low-voltage supervisory circuits implemented with SOI Technology*. Academic Press.

Sun, P., & Reano, R. M. (2010). Submilliwatt thermo-optic switches using free-standing silicon-on-insulator strip waveguides. *Optics Express, 18*(8), 8406–8411. doi:10.1364/OE.18.008406 PMID:20588686

Chapter 3

Low–Power High–Speed Eight–Bit Universal Shift Register Design Using Clock Gating Technique

Preeti Sahu
Poornima University, India

ABSTRACT

A register is basically known as a storage device for units in circuits. In data processing systems, they are used to immediately transfer data by using CPU. In digital electronics, shift registers are known as the sequential logic circuits that are used to store data temporally and transfer the data to its output for each and every clock pulse. Shift registers are found as digital memory unit storage in such devices as calculators, computers, etc. Based on shifting data, shift registers are classified in two types: universal shift register and bidirectional shift register. This chapter dealt with design and implementation of 8-bit universal shift register with CG scheme for minimizing power. Circuit operation is performed by Xilinx-14.7 software tool and simulated with I-SIM simulator tool using VHDL language. XPE tool is used to optimize power in the circuit. Results improved the power consumption in circuit by 40.65%. Also 4.76% of area was increased due to adding external circuitry and delay was reduced by 12.93% in the proposed design.

INTRODUCTION

At now days VLSI has important role in many applications such as DSP, RF,

DOI: 10.4018/978-1-6684-4974-5.ch003

communications network, microwave applications, MEMS, and Space application, Robotics etc. An electronic circuit that consist of elements, which may be a transistor, diodes or resistors combined in such a manner that they perform a logic operation called gate circuit which are known as basic building blocks of a digital system. There are 3 main parameters in VLSI digital design: Power, Area, and Delay. In this paper dynamic power reduction in 8-bit USR circuit is discussed. Generally power consumption is increased in the electronic system and the integrated circuits in particularly manner because of their complexity due to large number of circuits on a single chip. So, there is needed design a circuit with low power consumption. The optimization is described in terms of generating the best design according to goal. In VLSI system designing mostly three sources of power dissipation discussed namely as, Dynamic, or circuit-switching power, Static Power and Short- circuit Power. Dynamic power is a very simple approach to estimate energy consumption in a CMOS circuit (Sahu and Agrahari, n.d., 2021).

Dynamic power is caused by switching activities of the circuit. Increment of dynamic power in circuit is depends on the higher operating frequencies which leads more frequent switching activities in the circuit. Dynamic power of circuit is described as given expression, where energy consumption in CMOS circuits is estimated by the capacitance to be switched. The charging and discharging of capacitance is known as most significant source of dynamic power dissipation in VLSI circuits (Joshi and Jangir 2019).

$$Pd = C.V^2.f$$

Static power is related to the changes in states of the circuits, means that the static power is due to the changing of circuit states that are 0 to 1 or 1 to 0 rather than switching activities. In CMOS circuit leakage power is only source of static power dissipation. Low power consideration should be applied in digital CMOS technology at all levels of design abstraction and design activities. A low power design system also affects other features such as reliability, design cycle time, testability and design complexity. Chip area and speed are the major trade off considerations in designing of VLSI system (Niranjan, n.d.; Sivakumar and Sowmya, 2016; K. Anusha and Deepika, 2016).

Figure 1. Shift Register using D-Flip-flop Latches

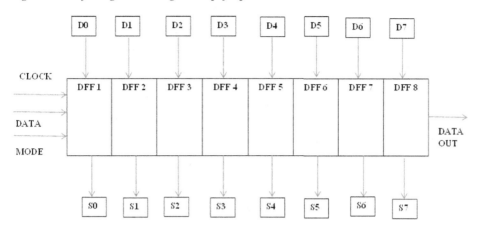

PROPOSED SYSTEM

Shift Register:

A digital circuit which formed by cascading the Flip- flop circuits is called **shift register**. In the shift register input of cascaded flip-flop is driven by output of previous flip-flop with a common clock signal. The data stored in the system by shifting from one location to next. A feedback connection is made by connecting the last flip-flop back to the first, where data is shifted by shifters for extending time period. They are used as a form of computer memory. Array is used to store the Data and read back. In a computer world, each bit is stored serially in the shift registers. In the way of serially, data can be fed one bit at a time and in the way of parallel fed the batch of data bits at the same time. Shift registers can be used as storing devices for data or movement/shifting devices for data. Therefore, these types of registers commonly implemented in calculators or computers for storing and processing the data with shifting of data bits. They can also use as temporary storage units for binary operations as addition or multiplication. Shift registers can have inputs and outputs as both parallel and serial way (K. Anusha and Deepika 2016; Saranya et al., 2013). A simple block diagram of 8-bit shift register using DFF is shown in Figure 1.

An 8-bit shift register shown as above Figure 1, where 8-bit data operation performed with using D- Latches. Clock signal is required for operating the data flow. This scheme described the parallel-In- Parallel-Out Shift register. Here data can be loaded as input in the parallel way and their corresponded output also gets in parallel way (Paliwal et al., 2020).

Figure 2. Block Diagram of 8-bit Shift Register

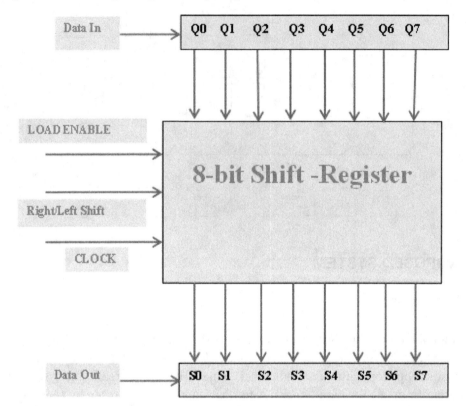

A Parallel-in-Parallel-Out (PIPO) shift operated as very fast and within a single clock pulse it provides the output as corresponded input. Figure 2 explained the workflow of 8-bit shift register. This is Bidirectional shift register which performed the data serially and parallel both. A clock signal is needed to start to data flow in register. Input different Parallel data and get their corresponded output after shifting process also in parallel mode (Daboul et al., 2018).

Universal Shift Register:

A USR (Universal Shift Register) is a register which perform the operation to store the data and shifts the data with having the parallel load capability. Due to parallel load capability USR is used to perform input & output operations in both serial and parallel modes. USR is a combination of unidirectional shift registers and bidirectional shift registers. It is also known as a shift register having parallel load or parallel-in- parallel-out shift register (Barman and Kumar, 2018; Tamil and Shanmugasundaram, 2018)

Figure 3. Block Diagram of Universal Shift Register Circuit

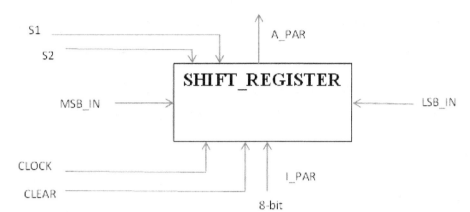

The USR circuit capable to perform following 3-operations, such as

Parallel Load Operation: stores the data in parallel as well as the data in parallel (Chandrakar and Roy, 2019).

Shift Left Operation – stores the data and transfers the data shifting towards left in the serial path (Sahu and Agrahari, 2021).

Shift Right Operation – stores the data and transfers the data by shifting towards right in the serial path (Bhattacharjee et al., 2016).

A simple block diagram of USR is shown in figure-3 as given below, which has serial inputs S1, S2 and a clock signal for controlling the shift-right operations (R. Keerthi Kiran and Kalpana, 2015).

All the parallel inputs and outputs are connected with the parallel load to transfer the data in parallel mode. When Clear pin is set at logic '0' it clears the register and CLK pin provides clock pulses to start the operations. The information or data in the register would not change when the circuit is in the control state. When the clock pulse is applied, the circuit will be out of their control state and ready to perform operations. The register circuit is performing as a USR if it operates with a parallel load and performs the shift operations towards the right and left. From the above figure can be explain the working of USR. Serial inputs S1 and S2 shift the data towards the right and left and store the data in the register (R Keerthi Kiran and Kalpana, 2015; Kakarala et al., 2015).

Working of the USR can be understood simply through table 1 as shown below. This table is also explained as truth table of USR system, which explained all the outputs to its corresponded inputs at S1 and S2 (Srinivasan et al., 2015).

If the selected pins S1= 0 and S2 = 0, then this register circuit is stays in control state and doesn't operate until the clock pulses is not applied. When S1 = 0 and

Table 1. Truth Table of USR

S1	S2	Mode of Operation
0	0	Locked State (No Change)
0	1	Shift Left
1	0	Shift Right
1	1	Parallel Load

S2 = 1, then circuit performs shift operation on the data to left and stores the data. When S1 = 1 and S2 = 0, then perform the right shift operation and shift the data to right. When S1 = 1 and S2 = 1, then register loads the data in parallel mode and performs the parallel loading operation and stores the data in same register.

Design Methodology Clock Gating

In Low power VLSI design Gating technique has proved better solution for power minimization. There are two types of gating technologies are studied for power minimizing in IC design i.e., Clock Gating and Power Gating. The Clock gating technique minimizes the power of circuit with reducing clock switching activities and the Power Gating scheme minimizing the power with adding sleep transistors as header and footer to the circuit. In this paper USR implemented using Clock Gating scheme which is shown in figure 4 & 5 as below.

A basic block diagram of CG scheme is shown in figure 4. Where D-FF circuit is used to generate Gated Clock signal and then it is provided to the main circuit as driver signal. When 'En' is 1, Clk signal gated with D-FF circuit otherwise the circuit is off (Anand et al., 2014).

AND gate-based CG implementation is easy to design and implementation for the circuit, a basic form of AND gate-based CG scheme is shown in figure as below where AND gate is used to generate gated clock signal and this gated clock signal provided to circuit as driver signal. AND gate-based CG scheme is shown in Figure 5.

Clock gating is widely used for dynamic power reduction of the circuit and little bit of static power also. Power Minimization also increases the speed.

EXPERIMENTAL ANALYSIS

In this section the circuit implementation and simulation work is defined for designing low power 8-bit Universal shift register. Experiment of circuit is performed by

Figure 4. Clock Gated Circuits

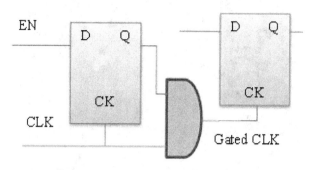

XILINX-14.7 software tool with VHDL coding. I-Sim simulator is used to perform the simulation process and got the waveforms as corresponded output.

An RTL schematic of 8-bit USR is shown in figure 6 as above. It consists of FF circuits as cascaded network and shifts the data from one to next. The technology view of 8-bit USR is shown in Figure 7, which consider the circuit diagram and their I/O paths for shifting the data. All network connections are shown by figure clearly which used to transfer the data to FFs. With this diagram can understand that data is loaded serially and output also gets in serially. Technology view of this design is shown as below. This technology schematic flow helps to understand the connections between latches and devices and its shifting process from one latch to another latch (Sahu and Agrahari, 2020).

In the next the simulated waveforms as shown in Figure 8 without CG and with CG implementation respectively. These waveforms are defined in terms of output as corresponding input data.

In Figure 8, simulated waveforms are shown for 8-bit USR design without CG implementation. Output waves are generated with the corresponded input data. Proposed USR performed outputs in four modes of operation i.e., SISO, SIPO,

Figure 5. CG implementation with Gate based Technology

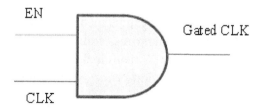

Figure 6. RTL schematic of 8-bit Universal Shift Register

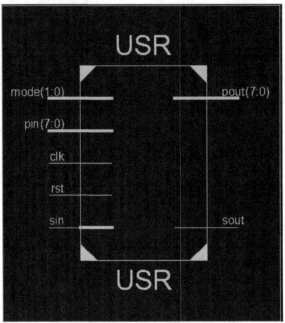

PISO and PIPO. When reset is '1', all the outputs received '0'. When reset is '1', apply different values on mode signal. When mode is '00', No change condition applies to signals and output received with serial input data on Sin and the register worked in SISO mode. When mode is '01', 1-bit shift left and output taken after shifted the corresponded input value, system worked as SIPO mode. When mode is '10' 1-bit of input shifted to right and corresponded output signal achieved and the register circuit worked in PISO mode. When mode is '11' then input shifted to parallel load and received with corresponding output, Here the proposed USR worked in PIPO mode.

RTL schematic of proposed USR with CG implementation design is shown in figure 9 as below. This schematic diagram explained the all-block connections and clarifies the design. Technology schematic of proposed design is shown in figure 10 which explained all the network connections briefly after CG implementation.

The CG implementation has been completed and output provided as the simulated waveforms. With Figure 11, CG scheme implementation on Universal shift register design can be shown. The simulation process with CG implementation starts after enabling the clock signal. If enable is '1' then clock signal provide as input to circuit otherwise circuit takes in OFF condition. Same as conventional design CG based

Figure 7. Technology Schematic of 8-bit Universal Shift Register

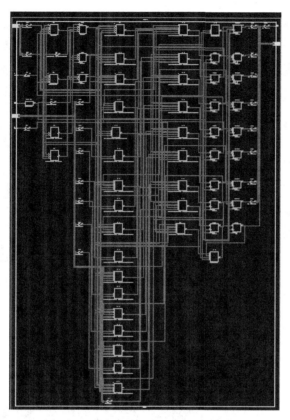

USR also performed operations for different operating modes like SISO, SIPO, PISO, and PIPO (Shinde and Salankar, 2011).

Area Optimization in proposed explained in table-3 as shown calculated with device utilization. USR has been below. Area is in both circuits has been reduced with proposed design as compared to conventional design. With analyzing the results static or leakage power reduction is achieved in proposed design using AND Gate based CG implementation. Clock is not used in longer path so dynamic switching or dynamic power is not placed at '0'. Different cases apply on mode respectively '00', '01', '10' and '11'. With CG implementation clock signal is controlled by Gated Clock signal which is generated after enabling the clock signal using AND gate. Due to this control signal delay and power is less than conventional USR design (Sahu and Agrahari, 2020). The compared results for different parameters in Conventional and Proposed designs are explained as tables in next section.

Figure 8. Simulation Results for 8-bit Universal Shift Register

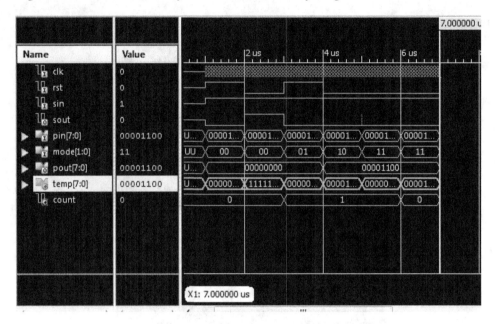

Figure 9. RTL schematic of 8-bit Universal Shift Register with CG implementation

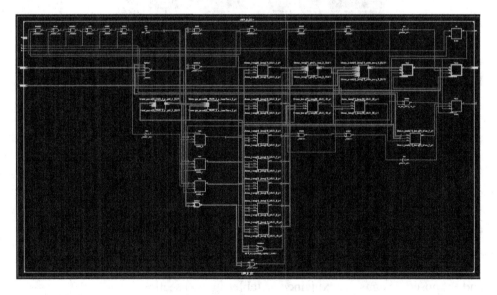

Figure 10. Technology schematic of 8-bit Universal Shift Register with CG implementation

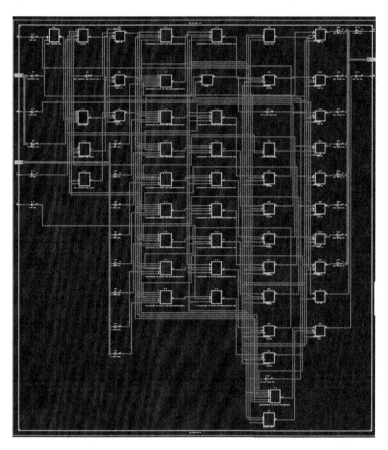

Figure 11. Simulation Results for 8-bit Universal Shift Register using CG technique

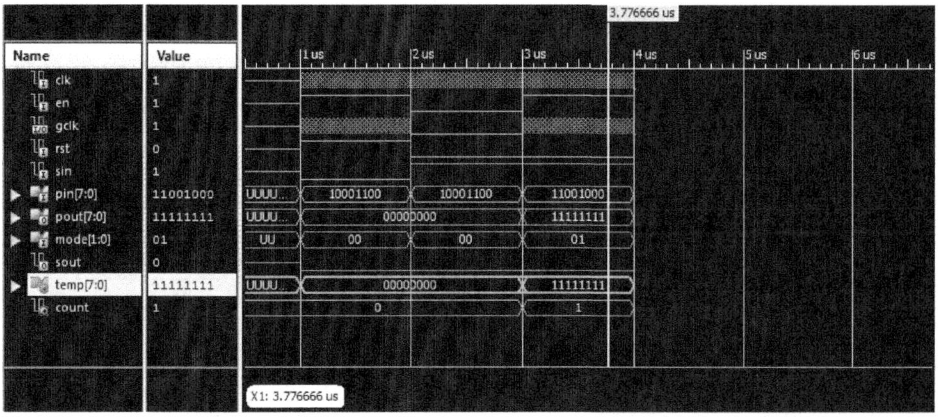

Table 2. Power Optimization in 8-bit Shift Register

Circuit	Frequency	Power
		Leakage Power
8-bit Universal Shift Register without CG implementation	523.478 MHz	82.16mW
8-bit Universal Shift Register with CG implementation	502.197 MHz	78.82mW

RESULTS AND DISCUSSION

This section has been defined the Results which provided after simulation part. In the proposed system Area, Delay and Power calculated with Xilinx-14.7 software. The power has been calculated with X-Power Estimator (XPE) tool. Results optimized area, delay, and power in proposed system at 502.197 MHz frequency and 1V supply voltage which explained in tabular form as given below.

Power Optimization in 8-bit USR with and without CG implementation is shown in Table-2 as below. There is leakage power optimized. Leakage power, Area reduction achieved with reduction in number of LUTs, I/O pins, Flip-flops. Proposed design achieved less area as compare to conventional USR design (Sharma 2012).

Delay Optimization has been explained in Table 4. Delay in the designed circuit has been achieved with calculate gate and net delay. Proposed design achieved with reduction in Gate delay.

CONCLUSION

As above results proved that CG scheme is better for minimizing power in the circuits. This scheme also has better solution for minimize the area utilization by devices and delay. With implementing power gating scheme shift register has been improved the power, delay, and area utilization. Power has been improved by 40.65% with implemented design. Area increased by 4.76% but delay has been improved by

Table 3. Area Optimization in 8-bit Shift Register

Circuit	Device Memory Utilization
Memory used in 8-bit Universal Shift Register without CG implementation	427252 KB
Memory used in 8-bit Universal Shift Register with CG implementation	447604 KB

Table 4. Delay Optimization in 8-bit Shift- Register

Circuit	Delay (in Nano-Second)		
	Delay before Clock	Delay after Clock	Delay to Clock to setup on destination
8-bit Universal Shift Register without CG implementation	1.539	0.647	1.910
8-bit Universal Shift Register with CG implementation	1.960	0.640	1.663

12.93%. Requirement of Low power devices are increases much more in IC world. So, PG scheme has better solution in this way. Proposed and implemented design considers less area with minimize the leakage in the circuit. The proposed universal shift register design achieved low power and improved performance of the circuit. Speed of the proposed USR system will be high than conventional USR system.

REFERENCES

Anand, Joseph, & Oommen. (2014). Performance Analysis and Implementation of Clock Gating Techniques for Low Power Applications. In *2014 International Conference on Science Engineering and Management Research (ICSEMR)*. IEEE.

Anusha & Deepika. (2016). Design and Implementation of Novel 8-Bit Universal Shift Register Using Reversible Logic Gates. *International Journal of Innovation in Engineering and Technology*.

Barman, J., & Kumar, V. (2018). Approximate Carry Look Ahead Adder (Cla) for Error Tolerant Applications. In *2018 2nd International Conference on Trends in Electronics and Informatics (ICOEI)*. IEEE. 10.1109/ICOEI.2018.8553739

Bhattacharjee, P., Majumder, A., & Das, T. D. (2016). A 90 Nm Leakage Control Transistor Based Clock Gating for Low Power Flip Flop Applications. In *2016 IEEE 59th International Midwest Symposium on Circuits and Systems (MWSCAS)*. IEEE. 10.1109/MWSCAS.2016.7870034

Chandrakar, K., & Roy, S. (2019). A SAT-Based Methodology for Effective Clock Gating for Power Minimization. *Journal of Circuits, Systems, and Computers*, 28(01), 1950011. doi:10.1142/S0218126619500117

Daboul, S., Hähnle, N., Held, S., & Schorr, U. (2018). Provably Fast and Near-Optimum Gate Sizing. *IEEE Transactions on Computer-Aided Design of Integrated Circuits and Systems, 37*(12), 3163–3176. doi:10.1109/TCAD.2018.2801231

Joshi & Jangir. (2019). Design of Low Power and High Speed Shift Register. *IOSR Journal of VLSI and Signal Processing, 9*(1), 28–33.

Keerthi Kiran, R., & Kalpana, A. B. (2015). Low Power 8, 16 & 32 Bit ALU Design Using Clock Gating. *International Journal of Scientific and Engineering Research, 6*(8).

Niranjan, V. (n.d.). *Low power and high performance shift registers using pulsed latch technique.* Academic Press.

Paliwal, P., Sharma, J. B., & Nath, V. (2020). Comparative Study of FFA Architectures Using Different Multiplier and Adder Topologies. *Microsystem Technologies, 26*(5), 1455–1462. doi:10.100700542-019-04678-8

Sahu & Agrahari. (2021). Low Power Design and Challenges in VLSI with IoT Systems. *International Conference on Advanced Computing and Communication Technology.*

Sahu & Agrahari. (n.d.). Optimization of Parameters in 8-Bit ALU Circuit With Clock Gating Technique. *PRATIBHA: International Journal of Science, Spirituality, Business and Technology, 34.*

Sahu, P., & Agrahari, S. K. (2020). Comparative Analysis of Different Clock Gating Techniques. In *2020 5th IEEE International Conference on Recent Advances and Innovations in Engineering (ICRAIE).* IEEE. 10.1109/ICRAIE51050.2020.9358375

Sahu, P., & Agrahari, S. K. (2021). Power and Performance Optimization in 16- Bit ALU Using Power Gating. *International Conference on Recent Trends in Electrical, Electronics & Computer Engineering for Environmental and Sustainable Development.*

Sandhya, Krishna, & Satamraju. (2015). *A Novel Approach for Auto Clock Gating of Flip-Flops.* IJSER.

Saranya, M., Vijayakumar, V., Ravi, T., & Kannan, V. (2013). Design of Low Power Universal Shift Register. *International Journal of Engineering Research & Technology.*

Sharma, D. K. (2012). Effects of Different Clock Gating Techinques on Design (Vol. 3). Academic Press.

Shinde, J., & Salankar, S. S. (2011). Clock Gating—A Power Optimizing Technique for VLSI Circuits. In *2011 annual IEEE India conference*. IEEE. 10.1109/INDCON.2011.6139440

Sivakumar, S. A., & Sowmya, R. (2016). Design of Low Power Universal Shift Register Using Pipe Logic Flip Flops. *International Journal of Advanced Research in Computer and Communication Engineering*, *5*(5), 55–59.

Srinivasan, N., Prakash, N. S., Shalakha D, Sivaranjani D, Sri Lakshmi G, S., & Sundari, B. B. T. (2015). Power Reduction by Clock Gating Technique. *Procedia Technology*, *21*, 631–635. doi:10.1016/j.protcy.2015.10.075

Tamil, S. C., & Shanmugasundaram, N. (2018). Clock Gating Techniques: An Overview. *2018 Conference on Emerging Devices and Smart Systems (ICEDSS)*.

Chapter 4

Dynamic Body Bias:
A Transistor-Level Technique for the Design of Low-Voltage CMOS Analog Circuits

Vandana Niranjan
Indira Gandhi Delhi Technical University for Women, India

ABSTRACT

The demand for portable device applications has grown immensely. For such applications, low voltage and low power operation is an essential prerequisite to prevent overheating and ensure reliable functioning. Low voltage operation curtails the total number, weight, and dimensions of batteries, and low power consumption extends battery life. The shrinking size of MOS transistors in CMOS processes necessitates the use of lower supply voltages. Since the threshold voltage of MOS transistor is not diminished at the same rate as the power supply voltage, analog designers face problems due to shrinking voltage headroom. One of the findings that can overcome the issues introduced by comparably high threshold voltages is based on the enactment of body bias approach. In such a solution, a relatively small potential is applied at body terminal of a MOS transistor to adjust its threshold voltage. This chapter discussed that body bias approach is an attractive opportunity for utilizing the body effect positively to improve the performance of low voltage-integrated circuits.

BACKGROUND

The demand for portable had held electronic and implantable device applications

DOI: 10.4018/978-1-6684-4974-5.ch004

has grown immensely in the last few decades. For such applications, low voltage and low power operation is an essential prerequisite to prevent overheating and ensure reliable functioning without failure. Low voltage operation is preferred for possible curtailment in the total number, weight and dimensions of batteries and low power consumption extends the operation period for battery powered devices. The shrinking size of MOS transistors in CMOS processes necessitates the use of lower supply voltages. Since the threshold voltage of MOS transistor is not diminished at the same rate as the power supply voltage, analog designers face problems due to shrinking voltage headroom. At reduced supply voltage, diminishing headroom sets new challenges to ameliorate or even prolong the circuit performance. One of the findings, which can overcome the issues introduced by comparably high threshold voltages, is based on the enactment of body bias approach. In such a solution, a relatively small potential is applied at body terminal of a MOS transistor to adjust its threshold voltage. Reconfiguring the MOS transistor in this way broaden the applicability of basic CMOS analog building blocks to low supply voltages. Body effect in a MOS transistor was contemplated in the past as an exclusive source of unwanted second order effects. The main objective of this chapter is to apprise about that how body bias approach is an attractive opportunity for utilizing the body effect positively to improve the performance of low voltage integrated circuits. Also, as the power supply voltage approaches the transistor threshold voltage, the circuit performance becomes extremely sensitive to process variations and temperature alterations. Body bias approach not only augments the performance, but also improves the circuit robustness against process and temperature variations. Body bias approach accredits a heterogeneity of effective body bias techniques. These bias techniques require triple-well CMOS technology for implementation, although at slightly higher cost but it's free from latch-up and is more immune to noise. In this chapter, the basic principle of dynamic body bias approach has been explained. The inspiration towards minimizing supply voltages and power consumption in mixed-signal integrated circuits is expanding demand for battery-operated portable electronic devices. However, this cutback in the supply voltage primarily affects the performance of CMOS analog circuits in terms of dynamic range, noise, speed, linearity, gain and bandwidth. The importance of dynamic body bias technique to circumvent these limitations and improve the performance of circuits capable of operating under low supply voltage is explained. Various reported applications of dynamic body bias technique in analog circuits have been reviewed. It is also emphasized that for implementation and fabrication of DBB technique, triple-well CMOS technology is necessary. This technology is explained with the help of a cross section figure. A small signal model is very important when it comes to design of analog circuits. Therefore small signal model of a dynamic body biased MOS transistor is provided. Author has also discussed various modification in small signal parameters due to dynamic body bias in a MOS transistor. This chapter is concluded in the end. It becomes quite clear that as now a days, most of the integrated circuits work at supply voltage less than 1 volt and therefore dynamic body bias approach is quite attractive option for utilizing both gate and body transconductance of a MOS transistor effectively. This will not only help to achieve higher output swing but also wide dynamic range and higher overdrive voltage. DBB technique is feasible solution in design of low voltage CMOS analog circuits.

INTRODUCTION

With the proliferation of portable battery powered electronic equipment, low power at low voltage design has become necessity in most of the VLSI circuits and systems. The demand for implantable medical devices such as pacemaker, neurostimulators, cardioverter defibrillator etc has also grown immensely in the last few decades. Further, to keep up large time interval between battery charge-recharge cycle all implantable electronic products need to have lower power consumption. This also ensures reduction in the weight, size and number of batteries used. Apart from this, low power at low voltage operation is imperative to prevent overheating of portable devices and ensure reliable functioning (Matej et al., 2018; Viera et al., 2018).

As MOS scaling is increasing, VLSI design and research group has estimated increased usage of digital logic circuits in contrast to RF/analog integrated circuits. Since this world we live in and all real signals are analog/continuous in nature therefore it is practically unrealistic to replace all analog circuits. Digital circuits fully benefit from advances in MOS scaling technology as compared to analog circuits. On the contrary, the functionality of analog circuits often suffers at low voltage levels due to degradation of MOS transistor's intrinsic gain i.e. $g_m r_o$. As channel length is decreasing, the maximum achievable intrinsic gain in deep submicron technologies is becoming unsatisfactorily low due to reduction in small signal output resistance r_o. Another important limitation that analog circuits face in a low voltage design scenario is that MOS transistor's threshold voltage is not able to reduce further at the same pace as the power supply node voltages. Due to this threshold voltage scaling limitation, it is very challenging to maintain the high performance of analog circuits at lower supply voltage. Various components constituting the threshold voltage of a MOS transistor is expressed using following expression:

$$V_{T0} = \varphi_{GC} - \psi_s - \frac{Q_{B0}}{C_{ox}} - \frac{Q_{ox}}{C_{ox}} \tag{1}$$

where, V_{T0} denotes threshold voltage of MOS transistor under the condition that both body/bulk and source terminals of transistor are at equal potential. Further, difference in work function between gate and channel region reflecting built-in potential of the MOS transistor is denoted as φG_C and surface potential in strong inversion region is ψs. Co_x denotes gate oxide capacitance per unit area whereas QB_0 and Qo_x represents the depletion region charge density and charge density at surface between the silicon substrate and gate oxide respectively.

The main components present in the V_{T0} that do not scale by the same amount as the supply voltage are depletion region charge and built-in-potential. Although scaling is progressing but transistor's threshold voltage is not anticipated to scale down much lower than that available today. Therefore, some other approaches must be investigated to find appropriate solutions to the challenges introduced by the un-scalable value of relatively high threshold voltages. There are many non-conventional techniques based on CMOS technology i.e. bulk-driven, floating-gate and quasi-floating-gate techniques have been proposed by researchers as another ways to reduce the design complexity and push the voltage supply towards threshold voltage of the MOS transistors. However due to various drawback associated with each of them, another techniques need to be explored (Chawla et al., 2010; Niranjan & Gupta, 2009; Niranjan & Gupta, 2011).

One of the most feasible solutions to design low voltage analog circuits with low V_{T0} is the application of dynamic body bias (DBB) technique. This technique is easily implemented at transistor level. In DBB technique, both the body and the gate terminals are connected together, and bias/potential applied at this terminal adjusts the threshold voltage of transistor. Reconfiguring a MOS transistor implementation in this way accredits the applicability of foundational analog building blocks to domain of low supply voltage. Henceforth throughout the chapter, short notation DBB technique has been used instead of dynamic body bias technique. In the next section we will study the operating principle of DBB technique.

WORKING PRINCIPLE OF DBB TECHNIQUE

Use of body terminal in a MOS transistor is less matured as contrast to the other three terminals. Body bias is basically the technique of putting a bias voltage at the body/bulk terminal to change the threshold voltage of a MOS transistor. When value of bias/potential of body terminal is different than the bias/potential of source terminal, then the body/bulk terminal would also influence the working of a MOS transistor and will modulate its threshold voltage. This change in the threshold voltage resulting due to difference of bias/potential between body and source terminal is "body effect".

Shichman Hodges model can be easily used to estimate the threshold voltage of an n type MOS transistor under body biased conditions. This model is linked with the emergence of a conducting inversion layer at the source end, expressed as

$$V_T = V_{T0} + \gamma \left(\sqrt{|\psi_s + V_{SB}|} - \sqrt{|\psi_s|} \right) \tag{2}$$

where, V_{SB} represents potential difference between bulk/body and source terminal, V_T denotes threshold voltage when bulk/body and source terminals are at different potentials and body effect factor is represented by γ. The body effect factor is an indication of the effect of variations in V_{SB} on threshold voltage and is expressed as

$$\gamma = \frac{\sqrt{2q\varepsilon_{si}N_A}}{C_{ox}} \tag{3}$$

where, ε_{si} denotes silicon's dielectric permittivity, q represents electronic charge and doping density of body/substrate is denoted by N_A.

It is interesting to note that Shichman Hodges model is not only a comprehensible first-order model appropriate for long-channel transistors but also suits the purpose to justify that threshold voltage is nonlinearly contingent on the applied potential/bias voltage at body terminal (Niranjan et al., 2012; Niranjan et al., 2014; Tsividis, 1999). Thus effectual threshold voltage control can be easily achieved at transistor level by using body bias strategy and the this alteration in the threshold voltage is expressed as

$$\Delta V_T = \gamma \left(\sqrt{|\psi_s + V_{SB}|} - \sqrt{|\psi_s|} \right) \tag{4}$$

It follows From Eq. (4) that body bias strategy not only take advantage of the body effect but also bestows an effective " virtual knob" to manage the threshold voltage of MOS transistors. Thus, threshold voltage of a MOS transistor can be lowered or increased using body bias approach.

Transistor level implementation of DBB technique in n type MOS transistor is shown in Fig.1. Here, gate and body terminals of a MOS transistor are tied in unison. Due to this connection, with every change in the input signal, the bias/potential voltage at body terminal will also change dynamically. This is the reason behind its name as dynamic body bias. When appropriate signal is applied at the input terminal, transistor starts conducting and body effect lowers its threshold voltage. However, in the off state, there is no change in threshold voltage.

Figure 1. DBB technique

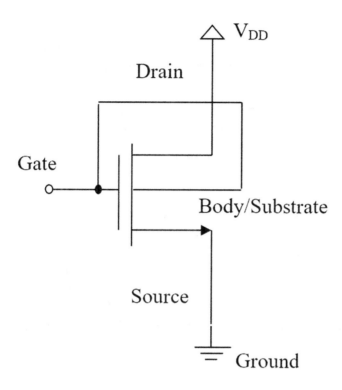

A MOS transistor using DBB technique is also perceived as dual-gate transistor due to identical bias voltage put in at gate and body terminals. Apart from the value of threshold voltage, gate, and body connection in DBB technique also affects the transconductance and capacitance parameters. This is by virtue of the fact that channel current depends functionally on threshold voltage and threshold voltage itself depends on body bias voltage. The drain current of transistor in Fig.1 is obtained as

$$I_D = \left(\frac{\mu_n C_{ox}}{2}\right)\frac{W}{L}\left[V_{GS} - \left\{V_{T0} + \gamma\left(\sqrt{|\psi_s + V_{SB}|} - \sqrt{|\psi_s|}\right)\right\}\right]^2 \tag{5}$$

It is observed from drain current equation that both body transconductance g_{mb} and gate transconductance g_m adds up to the channel current as expressed in following expressions.

$$\partial I_D = g_m \partial V_{GS} \tag{6}$$

and

$$\partial I_D = g_{mb} \partial V_{SB} \tag{7}$$

It is seen from Eq. (7) that the body potential/bias voltage also contributes to the channel current and thus body region of transistor poses as its second gate. The two transconductance are related to body bias voltage as

$$\frac{g_{mb}}{g_m} = \frac{\gamma}{2\sqrt{|\psi_s + V_{SB}|}} \tag{8}$$

The ration of both transconductance i.e., g_{mb}/g_m varies in the range 0.1 - 0.3 but sometimes increase significantly above those values for transistors with large body effect factor γ, operating with low V_{SB}.

Among various small-signal parameters influenced by DBB technique is the source to body junction capacitance. The source to body p-n junction has a parasitic capacitance due to depletion region and bias dependent. Source-body capacitance can be expressed using equation.9, assuming constant doping profile in p type and n type regions.

$$C_{sb} = \frac{C_{sbo}}{\sqrt{1 + \dfrac{V_{SB}}{\varphi_0}}} \tag{9}$$

where, C_{sb0} represents the value of body-source capacitance at zero bias voltage and the built-in-potential of the body-source p-n junction is denoted as φ_0.

The ratio of two transconductance i.e g_{mb}/g_m is associated to transistor parasitic capacitances as expressed in following expression

$$\frac{g_{mb}}{g_m} \approx \frac{C_{sb}}{C_{gs}} \tag{10}$$

Thus, DBB technique can fine-tune the magnitudes of body transconductance, threshold voltage and body capacitance of MOS transistor and thus can be utilized to improve the performance of RF/analog circuits in a low voltage design domain (Jhajharia & Niranjan, 2016; Niranjan et al., 2013a; Niranjan et al., 2014a; Niranjan et al., 2014b).

Due to DBB technique, body-source junction of MOS transistor gains slight forward bias. Therefore, magnitude of input signal voltages are kept low enough (< 0.6 V) to restrict high junction current. When voltage levels are low, the concentration of mobile charge carriers in the depletion region is also very low. As a result, traditional SPICE MOS models are still relevant for simulation/analysis of DBB technique based analog circuits. In the next section we will study about the effects of DBB technique on the small-signal parameters of MOS transistor.

EFFECT OF DBB TECHNIQUE ON SMALL SIGNAL MODEL OF MOS TRANSISTOR

Small signal modelling is premise of all analog circuit anatomy. Design of analog circuits to a great degree depends on using a precise MOS model to foresee the correct performance of the designed circuit. CAD tool stand in need of exceedingly error-free models which are very difficult to comprehend. Therefore, for simple estimations using pencil and paper design, it is more appropriate to use a customized simple small signal MOS model. This dedicated model assist insight into circuit operation. From Eq. (2), it may be observed that

$$V_T \propto \sqrt{|\psi_s| - V_{GS}} \tag{11}$$

It follows from Eq. (11) that for all numerical values of the gate input such that $V_{GS} > \psi_s$, circuit simulation program may be unsuccessful to converge. Therefore, DBB technique is more acceptable and applicable for very low voltage integrated circuits as sizable errors can be anticipated for higher gate input (V_{GS}) voltage.

Small signal model of n type MOS transistor using DBB technique is shown in Fig. 2. As both gate and body terminals are utilized as signal input, body terminal has significant impact in its behaviour.

From the model it is seen that there are two current sources/generators which are managed by the same applied voltage V_{gs} and the capacitances C_{gs}, C_{bd}, C_{gd}, represent gate-source, body-drain and gate-drain parasitic junction capacitances respectively. The body region of a MOS transistor has definable resistance R_{body}. Owing to gate and body interconnection, an added parasitic resistance and capacitance associated with the body region comes into picture and is shown as R_{body} and C_{body} respectively.

Thus, DBB technique results into following modified parameters:

• The effective input capacitance of transistor becomes.

Figure 2. Model of MOS transistor using DBB technique

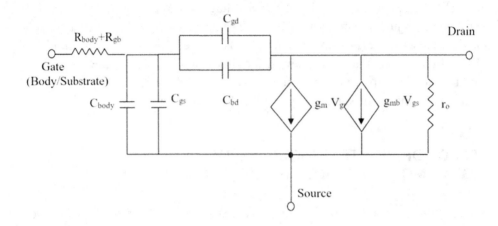

$$C_{DBB} = C_{gs} + C_{body} \tag{12}$$

- The effective input resistance of transistor becomes.

$$R_{DBB} = R_{gb} + R_{body} \tag{13}$$

where gate-body contact resistance is denoted as R_{gb} and it has ohmic nature having a very small value. The body contact provides a resistive path to the transistor for discharging/ charging of the body. Body resistance R_{body} have an effect on the frequency behaviour of RF/analog circuits as it forms a pole with body capacitance affecting parameters like gain, bandwidth etc.

- The effective transconductance of transistor becomes.

$$G_{m(DBB)} = g_m + g_{mb} \tag{14}$$

Thus, DBB technique strengthen the transconductance and increases driving current of MOS transistor due to dynamic scaling down of the threshold voltage. For every input, $V_{GS} = V_{BS}$ and drain current in strong inversion is given by

$$I_{DBB} = \frac{\mu_n C_{ox}}{2} \frac{W}{L} \left[V_{GS} - \left\{ V_{T0} + \gamma \left(\sqrt{|\psi_s - V_{GS}|} - \sqrt{|\psi_s|} \right) \right\} \right]^2 \tag{15}$$

where all symbols in the expression have their standard meaning.

- DBB technique modifies the unity gain frequency as

$$f_{T(DBB)} = \frac{g_{m(DBB)}}{2\pi C_{DBB}} = \frac{g_m + g_{mb}}{2\pi \left(C_{gs} + C_{body} \right)} \tag{16}$$

- DBB technique modifies the spectral density of input-referred noise power as

$$v_{noise(DBB)}^2 \left(f \right) = \frac{i_{ni}^2}{\left[g_{m(DBB)} \right]^2} = \frac{i_{ni}^2}{\left(g_m + g_{mb} \right)^2} \tag{17}$$

where i_{ni}^2 represents the total drain current produced by all noise sources. Thus, spectral density of input-referred noise power is lowered due to higher effective transconductance resulting from DBB technique (Chaudhry et al., 2015; Deo et al., 2014; Garg et al., 2014). In the next section we will study about the fabrication process used for implementation of DBB technique in a MOS transistor.

Effect of Mobile Charge Carriers in The Depletion Region

As deliberated in the previous section, the traditional MOS transistor model is used to predict the effect of DBB technique on various parameters of MOS transistor. However, the conventional model is based on the presumption that the channel is relieved of mobile charge carriers i.e., total depletion approximation thus neglecting any domination of mobile charge carriers present within the "depletion layer".

DBB technique slightly forward biases the body-source junction of MOS transistor. As a result, few mobile carriers are present in source-body depletion region. These charge carriers may affect the threshold voltage. To gauge the effect of mobile charge carriers, the Pao Sah equations were extended to model the effects of DBB technique as "parallel combination of the MOS transistor and a parasitic lateral bipolar junction transistor". This also justifies the enhanced current driving capability of DBB technique (Niranjan, 2017; Niranjan et al., 2017; Soni et al., 2017). Based on this assessment, it was concluded that for low forward biases (< 0.6 V) with negligible short channel effects, conventional SPICE MOS model is valid for simulating circuit design of DBB technique since the presence of minority charge carriers at low forward biases are negligible.

TRIPLE WELL CMOS PROCESS
TECHNOLOGY FOR DBB TECHNIQUE

In conventional bulk-CMOS process technology, the p type substrate is common for all n type i.e NMOS transistors. As a result, n type transistors cannot be designed to operate as four terminal devices because of their shared jointly shared connection with p type substrate. Therefore, implementation of DBB technique in MOS transistor requires supplementary implant layer in fabrication process to gain access to the body terminal of individual MOS transistor. This typical fabrication process is known as triple-well CMOS technology. Using triple-well structure, DBB technique can be put into effect in both n type and p type MOS transistors in a circuit. Thus, DBB technique offers many advantages but at the expenses of larger die area due to the usage of deep n-well process.

The working principle of triple well process is demonstrated in Fig.3 by a more intelligible cross-section of MOS transistors using DBB technique with terminals source S, shorted gate and body G (B) and drain D. A p well is formed in a layer represented as deep n well, which in turn is fabricated on a p type substrate. Triple well technology is slightly high budget in contrast to conventional CMOS process technology because an additional well is used for p type and n type MOS transistors which enables independent bias/potential control knob on the body/bulk terminal of each MOS transistors. The main primacy of fabricating both types of MOS transistors in distinct wells is that extra thick deep N well layer isolates each p well from the common p-substrate and thus stays clear of any surplus current arising due to conduction of parasitic diodes. The deep N well layer also separates the resistive path from any digital noise source on chip into the RF/analog circuits. This is particularly advantageous when using distinct V_{DD} and V_{SS} for the digital/analog circuits. It is interesting to mention here that triple well technology qualifies the prospect of biasing both the p substrate and the p well independently.

Although triple well process involves added fabrication cost, but it is more sturdy to process and parasitic junction capacitance alterations. This structure is also effective in reducing crosstalk and substrate noise in circuits. In addition, the p-n junction capacitance related with triple well anatomy works as a dissociating capacitance and adds to diminution of noise generated in the power supply. Thus, total noise alleviation in a triple well structure is far more superior than that of a twin well process. The triple well process technology is more reliable due to reduction in latch up conditions (Chaudhry et al., 2014; Niranjan et al., 2013b; Niranjan et al., 2014c).

Figure 3. Implementation of DBB technique utilizing triple well CMOS process technology

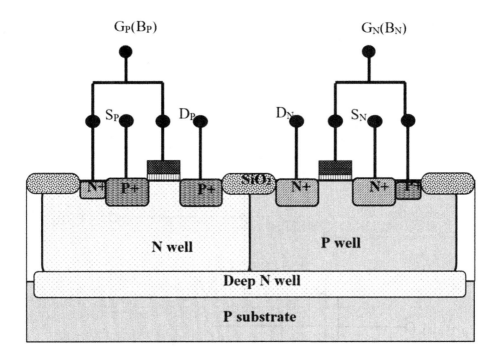

MERITS OF DBB TECHNIQUE FOR DESIGN OF LOW VOLTAGE ANALOG CIRCUITS

As already mentioned in earlier section, the threshold voltage of MOS transistor is not anticipated to scale down more than what is obtainable today. Therefore, analog designers face many challenges and limitations. This has paved way too many non-traditional low voltage realization techniques such as quasi floating gate, floating gate and body/bulk driven technique to push the power supply voltage in the vicinity of threshold voltage of the MOS transistor. But these techniques have drawbacks such as complex circuit structure, low transition frequency, low transconductance, higher noise and require more chip area. In this section, merits of dynamic body bias technique over bulk driven and gate driven techniques have been explained.

A traditional gate-driven technique is shown in Fig.4 showing transistor M_1 biased by employing power supply voltage V_{DD} and biasing current I_B with body terminal at ground potential. As the input signal is put in at the gate terminal therefore only gate transconductance (g_m) adds up to the drain current.

Figure 4. Gate-driven technique

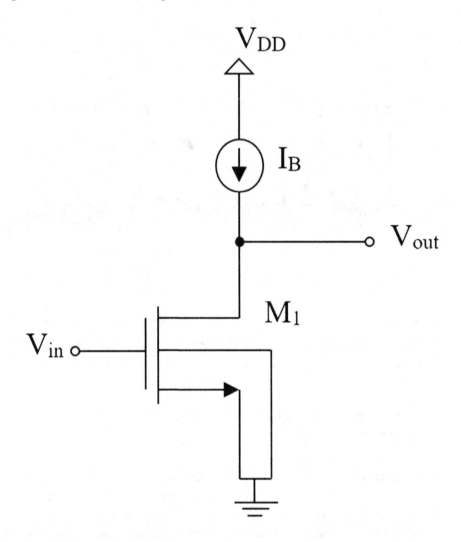

A body-driven technique suitable for low voltage design environment is shown in Fig.5. Here, input signal is put in at the body/bulk terminal of the MOS transistor and appropriate bias/DC voltage is also put in at the gate terminal to retain the transistor in the conducting state. As we see that the gate terminal is at ground potential for the input signal in body-driven transistor therefore, only bulk transconductance (g_{mb}) is able to contribute to conduction current.

In DBB technique both body and gate terminals are connected simultaneously and used as signal input terminal. As input signal gets applied to the body as well as gate terminal jointly at the same instance therefore both g_{mb} and g_m contributes

Figure 5. Body-driven technique

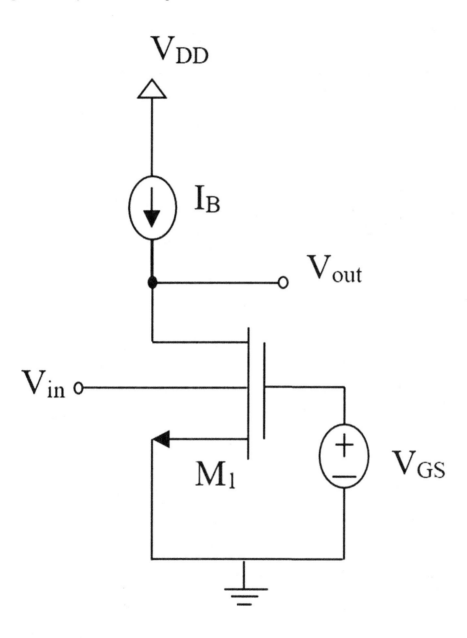

to the drain current in a MOS transistor. Thus dynamic body bias technique can be viewed as amalgamation of bulk/body driven and gate driven techniques for equitable comparison (Niranjan et al., 2014d; Niranjan et al., 2015; Singh, Chaudhry, Niranjan et al, 2015), summarized in Table.1.

Table 1. Comparison of DBB technique with gate driven and bulk driven techniques

In summary, DBB technique has lower input referred noise, higher effective transconductance, lower threshold voltage and higher transition frequency as compared to gate driven and bulk driven techniques. For analog circuit design, DBB technique offers various advantages such as	• Voltage headroom is a term used to quantify about how close the input signal and output signal of an amplifier can swing to the supply rails. In low voltage design environment there is not enough headroom voltage for signal swing and thereby decreasing the threshold voltage can be beneficial. Therefore, using the DBB technique, it is feasible to enlarge the range of input voltage in a circuit.	• As source-body junction is forward biased, transversal electric field reduces thereby decreasing the degeneration of the carrier mobility. It is quite evident that higher mobility leads to an enhanced current drive capability which is also helpful in designing circuits for high speed applications.	• DBB technique clearly manifest higher cutoff frequency and lower noise which is helpful in designing circuits for low noise and wide band applications.	• Generally the MOS transistor is biased in the saturation region for analog circuit designed to process incremental input signals. DBB technique saturates drain current rapidly due to minute channel length modulation effects.
• DBB technique improves the intrinsic gain of MOS transistor. This is mainly due to fact the output resistance improves due to repression of short channel effects resulting in efficacious gate control of the channel charge.	• Implementation of DBB technique in analog circuits is easily feasible since it out-turn in a simpler circuit without requirement of any auxiliary circuitry for generation of bias voltage.	• DBB technique is also suitable for design of subthreshold circuits since it has ideal subthreshold characteristics. The subthreshold swing is significant criterion in weak inversion modeling of MOS transistor, principally for low voltage high gain analog applications and imaging circuits.	• Turn ON performance of MOS transistor improves as the value of the swing decreases. Using DBB technique, the ideal subthreshold swing of 60 mV/ dec can be realized.	Although DBB technique has several advantages, it also has certain limitations, namely.
• The body capacitance (C_{body}) slightly degrades the unity gain frequency of transistor but higher transconductance value compensates to small extent.	• Drain-body capacitor forms a miller capacitor therefore use of DBB technique is an important issue in the design of RF/ analog circuits.	• Body resistance can be lowered further by using large amount of contacts since body resistance deteriorate the frequency response and adds up to noise.	• The available range for variations in input signal (generally less than 600 mV) is inadequate to limit large parasitic currents inside body region of the transistor would also degenerate the circuit performance.	• Hot carrier reliability is an important issue in DBB technique. 'Hot carriers' make reference to mobile charge carriers in the substrate/body of a MOS transistor that have energies sufficiently above mean. Hot carriers are present due to high intensity fields in the drain region of transistor. They compromises the working of transistor by degrading the oxide and the Si-SiO2 interface and creating charged defects in the oxide layer. These effects add up to reliability problems in DBB technique.
Thus, DBB technique is suitable in the design of low power and low voltage analog circuits to acquire higher output swing, large dynamic range such as rail-to-rail voltage range and higher overdrive voltage.	**Applications of DBB Technique in Analog Circuits**	DBB technique has been extensively used in digital circuits since 1994 but only recently it has attracted attention in analog circuits. In the following section we will study about various analog applications using DBB technique reported in published literature. In the published literature many authors have used the term dynamic threshold MOS (DTMOS) as short notation for MOS transistor implemented with dynamic body bias (DBB) technique.	Analog circuits such as analog/ digital converters necessitate voltage references. The bandgap voltage reference must exhibit both low temperature coefficient and high-power supply rejection being the most accepted high quality voltage reference used in IC industry today. However, chip design is now mostly dominated by low voltage and low power intents. With this endeavour to achieve, a low voltage low power bandgap reference circuit has been first reported using DBB technique in the year 1999 by authors Anne-Johan. This circuit operates at low supply voltage of 0.85 V and at the same time-consuming small power about 1 µW.	In the year 2006, authors Maymandi nejad et al. (Maymandi-Nejad & Sachdev, 2006) proposed a low voltage differential amplifier with a common mode feedback (CMFB) circuit and a low voltage comparator using DBB technique. In the case of feedback circuit, this technique reduces circuit intricacy without consuming any additional power. Similarly, the comparator achieves rail to rail input range.

Authors Achigui et al. in the year 2007, designed a low voltage differential class

AB opamp using DBB technique. The proposed operational amplifier makes use of this technique for minimizing the threshold voltage requirements of transistors and to get larger overdrive voltage. DBB technique has strengthen the input common mode range and further resulting in better matching of transistors in the input stage of differential. The proposed operational amplifier is able to operate at low supply voltage of 1 V and therefore is more suitable for low noise and low power applications (Achigui et al., 2007).

A second-generation current conveyor (CCII) circuit working at supply voltage of 0.4 V in the subthreshold region has been proposed by the author Uygur. This circuit consumes total power of only 214 nW and the outcome of using DBB technique is a very compact circuit topology. Recently in the year 2014, a voltage differencing transconductance amplifier been reported in the literature, operating in the subthreshold region. DBB technique in the amplifier entitles ultra-low voltage operation under very low symmetric supply voltage of $\pm$ 0.2 V (Uygur & Kuntman, 2014).

The voltage follower using DBB technique, proposed in (Kanjanop et al., 2011) can drive $\pm$ 0.25 V to the resistive load of 500 Ω. A low harmonic distortion of 0.4% is achieved at the operating frequency of 1 MHz. The proposed circuit is designed using complementary source follower with a common-source output stage. The power dissipation and bandwidth of the voltage follower are 103 μW and 288 MHz respectively. Similarly, a recent reported application reported in literature (Kanjanop & Kasemsuwan, 2011) is class AB current differencing buffered amplifier. Two current mirrors form the basis of this circuit and voltage follower is also included in which DBB technique allows low voltage operation at supply voltage of 0.7 V. Another circuit utilizing class AB current mirror configuration with a common source output stage is reported. The proposed circuit has been designed using a 0.13 μm CMOS process technology and use of DBB technique enables operation at low supply voltage of 0.7 V.

Authors Ehsan et al. (Kargaran et al., 2012) have proposed Operational Transconductance Amplifier (OTA) utilizing weak inversion operating region of MOS transistors, for biomedical applications. This circuit consumes only 386 nW of power. The reported OTA uses DBB technique to ensure low voltage operation at supply voltage of 0.4 V. Further, right half plane (RHP) zero controlling techniques have improved its unity gain-bandwidth. Many OTA amplifiers and operational amplifiers have been proposed in literature (Khosrojerdi et al., 2013; Liu et al., 2014; Suadet et al., 2011). It is interesting to note that in most of the applications, DBB technique has been used in the differential input stage to enable low power and low voltage operation.

Recently in the year 2015, authors V. Niranjan et.al have enhanced the performance parameters of many basic building blocks of analog integrated circuits at low power voltage of 1V. The performance parameters such as intrinsic gain and output impedance of self cascode structure has been improved using DBB technique. Further, the bandwidth of current mirrors for cascode and self-biased cascode configuration has been increased using DBB technique by a factor of 2. Also, a new four quadrant analog multiplier has been reported using dynamic body biased MOS transistors. The improved multiplier is able to operate in a wide frequency range of about 250 MHz. High linearity at low supply voltage that too consuming very small power is achieved using DBB technique. This technique has slightest overhead in terms of circuit complexity and power consumption i.e. required circuitry for bias voltage generation. The performance parameters of flipped voltage follower (FVF) has been enhanced using dynamic body bias technique. This bandwidth extension ration (BWER) of 2.6 is achieved without degrading the input and output impedance (Garg & Niranjan, 2015; Niranjan, 2015; Prerna & Niranjan, 2015; Singh, Niranjan, & Kumar, 2015).

Authors Basak et.al in the year 2016 and year 2017, proposed voltage differencing current conveyor and voltage differencing buffered amplifier using DBB technique. The reported circuits can operate at an ultra-low supply voltage 0.4 V (Basak & Kaçar, 2016; Basak & Kaçar, 2017). In the year 2017, authors Jooq et.al reported two stage OTA in 0.18µm CMOS technology. This amplifier utilizes DBB technique to provide a low voltage power supply and low power consumption amplifier. The proposed amplifier operates with 0.4V power supply and consumes only 163nW with 82.77dB open loop gain and 243 KHz unity-gain bandwidth (Jooq et al., 2017).

Authors Asli et.al in the year 2018, designed rectifier for energy scavenging to convert RF signals into electricity in 130 µm process technology. Due to DBB technique, the reported rectifier attains dynamically controlled threshold voltage. This technique reduces leakage current in the transistors in differential drive topology and at the same time also decreases the threshold voltage resulting in improved sensitivity of the rectifier (Asli & Wong, 2018). Recently many other applications of DBB technique in Analog domain such as filter, opamps, analog squarer, single even mitigation etc have also been reported by researchers (Alaybeyoglu & Ozenli, 2022; Keleş & Keleş, 2023; Konal & Kacar, 2021; Liu, 2020; Yildirim, 2021).

CONCLUSION

DBB technique proffers good panorama as a low voltage analog technique as it furnishes many improvements in small signal parameters for analog accomplishment. In spite of the fact that body-source junction parasitic diode get marginally forward biased but any considerable conducting p-n junction current can be banished by ceiling the gate input to approximately one diode voltage. DBB technique reduces the threshold voltage by more than 25%. Due to the decrease in lower effective normal field and depletion region charge in the channel, the mobility of charge carriers in DBB technique is higher than in gate-driven technique. Further, as the potential in the channel region is forcibly dominated by two gates i.e., the gate and body terminals, it results in higher transconductance due to speedy current transport. The DBB technique improves the transconductance of MOS transistor by 30% to 40%.

Apart from this, DBB technique offers predominant command over the short channel effects such as hot carrier effects, channel length modulation and drain induced barrier lowering. These phenomena affects the reliability and stability of small geometry MOS transistors. Therefore, output resistance and the intrinsic gain of a MOS transistor can be enhanced using DBB technique owing to curtailed channel length modulation effects. As input signa/voltage is applied directly to the body terminal, DBB technique has preferred attributes for design of subthreshold analog circuits.

DBB technique has succeeded in dealing with the variations in transconductance and drain current due to the process, temperature and voltage deviations. The performance deviations across the process corners has been reduced to as low as 12%. Therefore DBB technique is very effective in desensitizing the circuit against the process, temperature and voltage variations. The slow and steady increase of environmental electromagnetic pollution has led to constant rise in the level of RF interference. In present day integrated circuits, these interfering signals superimpose onto nominal desired signals. As a result, a high immunity to electromagnetic interference has accordingly became an obligatory requirement for low voltage integrated circuits. DBB technique results in electromagnetic interference immune designs. DBB technique enhances the maximum oscillation and cutoff frequency of a MOS transistor. Noise has been reduced to more than 30 times, thus making it ideal contender for low noise RF applications. There is not much temperature dependence of parameters in DBB technique. The cutoff frequency has positive temperature coefficient for low voltage applications (< 0.6 V) and transconductance improves with increase in temperature.

It is possible to implement DBB technique in n type and p type MOS transistors by using triple well option in CMOS process technology. Possibility of Triple well option contributes to isolation of noise sensitive n type MOS transistors in RF/ analog circuits from digital n-type MOS transistors that altogether release significant current to the substrate and thus generate "body/substrate noise". Thus, triple well alternative is instrumental for mixed signal circuits as accelerating integration densities can result in more complicated systems on chip. It is somewhat safe to use DBB technique for implementation of low voltage applications more than other applications. At low voltage levels, concentrations of mobile carrier do not increase to high levels in modern heavily doped substrates, resulting in high turn on voltage of parasitic diodes. It is of prime importance to record that compact model such as BSIM3V3 is yet logical and valid for forward bias less than 0.6 V at body source junctions, surprisingly those models presume total depletion approximation. Under these low voltage levels, free charge carriers' concentration in the channel are so inconsiderable in number that their consequences need no attention. Therefore, BSIM3V3 model can be agreeably used for circuit simulation. Thus, it is concluded that in order to get higher output swing, wide dynamic range and higher overdrive voltage, DBB technique is feasible solution in design of low voltage analog circuits.

REFERENCES

Achigui, H. F., Sawan, M., & Fayomi, C. J. B. (2007). 1 V fully balanced differential amplifiers: Implementation and experimental results. *Analog Integrated Circuits and Signal Processing*, *53*(1), 19–25. doi:10.100710470-006-9002-z

Alaybeyoglu, E., & Ozenli, D. (2022). Operational Amplifier Design Employing DTMOS Technique with Dual Supply Voltages. *Journal of Circuits, Systems and ComputersVol.*, *31*(02), 2250035. doi:10.1142/S0218126622500359

Asli, A., & Wong, Y. C. (2018). −31 dBm sensitivity high efficiency rectifier for energy scavenging. *International Journal of Electronics and Communication*, *91*, 44–54. doi:10.1016/j.aeue.2018.04.019

Basak, M. E., & Kaçar, F. (2016). Ultra-Low Voltage VDCC Design by Using DTMOS. *Acta Physica Polonica*, *130*(1), 223–225. doi:10.12693/APhysPolA.130.223

Basak, M. E., & Kaçar, F. (2017). Ultra-Low Voltage VDBA Design By Using PMOS DTMOS Transistors. *Istanbul University-Journal of Electrical & Electronics Engineering*, *17*(2), 3463–3469.

Chaudhry, A., Niranjan, V., & Kumar, A. (2014). Bandwidth Extension of Analog Multiplier using Dynamic Threshold MOS Transistor. *Proceeding IEEE International Conference on Reliability, Infocom technologies and optimization*. 10.1109/ICRITO.2014.7014682

Chaudhry, A., Niranjan, V., & Kumar, A. (2015). Wideband Analog Multiplier Using DTMOS And Self Cascode Current Mirror. *Proceeding IEEE International Conference on Reliability, Infocom technologies and optimization*.

Chawla, C., Niranjan, V., & Chopra, V. (2010). Comparative study between Dynamic threshold MOSFET and Conventional MOSFET. *Proceeding National conference on design &. communication technology*.

Deo, A., Niranjan, V., & Kumar, A. (2014). Improving gain of Class-E amplifier using DTMOS for biomedical devices. *Proceeding IEEE International Conference on Medical Imaging, m-health & Emerging Communications Systems.* 10.1109/MedCom.2014.7005564

Garg, S., Chaudhary, G., Niranjan, V., & Kumar, A. (2014). Bandwidth Extension of Voltage Follower using DTMOS transistor. *Proceeding IEEE International Conference on Innovative Applications of Computational Intelligence on Power, Energy and Controls with their impact on Humanity.* 10.1109/CIPECH.2014.7018211

Garg, S., & Niranjan, V. (2015). DTMOS transistor with self-cascode subcircuit for achieving high bandwidth in analog applications. *International Journal of Computers and Applications, 127*(11), 19–31. doi:10.5120/ijca2015906538

Jhajharia, H., & Niranjan, V. (2016). Exploiting Body Effect to Improve the Performance of Amplifier. *International Journal of Electronics. Electrical and Computational System, 5*(11), 9–14.

Jooq, M. K. Q., Miralaei, M., & Ramezani, A. (2017). Post-Layout Simulation of an Ultra-Low-Power OTA Using DTMOS Input Differential Pair. *International Journal of Electronics Letters., 6*(2), 168–180. doi:10.1080/21681724.2017.1335782

Kanjanop, A., & Kasemsuwan, V. (2011). Low voltage class AB current differencing buffered amplifier (CDBA). *International Symposium on Intelligent Signal Processing and Communications Systems*, 1-5. 10.1109/ISPACS.2011.6146107

Kanjanop, A., Suadet, A., Singhanath, P., Thongleam, T., Kuankid, S., & Kasemsuwan, V. (2011). An ultra low voltage rail-to-rail DTMOS voltage follower. *Proceedings International Conference on Modeling, Simulation and Applied Optimization*, 1-5. 10.1109/ICMSAO.2011.5775534

Kargaran, E., Sawan, M., Mafinezhad, K., & Nabovati, H. (2012). Design of 0.4V, 386nW OTA using DTMOS technique for biomedical applications. *Proceedings IEEE International Midwest Symposium on Circuits and Systems*, 270-273. 10.1109/MWSCAS.2012.6292009

Keleş, S., & Keleş, F. (2023). Low Voltage-Low Power Wide Range FGMOS Fully Differential Difference Current Conveyor And Application Examples. *International Journal of Electronics*, 1–16. doi:10.1080/00207217.2022.2164079

Khosrojerdi, ARezvani, RPourandoost, A. (2013). 0.8 V 191.9 nW DTMOS Current Mirror OTA in 0.18 μm CMOS Process. *Majlesi Journal of Telecommunication Devices, 2*(3), 251–254.

Konal, M., & Kacar, F. (2021). DTMOS based low-voltage low-power all-pass filter. *Analog Integr Circ Sig Process*, *108*(1), 173–179. doi:10.100710470-021-01878-z

Liu, J. (2020). Current mirror featuring DTMOS for analog single-event transient mitigation in space application. *Semiconductor Science and Technology, 35*(8).

Liu, J., Han, Y., Xie, L., Wang, Y., & Wen, G. (2014). A 1-V DTMOS-Based fully differential telescopic OTA. *IEEE Asia Pacific Conference on Circuits and Systems (APCCAS)*, 49-52. 10.1109/APCCAS.2014.7032716

Matej, R., Viera, S., & Daniel, A. (2018). Design techniques for low-voltage analog integrated circuits. *Journal of Electrical Engineering*, *68*(4), 245–255.

Maymandi-Nejad & Sachdev. (2006). DTMOS technique for low-voltage analog circuits. *IEEE Transactions on VLSI Systems*, *14*(10), 1151-1156.

Niranjan, V. (2015). *Performance Improvement Of Low Voltage CMOS Circuits Using Body Bias Approach*. PhD Thesis.

Niranjan, V. (2017). Wideband current mirror using transconductance boosting technique. *International Journal of Advance Research in Science and Engineering*, *6*(12), 1159–1171.

Niranjan, V., & Gupta, M. (2009). Low voltage four quadrant analog multiplier using dynamic threshold MOS transistors. *Microelectronics International*, *26*(1), 47–52. doi:10.1108/13565360910923179

Niranjan, V., & Gupta, M. (2011). Body Biasing-A circuit level approach to reduce leakage in Low power CMOS circuits. *Journal of Active and Passive Electronic Devices.*, *6*(1-2), 89–99.

Niranjan, V., Gupta, M., & Jain, S. B. (2012). A Novel 0.5 Volt Analog Multiplier using dynamic body bias technique. *Proceeding National Conference on Advanced VLSI and Embedded Technology*.

Niranjan, V., Kumar, A., & Jain, S. B. (2013a). Low Voltage Flipped Voltage Follower based Current Mirror using DTMOS Technique. *Proceeding IEEE International Conference on Multimedia, Signal Processing and Communication Technologies*, 250-254. 10.1109/MSPCT.2013.6782129

Niranjan, V., Kumar, A., & Jain, S. B. (2013b). Triple Well Subthreshold CMOS Logic Using Body-bias Technique. *Proceeding IEEE International Conference on Signal Processing, Computing and Control*, 1-6. 10.1109/ISPCC.2013.6663447

Niranjan, V., Kumar, A., & Jain, S. B. (2014a). Maximum bandwidth enhancement of current mirror using series-resistor and dynamic body bias technique. *Wuxiandian Gongcheng, 23*(3), 922–930.

Niranjan, V., Kumar, A., & Jain, S.B. (2014b). Composite transistor cell using dynamic body bias for high gain and low-voltage applications. *Journal of Circuits, Systems, and Computers, 23*(8), 1-18.

Niranjan, V., Kumar, A., & Jain, S. B. (2014c). Low Voltage Self cascode amplifier using dynamic body bias technique. *Proceeding International Conference on VLSI and Signal Processing.*

Niranjan, V., Kumar, A., & Jain, S. B. (2014d). Low-voltage and High-speed Flipped Voltage Follower Using DTMOS transistor. *Proceeding IEEE International Conference on Signal Propagation and Computer technology*, 145-150. 10.1109/ICSPCT.2014.6884882

Niranjan, V., Kumar, A., & Jain, S. B. (2015). Low-voltage gate and body driven self-biased cascode current mirror with enhanced bandwidth. *International Journal of Circuits and Architecture Design, 1*(4), 320–342. doi:10.1504/IJCAD.2015.072615

Niranjan, V., Kumar, A., & Jain, S. B. (2017). Improving Bandwidth of Flipped Voltage Follower Using Gate-Body Driven Technique. *Journal of Engineering Science and Technology, 12*(1), 83–102.

Niranjan, V., Singh, A., & Kumar, A. (2014). Dynamic Threshold MOS transistor for Low Voltage Analog Circuits. *Proceeding International Conference on Recent Trends & Issues in Engineering and Technology.*

Prerna & Niranjan, V. (2015). Analog Multiplier Using DTMOS-CCII Suitable for Biomedical Application. *Proceeding IEEE International Conference on Computing, Communication and Automation.*

Singh, A., Chaudhry, A., Niranjan, V., & Kumar, A. (2015). Improving Gain Bandwidth Product Using Negative Resistance And DTMOS Technique. *Proceeding 39th National Systems Conference.* 10.1109/NATSYS.2015.7489087

Singh, A., Niranjan, V., & Kumar, A. (2015). A novel technique to achieve high bandwidth at low supply voltage. *Proceeding IEEE International Conference on Computational intelligence and communication technology.* 10.1109/CICT.2015.48

Soni, S., Niranjan, V., & Kumar, A. (2017). High gain analog cell using biasing technique via gate and body terminals. *Proceeding IEEE International conference on Recent Innovations in signal processing and Embedded Systems.* 10.1109/RISE.2017.8378187

Suadet, A., Thongleam, T., Kanjanop, A., Singhanath, P., Hirunsing, B., Chuenta, W., & Kasemsuwan, V. (2011). A 0.8 V class-AB linear OTA using DTMOS for high-frequency applications. *Proceedings International Conference on Modeling, Simulation and Applied Optimization,* 1-5. 10.1109/ICMSAO.2011.5775478

Tsividis, Y. (1999). *Operation and Modelling of the MOS Transistor* (2nd ed.). McGraw-Hill.

Uygur, A., & Kuntman, H. (2014). A very compact, 0.4 V DTMOS CCII employed in an audio-frequency filter. *Analog Integrated Circuits and Signal Processing,* *81*(1), 89–98. doi:10.100710470-014-0365-2

Viera, S., Matej, R., Martin, K., Daniel, A., Lukas, N., Michal, S., & Miroslav, P. (2018). Ultra-Low Voltage Analog IC Design: Challenges, Methods and Examples. *Wuxiandian Gongcheng,* 27(1), 171–185.

Yildirim, M. (2021). Design of Low-Voltage and Low-Power DTMOS Based Analog Multiplier Utilizing Current Squarer. *International Journal of Electronics Letters,* 9(1), 1–13. doi:10.1080/21681724.2021.1889041

Chapter 5
A Detailed Study on Single Electron Transistors in Nano Device Technologies

S. Darwin

https://orcid.org/0000-0002-4316-1459
Dr. Sivanthi Aditanar College of Engineering, India

E. Fantin Irudaya Raj

https://orcid.org/0000-0003-2051-3383
Dr. Sivanthi Aditanar College of Engineering, India

M. Appadurai
Dr. Sivanthi Aditanar College of Engineering, India

M. Chithambara Thanu
Dr. Sivanthi Aditanar College of Engineering, India

ABSTRACT

The rapid advancement of integrated circuit (IC) technology in the recent decades paved the path for miniaturization of electronic devices. Nowadays all the handheld devices are battery operated, which moves the researchers to develop the devices with low power utilization, high-speed operating capability, and low cost. The advancement in technology scaling is crucial for enhancing the effectiveness of IC in the areas of latency, power dissipation, and signal processing. The chapter provides an outline of the history of nano electronic device development and emphasizes the potential of the single electron transistor (SET) as a new nano device that will eventually replace more traditional ones.

DOI: 10.4018/978-1-6684-4974-5.ch005

INTRODUCTION

Metal Oxide Semiconductor Field Effect Transistor (MOSFET) has been a fundamental component of all devices with computing platform and memory elements over the past 50 years. Continuous device geometry minimization, nowadays dipped into the range of micron, is mostly to enhance the efficiency of the device. In addition, limitations enforced by manufacturing processes and principles governing quantum physics prevent feature size decrease. The researchers are compelled by these limitations to investigate novel transistor substitutes for ultra-dense circuits. These novel technologies are referred to as nano-devices, and the science underlying them is referred to as nanotechnology (Khan, 2014; Zhang et al., 2018). The basic and fundamental element of Very Large-Scale Integration (VLSI) is MOSFET. The enhance speed, large package density of transistors and minimum power consumption are the requirements for an effective VLSI design. This can be achieved by already mentioned device geometry minimization but this leads to various shorter channel length effects in the device (Jacob et al., 2017; Krautschneider et al., 1997; Kuhn et al., 2008). Many nano-devices have been investigated to reduce these drawbacks and improve performance (Parekh, 2019). Depending on the operating concepts and fabrication methods, the devices are divided into three major types. They are carbon nanotube-based transistors, quantum solid state devices and molecular principle. Devices within the first segment resemble typical MOSFETs but differ from them in terms of size and composition because they are constructed of carbon nanotubes. The remaining categories are constructed independently but both involve quantum effects. The solid-state transistors utilize fabrication techniques comparable to, those utilized for MOSFETs. A cutting-edge strategy, molecular electronics needs new components and a novel operating concept. Quantum effects are taken advantage of by solid state quantum effect devices. A tiny island that confines the conducting charge in the sort of electrons is a crucial component shared by all of these circuits. This island is comparable to a MOSFET's channel. These electron confinement-based devices can be divided into two groups. One is quantum dots and the other single electron transistors (SET). The SET is a special kind of semiconductor switches that amplifies the current by using the process of controlled electron tunneling. The typical MOSFET works by allowing charge to move between both the drain and source electrodes under the supervision of the gate electrode. Millions of electric charge flow through the channel for current conduction in MOSFETs, which is detrimental since it results in thermal dissipation and power outages (Durrani, 2010; Mahapatra & Ionescu, 2005). The nanotechnology-based devices are divided into three major types as shown in fig.1. The solid-state mechanism utilizing quantum effects that confines the charge in the form of electrons as islands is a common essential future of this kind of nano-devices. The bifurcated device classification

Figure 1. Nanotechnology device classification

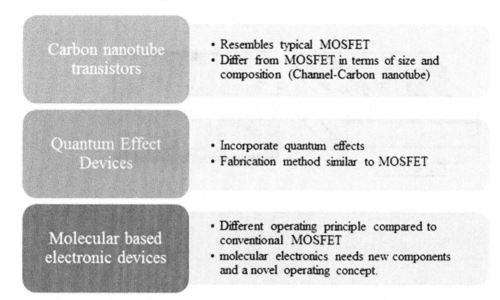

based on confinement of electrons are of quantum dots and single electron. The operation strategy of SET is evident by many researchers (Bounouar et al., 2012; Castro-Gonzalez & Sarmiento-Reyes, 2014; Chi et al., 2010; Gonza'lez et al., 2012; Neelakandan et al., 2022). The SET and CMOS technologies work best together. Hybrid SET-MOSFETs, which combine the best features among both devices, offer low power consumption, great density, and scalability potential.

The innovative SET device is extremely effective and can be utilized for a wide range of best to achieve applications, including charge sensors, infrared and detecting methodologies based on microwave, electrometers, memories, and logic architectures. Numerous studies linked to SET have been conducted until now (). The SET has the ability to function as a voltage-biased switch that turns on when the tunnelling current travels across it and come to off condition if the current not pass through. Classical physics implies that if a capacitor is specifically linked to a source of fixed voltage, no current can flow through it. But in accordance with quantum physics, the likelihood that an electron will travel through a small dielectric barrier, or a small thin capacitor is not zero. This implies the concept of quantum tunneling. The amount of current flow caused by electrons tunnelling is quite small. Consequently, tunnelling current is fairly low (in nanoampere). As a result, SET-based electronic circuits' voltage level demand is likewise relatively low. High chip compactness, fast throughput, and low power are required for VLSI designs. Scaling is an efficient way to achieve these. However, constant device scaling results in a

Figure 2. General schematic of SET

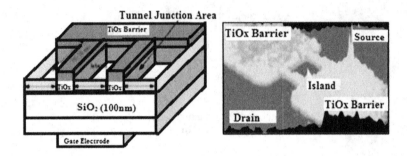

number of undesirable problems, including the short channel effect, increased power dissipation, significant leakage current, manufacturing uncertainty, and design dependability. By combining technologically sophisticated devices like SET, these can be alleviated.

In this paper, the theoretical investigation of SET is discussed in section 2. In section 3 the modeling aspects of SET is analyzed. The information about the fabrication process involved in SET is in section 4. The current applications of SET are discussed in section 5. Finally, in section 6 the conclusion is presented.

THEORETICAL POINT OF SET

The SET is a 3-point Single Electron Device (SED) that has a maximum operating speed and low consumption of power. Additionally, as size approaches the nanoscale, quantum mechanical processes begin to operate, which improves the effectiveness of SETs (Singh et al., 2012). The SET schematic is of a metal island in between two tunneling junctions, connecting the drain, source, and gate electrodes, just like a regular FET. The general schematic and the circuit (equivalent) diagram of SET is depicted in fig. 2 & 3.

The SET works by having electrons travel from the source (origin) to the island and then from the island to the destination that is drain through two junctions. The island creates tunnel connections at the island and uses a small insulator to isolate the source and drain contacts. As electrons traverse through the ultra-thin insulator tunneling contacts, they can be thought of as leaky capacitors. The gate controls the

Figure 3. SET Equivalent Circuit

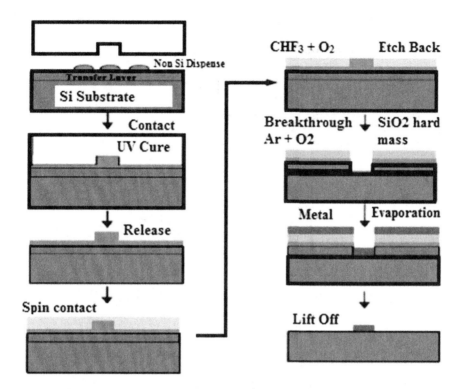

island's potential. Numerous gates could be used to realize the SET. The controlling gate and the adjusting gate are the two gates found in the most widely used SET device, although. These gates are connected to the island by means of a substantial dielectric layer. Either by eliminating or inserting an energy of electron larger than the energy induced by heat, the change in electrostatic energy is achieved, which then tracks the movement of electrons in and out of quantum dots. To put it another way, a quantum dot is a tiny conductive island with a variable number of electrons in specific orbitals. With the help of the gate electrode, the SET keeps track of the charge transfer between the drain/ source (origin). The electrical functionality of the device—whether it allows current to pass between the source and drain easily or not—depends on the movement of individual elementary charges and the voltage at the gate terminal. The source will send one electron tunneling to the island after the Coulomb blockade is broken, adding one extra surplus electron. According to that, there will be a tunneling procedure from the island to the drain.

According to the principles of quantum physics, the energy levels of SET devices are quantized and have electrons with finite number as dimension approaches zero.

Figure 4. SET Schematic

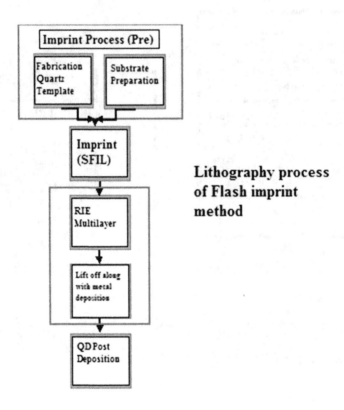

Every band level has a unique quantity of energy, and as an electron moves from one band levels to the other, each of which contains a different amount of energy, this either earns or destroys energy dependent upon whether it is travelling from a greater to a lesser or less to a high energy state. It has been demonstrated that this electron motion is caused by the tunnelling mechanism through a tunnel junction (Buxboim & Schiller, 2003; Goldhaber-Gordon et al., 1998; Kastner & Goldhaber-Gordon, 2001; Pushparaj et al., 2022). The SET is constructed with semiconductors, nanotubes (carbon) and metals in the size dimensions of nanometer level in order to provide better power consumption compared to other transistors. In SET, The conductive island is situated between two tunnel junctions that provide capacitance for a stored charge. The gate capacitance steers out the capacitance value of SET. A single transistor makes use of the feature that charging is defined by the barrier when the junction is adequately resistive. Fig. 4 shows the schematic structure of the SET.

The set of work associated with the construction of SET based on Coulomb energy, resistance in the tunnel region and excess electrons in the island region. A basic tunnel junction really acts as a barrier between electrodes that are naturally conductive by way of a thin strip of insulation. Insulating hurdles inhibit resulting from electron transport tunneling from the source to the island in compliance with the electrodynamics concept. This causes a rise in the electrostatic energy of the island, which is provided by

$$E_c = \frac{e^2}{2C} \tag{1}$$

Where the effective island capacitance is C. Coulomb blockade energy is another name for the electrostatic energy indicated by Ec. This energy is needed to make an island's existing electrons resist an incoming electron. The capacitance C of the island is quite low in the scenario of a small setup. Because of the extraordinarily high E_c predicted by the equation above, electrons cannot move simultaneously but rather pass one at a time. This condition is referred to as a "Coulomb blockage". Each of these two scenarios can reverse the restriction of transfer of electrons:

a)   The charge energy of Coulomb surpass $T \sim T_o = \dfrac{E_c}{K_B} = \dfrac{e}{2c}$ (2)

b)   the external force overcomes the charge of Coulomb that is $V \sim V_t = \dfrac{Ec}{e} = \dfrac{e}{2c}$

   (3)

Threshold voltage, or V_t, is referred to as a supplied voltage just high enough to push electron energy past the coulomb blockage of tunneling, allowing current to begin to flow via the tunnel junction. The fig. 5 depicts the circuit diagram of dual tunnel SET junction transistor. The degree to which the barrier conducts electron waves, the thickness of the barrier, and the ratio of the amount of wave modes that interact with the barrier at the tunnel juncture to the electrons' wavelength all affect the tunnel juncture electrical comparability.

Due to the fact that the tunnel junction consists of two number of conductors and an insulated coat between both the electrode (source or drain) and the island. As a result, the tunnel resistance R and tunnel capacitance C are used to define tunnel junctions. In this scenario, the junction (tunnel) serves as a capacitive and the insulation layer serves as the tunnel capacitor C's insulating medium. As an electron passes through the junction, a fundamental charge is added to the tunnel capacitance, creating a potential. The voltage created in the junction (tunnel) could be adequate to prohibit additional electron from tunneling with relatively low capacitance. If the voltage level in this case is less than the potential created in the junction, the electric

Figure 5. Dual tunnel SET junction circuit

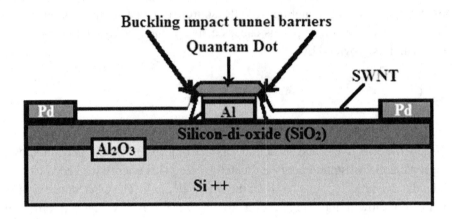

current will be inhibited. Thus, the coulomb blockade is defined as the increase in tunnel junction resistivity at zero voltage. Therefore, coulomb blockage may be described as a rise in junction resistance of an electronic system with at minimum one low capacitance tunnel junction at really small voltage levels. The Coulomb blockage is meet out only the following conditions are satisfied (Darwin et al., 2022; Joyez & Esteve, 1997; Korotkov, 1994; Lee et al., 2010; Zhu et al., 2009)

i)   The voltage level ought to be less than the expectation value multiplied by the island's self-capacitance.
ii)   The KBT temperature must be lower than the KBT charging energy.
iii) Heisenberg's uncertainty principle dictates that the tunneling resistance (R_T) must be larger than $\dfrac{h}{2 \blacklozenge e^2}$.

In case of excess electrons, an electron moving toward a sphere while experiencing a slight attractive force. When a single electron charges a sphere, additional electrons will experience a strong repulsive force. Through tunneling, electrons from the island are added to or subtracted from, positive or negative charging it. The island charged by additional electrons are called excess electrons (n). The system's electrostatic energy, which is reliant on the SET's charge energy, is impacted by the existence of extra electrons. Tunneling will be effectively prohibited and the Coulomb charges potential will function as a blockade if an additional surplus electron makes the system's energy rise. The defines the Coulomb blockade. Numerous intriguing devices are possible based on the Coulomb blockade effects, including highly accurate standards, extremely precise electrometers, logic circuits based on digital concepts

Figure 6. SET with current flow directions

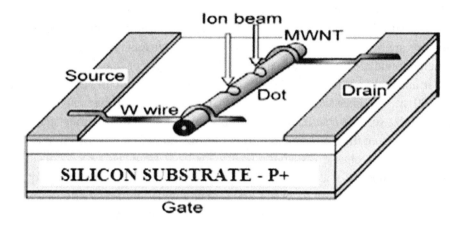

and memory with extremely low power usage. An electron will transit from the origin (source) to the island once the Coulomb blockade has been broken, providing one extra surplus electron. Similar tunneling will take place between the island and the drain. Fig. 6 depicts the circuit SET with the current flow direction. When compared to MOSFETs, where the drain current is produced by numerous electrons moving at once, the flow of electrons in dot/Island is due to Coulomb Blockade. A quantum dot is a mesoscopic setup in which the introduction or elimination of a single electron can result in an adjustment to the Coulomb energy larger than the thermal energy and govern the flow of electrons in and out the quantum dot.

The electrons' confined energy significantly contributes to the coulomb energy since their wavelength is directly proportional to the size of the dot. The island must be responsible for about 10 nm in size and have a capacitance with less than 10^{17} Farads.

VARIOUS MODELING ASPECTS OF SET

Substantial modelling is required for SET-based systems, which is done primarily using three primary methods. These include macro, Monte-Carlo and analytical modelling. Randomized tunnel timings are calculated for all potential occurrences and correlations in Monte-Carlo model. The generator of random numbers serves as the method's primary engine. Through a blocking layer that is subject to stochastic process, electrons are transferred. The simulation of big circuits takes a lot of time. The simulators based on Monte-Carlo model are SENECA, MOSES, KOSEC and SIMON (Fonseca et al., 1995; Miralaie et al., 2014; Wasshuber, 2001; Yu et al., 2000).

Macro modelling is another popular method. This method substitutes SET with a circuit made up of a combination of diodes, resistors, various sources, etc. In SPICE simulator settings, these can be effectively applied. The current-voltage equations are resolved by incorporating Kirchhoff's current and voltage rules. This method's foundation is entirely empirical and possibly not scalable. The model is further updated and modified to meet out the real world circuit simulations (Jain et al., 2015). The characteristics curves of SET are derived using analytical model technique by solving a series of equations. The circuit enters various states once the electron from the source-island-drain through tunneling. The exterior voltages and the circuit's charge distribution vary for each state. However, considering a limited number of states is the best approach for solving the master equation (Gnanasekar et al., 2021; Jia et al., 2004). The temperature restriction of SET is solved by the fullerene SET utilizing Quantum Dot arrays, which can function at ambient temperature. Analyzing the electrical properties of SET with aluminum island utilizing neural network. Modeling and analysis are done on the resistivity and quantum capacitance for SET utilizing carbon nanotubes (Abdelkrim, 2019; Khadem Hosseini et al., 2018).

The flexibility of the structure configuration and amazing accuracy of the Monte-Carlo model are provided. High computations, calculating times, and compatibility concerns with the well-known SPICE simulator are where this fall short. Therefore, it is only appropriate for SET structure research and the simulation of SET circuits. The macro approach offers quick calculations and has exceptional flexibility with current circuit simulators. This is appropriate for contextual examination of SET (large-scale) / CMOS-SET circuits of hybrid nature, although its accuracy and flexibility are rather constrained. Strong compatibility, great flexibility, and accuracy are supported by the analytical approach; however, the computation speed is constrained for complicated designs. This model has a balanced configuration of flexibility and speed and is appropriate for small- to moderate circuit simulation. Mahapatra - Ionescu Banerjee (MIB) model circuit design has the integration of CAD and SPICE to enhance the compilation process simpler.

In conclusion, it is believed that the best SET modelling may be achieved by combining master equations with macro models, in which the formula (empirical) of the macro model is helpful to simplify the principal equations. Additionally, a physical parameter driven, precise, and concise analytical model that facilitates development and design utilizing CAD tools is required for SET based complicated circuit designs.

Figure 7. SOI based SET fabrication procedure

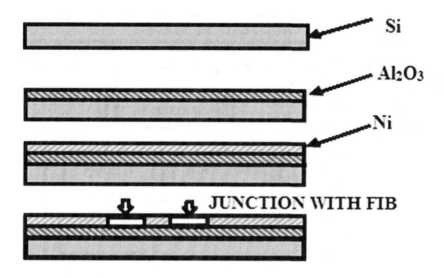

FABRICATION PROCESS IN SET

The fabrication processes can be divided into two categories: physical methods and chemical approaches. Thin film and lithographic technology are frequently combined in the physical approaches. Devices are available with carefully crafted electron densities and geometries. Recently, numerous wet chemical processes have been employed to create a variety of nanostructures. This method is advantageous since it is inexpensive and has good control over the size of the islands, and it may prove to be a useful one in the future. Although this method has not reached industrial maturity. There are numerous methods for creating SET, including STM/AFM, Electron Beam Lithography, and the creation of SET utilizing carbon nanotubes (CNT).

Lithography

Electron Beam Lithography (EBL) has often been utilized in the semiconductor sector to create master's masks and recreates from computer-generated image files for computer-aided design. The current semiconductor industry needs extremely tiny patterns (far finer than can be seen with the human eye) for integrated electronic circuits, and EBL is a specialized method for producing these patterns. The fabrication procedure for SOI SET transistors is shown in fig. 7. The SOI substrate in question has a submerged oxide layer that is 300nanometers and is situated between a capping with n-type Si layer that is of 50nanometers. The carrier density with Si capping is

Figure 8. Lithography (Electron beam) utilized SOI SET under Electron microscope

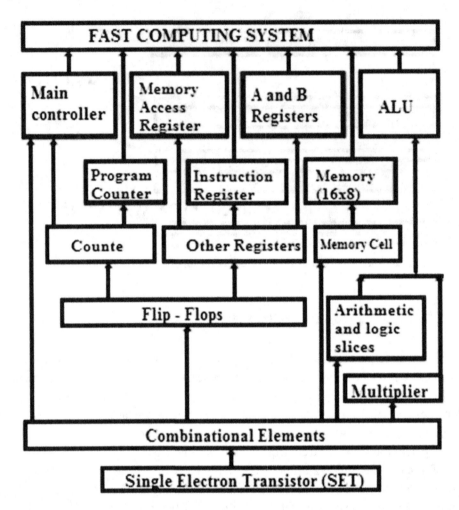

of 2.10^{19} cm³, while its resistance is 3.102 cm. Spin coating the SOI substrate with PMMA resists with bottom weight is of 100k along with top of 350k was done before the substrate was roasted for an hour at 180 degrees Celsius. The SET patterns were written using a highly focused lithography machine for electron beam.

Development was completed using a normal MIBK: IPA (1:3) at ambient temperature, accompanied by a 30-second IPA rinsing and air compressor drying. Reactive ion etching with fluorine-based plasma is the etching method is used. Al exhibits considerably higher lift off than Cr when Cr and Al are metalized, which involves in the metallization process of SOI SET.

Process Involved in Stm/Afm Nano-Oxidation

Recent advances in nanostructure production techniques employing the atomic force microscope (AFM) and Scanning Tunneling Microscope (STM) make them very appealing for use in electron and optical systems. The following are some instances of these nanostructure creation procedures. To use the air ambient STM, pattern and oxidizing of the hydro passivated Si and an approximate value of 35nm line is obtained. Due to its potential for extremely low functioning with just a few electrons, a single electron transistor (SET) is seen as a potential contender for a component of a potentially low-power, increased integrated circuit. Operating the SET at room temperature is essential for practical use. To achieve this, the SET island's area must be as tiny as 10 nm in order to lower the SET's overall capacitance and solve the thermal instability issues. However, the current conventional microfabrication technology cannot handle the 10-nm scale. With the use of a novel pattern generation technique based on the STM/AFM nano-oxidation process, we demonstrate the manufacturing of a SET. STM technology is used by SET in order to get around the control issues in self-organized systems. This method can be used to build a SET that works at room temperature. On a 100nm thermally oxidized SiO_2 per substrate is of n-type, a titanium (Ti) metal sheet with a thickness of 3 nm is formed. The Ti surface oxidizes as a result of the environment and moisture on the top (Matsumoto, 1998; Matsumoto et al., 1996; Sahatiya, 2013).

Step and Flash Imprint Lithography (SFIL)

A dual-layer imprint scheme-based technique for lithography is currently being researched. This method offers the ability to produce high aspect ratio, high-resolution designs at maximum throughput without the use of projection lenses. This method produces high-resolution relief pictures in the quartz by employing a normal mask blank as a design and patterned chromium as an etch mask. Spin-coating an organic transfer layer onto a silicon substrate is the SFIL technique. Over the coated silicon substrate, a surface-treated, transparent template holding relief structures of a circuit layout is precisely aligned. Reduced pressure, temperature, and Nanoimprint Lithography based on Ultra Violet technology is referred to as SFIL (UV-NIL). It has the advantage of preventing difficult modifications to the process parameters for multi-layer imprints caused by significant shifts in temperature and pressure. The capability of this method to be employed for large scale production is the second key benefit. The fig. 11 (a) and (b) depicts the clear view of the fabrication process and steps involved to make the SET with the help of process flow and chart representation.

Figure 9. (a) Nano oxidation process based on STM,(b) Atomic Force Microscopy (AFM) process of SET

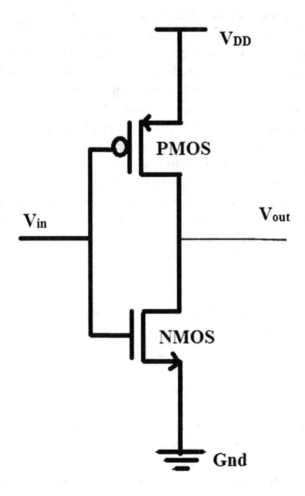

Carbon Nanotube with Single Walled Set Fabrication Process

With SET working temperature of about 100K, metallic grains and nanocrystals of colloidal nature produce smaller, more uniform dots. Furthermore, for SET functioning, these materials must be positioned in nanoscale gaps, which is very challenging to accomplish and results in a very low system yield. Due to their small diameters Single Walled Carbon Nanotubes (SWNTs) have recently been seen as viable options for SET manufacturing and connectivity.

The inclusion of tunnel barriers is necessary for the fabrication of SET utilizing SWNT. A SWNT has been proven to behave as a tunnel barrier of nanometer-sized

Figure 10. SET with titanium and ATM utilized SET

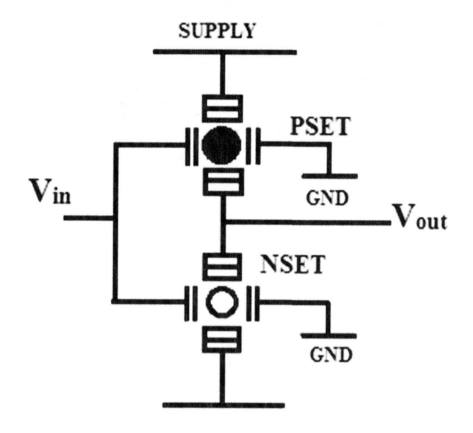

Figure 11. Detailed process flow of SFIL

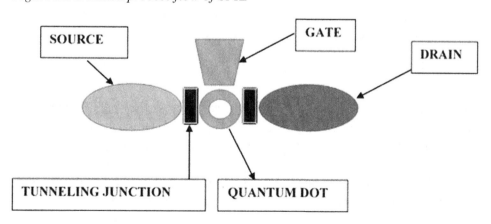

Figure 12. Flow diagram of SFIL

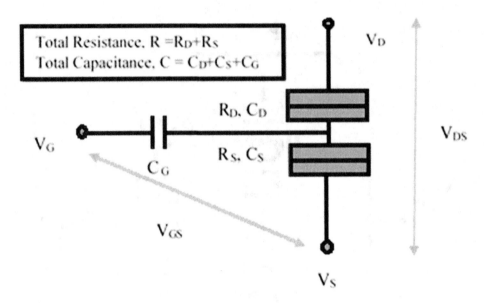

when bend at a certain location. A 100nanometers broad localized Al/Al2O3 surface (bottom) is coated with a SWNT. Pd source /drain electrodes with a 1 m spacing were placed in contact well with Si/Sio2 substrate's. The aluminum gate has three functions: it functions as a local bottom gate to regulate the behavior of the SET device, (ii) The quantum dot's size (L) is determined by the gate width and (iii) it specifies barrier with of the two tunnel at the corners by instinctually bend the nanotube to meet out the van der Walls relations with substrate. The oscillations are visible in low temperature electrical transport experiments up to 125K. The stability diagram displays energy level spacing of about 5 meV and charging energies of 12 to 15 meV. These energies confirm that the quantum dot is defined and governed by the localized gate and are consistent with a quantum dot dimension of around 100 nm. The fig. 12 depicts the nanotube set schematic structure.

Focused Ion Beam Based Cn-Set

The fig. 13 (a) and (b) depicts the schematic and process flow of FIB based SET fabrication process. The voltage is of 30kV and the Ga+ ion beam of reduced dimension value approximately 7nm. Ion-beam aided tungsten (W) deposition was carried out using $W(CO)_6$ gas for the nano interconnection to the CNT. The substrate was thermally generated oxide on a (100nm) highly doped p-silicon wafer. The fabricating procedure went like this. On the wafer, the 100nm Ti and Au of

Figure 13. Nanotube SET with Al/Al$_2$O$_3$ as gate to develop two tunnel barriers

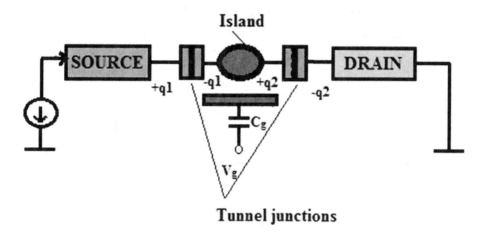

200nm pads associated with contact were first created using photolithography, then the deposition of metal layer and the upward lift procedure. As a gate electrode, the bottom metal surface was employed. Second, arc discharge-produced MWNTs were spread across the wafer. The Scanning Electron Microscopy was used to determine

Figure 14. Nanotube SET FIB based schematic

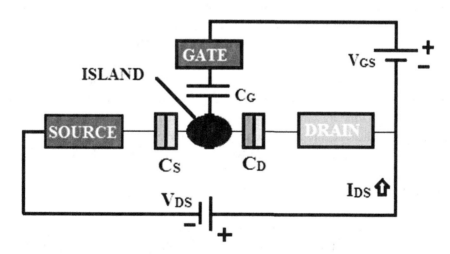

Figure 15. FIB based SET fabrication process

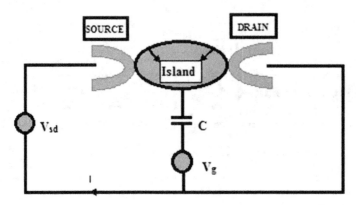

the locations of the MWNTs (SEM). After that, FIB-assisted deposition was used to deposit W nanowires from the MWNT's two ends to the contact pads. Finally, FIB etching created two trenches on the MWNT. The dimension aspect regarding ion beam of Ga+ and current factor is given as 18nm, 1.3×10^{18} cm^{-2}. Single electron transistors are created using both the high/low beam currents accessible on the FIB in a sequence of manufacturing stages. Higher currents are utilized to produce micro dimension structures, whilst lower currents are employed to produce nano dimension frameworks, which involve minimal energy doses.

Various research regarding the fabrication of SET device is analyzed and published. a SET fabrication techniques demonstration using a single, extremely tiny silicon QD linked to a gold broken interface with a nano - gap is exhibited. A SET that is integrated into a nano mechanical resonator and showed the maximum possible electron-phonon coupling. The construction of the SET using niobium and germanium quantum dots have introduced in the fabrication process. The nano damascene method for SET production is provided, along with information on how it integrates with high dielectric constant as a dielectric as gate stack MOSFETs and operates at room temperature (at V_{dd}=0.9V with cutting-edge nanowire enabling technology). Utilizing CMOS back-end-of-line (BEOL) appropriate technology, metal SET transistors are manufactured inside the chip interconnecting layers (Bai et al., 2018; Karbasian et al., 2015; Kim et al., 2003; Wen et al., 2020).

Design-Based Circuits in Set

The use of (FETs) is replaced by SETs in logic gates of the complementary type. It is noticed that the SET-only designs are substantially smaller in size and calls for comparatively fewer transistors (Dubuc et al., 2008). The basic logic gates with the

Table 1. Circuit implementation based on SET

Design	Number of transistors	Temperature	Power	Delay	References
Inverter	2		0.75nW	0.34nS	
NAND 2	10		1.3nW	21.9nS	
NOR 2	10	300 K	1.12nW	21.5nS	(Neelakandan et al., 2022)
DFF	84		11.8nW	12.3nS	
SR (8 bit)	672		92.1nW	1.56nS	
Inverter	2		0.73nW	1.02ps	
NAND 2	4	300K	0.1nW	7.37ps	(Sui et al., 2011)
NOR 2	4		0.1nW	7.13ps	
DFF	36		7.19nW	38.87ps	
OR and NOR gates	6	4K	-	-	(Eskandarian et al., 2018)
Frequency Doubler	2	15-300K	-	-	(Patel et al., 2019)
Random number generator	1	300K			(Tannu & Sharma, 2012)

utilization of SET have been proposed in many papers (Eskandarian et al., 2018; Lageweg et al., 2002; Maeda et al., 2012; Patel et al., 2019; Sahaÿ et al., 2013; Sui et al., 2011; Tannu & Sharma, 2012; Tsiolakis et al., 2010; Tucker, 1992). The different types of circuits implemented with SET device including inverters, multiplexers, flip-flops, frequency doubles, Analog to digital converters, random number generator and memory elements have been constructed. The SET can function reliably at ultra-high frequencies. The complexity of the devices increases when the transistor count reaches the maximum to meet out the feature needs of the society. The influence of CMOS fabrication process is sophisticated to place the SET transistors formed a stack like structure above the CMOS platform. This increases the efficiency in terms of power, delay, flexible in terms of temperature variations and low cost. Hence it also increases the package density of transistors in the integrated circuits. Table 1 shows the various circuit level enhancement of SET transistors.

SET FIELD OF APPLICATIONS

Because of the Coulomb domain, SET can be utilized as a memory cell. If an electron is present, the island can alter. In this manner, SET expands the amount of space that information may be stored on memory devices, such as whether or not a single electron is present on the SET island. Additionally, SETs are more useful than CMOS due to its low power consumption factor. However, making such a gadget is quite challenging. Nevertheless, researchers are still trying to figure it out, and if the fabrication if it does, there will be a revolutionary shift in the field of quantum technology (Aassime et al., 2001; Guo et al., 1997; Matsumoto, 2000).

It is claimed that SET-based electrometers operate capacitive connecting the gate node to an exterior charge source that will be estimated then the change in source-drain current is computed. Due to this device's large amplification coefficient can be used for computing a little current change. Research indicated that if there is a $e/2$ charge shift on the gate, and the Coulomb current island moves at 10^9 e/sec. This responsiveness is much better than typical MOSFIT electrometers. SETs have already been used as a tool for imaging and in metrological applications localized single transistor alterations (Hergenrother et al., 1993). The fundamental unit of dc current standards is single electron tunneling. The SET oscillations or Bloch oscillations for such a standard consider a straight forward oscillator that has an f-frequency RF source outside of it. Phase locking would allow for the transfer of a particular amount of electrons each cycle of an exterior RF signal, as a result, produce dc current, which has I=mef as its fundamental relationship to frequency. The coherent oscillations that can be produced by this setup are be eliminated by using such a reliable RF source to power gadgets such as SET turnstiles and pumps, but the oscillations are not in automated mode that are coherent (Kumar & Kaur, 2010). Benefits of the Single Electron Array include less shot noise and practical threshold voltage modification that would enabling it to recognize the frequency in the tetra hertz range of frequencies, where there is currently no backdrop limited radiation detectors offered. The SET devices operate with the support from Voltage State Mode (VSM). Here, in VSM, the gate electrode's supplied voltage V_G controls the source draining the current. Thus, the impacts of a single electron charging are inside the transistor, despite the fact that it seems to have a typical outside the electronic device that changes the current of many electronic charge sources, dc HI/low voltage states are represented by binary logic 1 and 0, respectively. The one drawback of VS circuits is that hardly any of the complimentary pair is almost too well, causing these circuitry' leakage current (static) in the quantity on the order of 10^{-4} e/RC exists (Beaumont et al., 2009; Koppinen et al., 2013). Applications for SET with nonvolatile memory capabilities making programmable logic. The halfway through the menstrual phase, hence the SET's functional logic (binary) is not or equivalent to conventional SET, as a result of which SET can be used as complementing p-MOS SETs or straightforward n-MOS SETs. In actuality, SET circuit function can be programmed depending on the function (Basanta Singh, 2013; Rai et al., 2019; Ubbelohde et al., 2012). The table 1 shows most of the applications are related to small scale logic circuits not meet out the large computing high speed applications.

The high performance SET based computing systems have executed instructions up to 14 numbers with the memory architecture based on Von-Neumann (Patel et al., 2020). The computing system's design includes the following components: realistic SET conditions for operation at room temperature parasitic on interconnects that are useful. Another crucial component of a computer system is its memory an element of the plan. The computing system's ALU is created using a cutting-edge slice-based methodology that realizes greater performance with minimal hardware design. The suggested SET-based system is based on the computing system may carry out 14 different instruction kinds. It runs numerous programs that is confirmed using various vector files. it automatically the tool for creating vector files is fully conceived and designed (Patel et al., 2018). During the evaluation, the SET's efficiency was based on computer system and computation using CMOS technology (16 nm) systems are validated and thoroughly analyzed. The variation analysis is done to assess design resilience to voltage, temperature fluctuations and process. Fig 14 shows the computing system based on SET

Low Power Consumption of Set

The inverter circuit on the PSET island mentioned above is black. It follows that negative background charge White dominates the entire NSET island. The counterintuitive tariff markings are seen on the backs of both islands, although it is assumed that all total values are equal. The detailed information for the back features of the SET inverter and CMOS, as well as the ground charges inverter (Kang, 2003; Mishra et al., 2021). Fig. 15(a) and (b) depict the conventional CMOS inverter and the SET inverter where PSET is represented as black.

Fig. 16 (a), (b), (c) and (d) illustrate the power consumption details in the conventional PMOS and NMOS devices. The inverter based on CMOS has an average power consumption related to PMOS of 3.3708 x 10^{-5}W and for NMOS 3.1701x10^{-5}W. The value of 6.5409x10^{-5}W is the total average power consumption of the CMOS inverter. At the same time, the consumption of the power values for the SET-based PSET is 6.061x10^{-12} W and NSET 6.722x10^{-12} W. Hence, the value of 13.3284x10^{-12}W is obtained as the total average power consumption of SET model transistors. From the observed values, the power consumption factor for the PSET inverter is much less than the CMOS inverter circuit, which indicates that SET-based transistors have a better scope in future semiconductor device technologies.

SET is a non-volatile memory-based approach for programmable SET logic. In addition to typical SETs, the semi-phase shift enhances their SET functionality. SET circuit operation is dependent on the function of the stored memory. The coulomb

Figure 16. Computing system based on SET

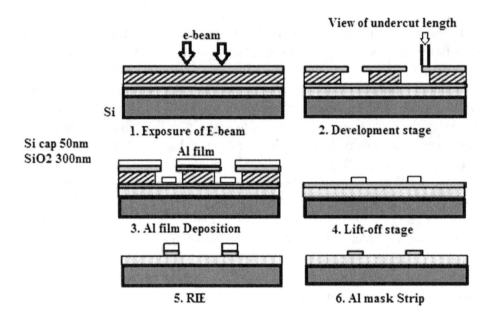

Figure 17. Conventional CMOS inverter

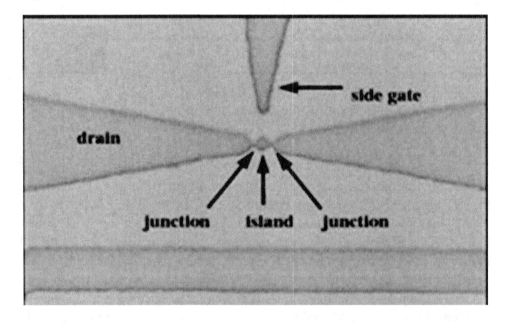

Figure 18. SET based inverter

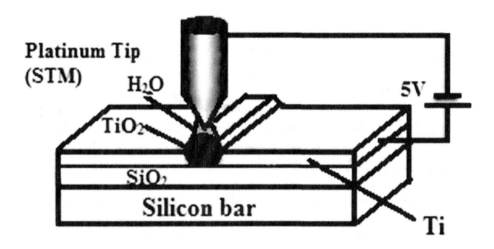

oscillation phase is changed by the quantum dot motion of the SET Island, which involves writing to or deleting from a memory function that imports data into or from the memory unit on the SET island to shift the coulomb oscillation phase. The phase shift is equivalent to 1/2 the oscillation's duration if the supplied load is sufficient.

SET is also used in microwave detectors. It's possible that a quasi-particle is introduced into the island at a comfortable distance. Just two electrons can be transferred simultaneously throughout the lengthy tunnelling process. The device is hence extremely sensitive to microwave radiation. Due to high sensitivities, single electron transistors are especially appropriate for use as electrometers. Two-gate electrodes are frequently utilized in applications such as electrometry.

Due to the growing number of semiconductor switches used to regulate electron tunnelling to enhance current, the share of the market of single electron transistors is

Table 2. CMOS and SET Comparative analysis

CMOS	SET
Physical limitation beyond 10nm technology	Better nanoscale device performance
Dissipation in terms of power is high	Power dissipation is in ultra-low range
Oscillation characteristics is not good	Better oscillation characteristics
Complex circuit	Simplified circuit
Power consumption is in the range of CMOS based inverter is 6.5409×10^{-5}W	Power consumption is in the range of CMOS based inverter is 13.328×10^{-12}W

Figure 19. Consumption of power in PMOS and NMOS in CMOS devices and Consumption of Power details in SET PMOS and NMOS devices

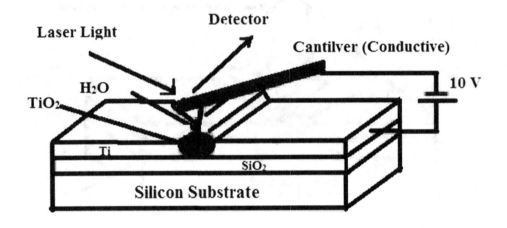

Figure 20.

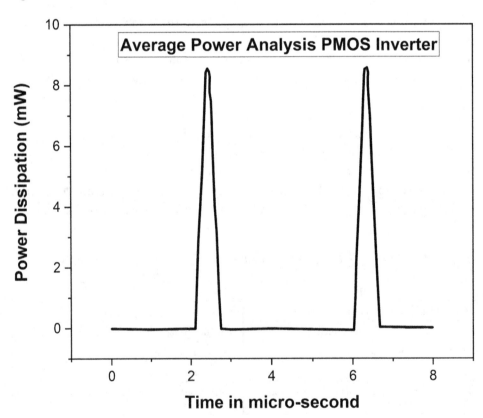

projected to rise. Single electron spectroscopy, ultrasensitive microwave detectors, infrared radiation detection, and other digital and analogue fields are expected to see an increase in the need for SET, which is expected to propel the growth of the single electron transistor business. Due to its energy-saving qualities and compatibility with CMOS technology, which increases operating efficiency, SET use is on the rise. Additionally, memory cells also employ single electron transistors, opening up additional opportunities for adoption. The usage of single electron transistors is expanding as a result of their ability to conserve energy and their compatibility with CMOS technology, which improves operational efficiency. Single electron transistors are also used by memory cells, creating further adoption possibilities. The SET market is expected to increase favorably as the automotive sector expands. The single electron transistor, however, is a novel technology with a small number of end users, which could hinder the market's expansion.

CONCLUSION

This article emphasis the overall introduction of SET, theoretical points behind SET devices, the modeling aspects involved in the SET, fabrication steps and the SET's field of applications. The SET technology uses extremely small amount of energy and power for various operations, carries out processes more quickly, and offers scalability all the way to the regime with dimensions of sub nanometer scale. SET innovation has a subject of extensive study due to their distinctive multi-functionality because of its multi-bit and multi-gate features. From the review analysis, out of the various SET models the MIB based model is efficient due to the integration of SPICE and CAD tools to enhance the simulation process. The study of SET manufacturing demonstrates that it is feasible and implementable in real technology. Additionally, it is possible to make the range of SET capacitance around 10^{-18} that makes it possible to operate at ambient temperature. The power consumption terms are examined for various circuits including basic gates, inverters, random number generator which shows the result values suitable for future integrated technologies. Finally, a comparison in terms of average power consumption for CMOS and SET based inverters is analyzed and the SET inverters have minimum amount of power utilization in the range pico-Watt range which implies SET based devices are much sophisticated to implement complex systems.

REFERENCES

Aassime, A., Johansson, G., Wendin, G., Schoelkopf, R. J., & Delsing, P. (2001). Radio Frequency Single-Electron Transistor as Readout Device for Qubits: Charge Sensitivity and Backaction. *Physical Review Letters*, *86*(15), 86. doi:10.1103/PhysRevLett.86.3376 PMID:11327974

Abdelkrim, M. (2019). Modeling and simulation of single-electron transistor (SET) with aluminum island using neural network. *Carpath J Electron Comput Eng*, *12*(1), 23–28. doi:10.2478/cjece-2019-0005

Ahsan, M. (2018). Single electron transistor (SET): Operation and application perspectives. *MIST Int J Sci Technol*, 6, 1.

Albert, J. R., Kaliannan, T., Singaram, G., Sehar, F. I. R. E., Periasamy, M., & Kuppusamy, S. (2022). A remote diagnosis using variable fractional order with reinforcement controller for solar-MPPT intelligent system. In *Photovoltaic Systems* (pp. 45–64). CRC Press. doi:10.1201/9781003202288-3

Bai, Z., Liu, X., Lian, Z., Zhang, K., Wang, G., Shi, S. F., Pi, X., & Song, F. (2018). A silicon cluster based single electron transistor with potential room-temperature switching. *Chinese Physics Letters*, *35*(3), 037301. doi:10.1088/0256-307X/35/3/037301

Basanta Singh, N. (2013). Design and Implementation of Hybrid SETCMOS 4-to1 MUX and 2-to-4 Decoder Circuits. *International Journal of Advanced Research in Electrical, Electronics and Instrumentation Engineering, 2.*

Beaumont, A., Dubuc, C., Beauvais, J., & Drouin, D. (2009). Room Temperature Single-Electron Transistor Featuring Gate-Enhanced ON-State Current. *IEEE Electron Device Letters*, *30*(7), 30. doi:10.1109/LED.2009.2021493

Bounouar, M. A., Beaumont, A., El Hajjam, K., Calmon, F., & Drouin, D. (2012). *Room temperature double gate single electron transistor based standard cell library. In 2012 IEEE/ACM international symposium on nanoscale architectures (NANOARCH).* IEEE. doi:10.1145/2765491.2765518

Buxboim, A., & Schiller, A. (2003). Current characteristics of the single electron transistor at the degeneracy point. *Physical Review. B*, *67*(16), 165320. doi:10.1103/PhysRevB.67.165320

Castro-Gonzalez, F., & Sarmiento-Reyes, A. (2014). *Development of a behavioral model of the single-electron transistor for hybrid circuit simulation. In 2014 international Caribbean conference on devices. Circuits and systems (ICCDCS).* IEEE. doi:10.1109/ICCDCS.2014.7016179

Chi, Y., Sui, B., Yi, X., Fang, L., & Zhou, H. (2010). Advances in the modeling of single electron transistors for the design of integrated circuit. *Journal of Nanoscience and Nanotechnology*, *10*(9), 6131–6135. doi:10.1166/jnn.2010.2560 PMID:21133161

Darwin, S., Rani, E. F. I., Raj, E. F. I., Appadurai, M., & Balaji, M. (2022, April). Performance Analysis of Carbon Nanotube Transistors-A Review. In *2022 6th International Conference on Trends in Electronics and Informatics (ICOEI)* (pp. 25-31). IEEE. 10.1109/ICOEI53556.2022.9776858

Deivakani, M., Kumar, S. S., Kumar, N. U., Raj, E. F. I., & Ramakrishna, V. (2021, March). VLSI implementation of discrete cosine transform approximation recursive algorithm. *Journal of Physics: Conference Series*, *1817*(1), 012017. doi:10.1088/1742-6596/1817/1/012017

Deng, G., & Chen, C. (2012). A SET/MOS hybrid multiplier using frequency synthesis. *IEEE Transactions on Very Large Scale Integration (VLSI) Systems*, *21*(9), 1738–1742. doi:10.1109/TVLSI.2012.2218264

Deyasi, A., & Sarkar, A. (2019). Effect of temperature on electrical characteristics of single electron transistor. *Microsystem Technologies*, *25*(5), 1875–1880. doi:10.100700542-018-3725-5

Dubuc, C., Beauvais, J., & Drouin, D. (2008). A nanodamascene process for advanced single-electron transistor fabrication. *IEEE Transactions on Nanotechnology*, *7*(1), 68–73. doi:10.1109/TNANO.2007.913430

Durrani, Z. A. K. (2010). *Single-electron devices and circuits in silicon.* World Scientiðc.

Eskandarian, A., Rajeyan, Z., & Ebrahimnezhad, H. (2018). Analysis and simulation of single electron transistor as an analogue frequency doubler. *Microelectronics Journal*, *75*, 52–60. doi:10.1016/j.mejo.2018.02.008

Fonseca, L., Korotkov, A., Likharev, K., & Odintsov, A. (1995). A numerical study of the dynamics and statistics of single electron systems. *Journal of Applied Physics*, *78*(5), 3238–3251. doi:10.1063/1.360752

Gnanasekar, A. K., Deivakani, M., Bathala, N., Raj, E., & Ramakrishna, V. (2021). Novel Low-Noise CMOS Bioamplifier for the Characterization of Neurodegenerative Diseases. In *GeNeDis 2020* (pp. 221–226). Springer. doi:10.1007/978-3-030-78787-5_27

Goldhaber-Gordon, D., Shtrikman, H., Mahalu, D., Abusch-Magder, D., Meirav, U., & Kastner, M.A. (1998). *Kondo effect in a single-electron transistor.* Academic Press.

Gonza'lez, F. J. C., Reyes, A. S., & Saenz, F. J. Z. (2012). Effects of single-electron transistor parameter variations on hybrid circuit design. In *2012 IEEE 3rd Latin American symposium on circuits and systems (LASCAS).* IEEE. 10.1109/LASCAS.2012.6180354

Guo, L., Leobandung, E., & Chou, S. Y. (1997). A Silicon Single-Electron Transistor Memory Operating at Room Temperature. *Science, 275*(5300), 275. doi:10.1126cience.275.5300.649 PMID:9005847

Hergenrother, J. M., Tuominen, M. T., Tighe, T. S., & Tinkham, M. (1993). Fabrication and characterization of Single-Electron tunneling transistors in the superconducting state. *IEEE Transactions on Applied Superconductivity, 3*(1), 3. doi:10.1109/77.233570

Inokawa, H., Nishimura, T., Singh, A., Satoh, H., & Takahashi, Y. (2018). *Ultrahigh-frequency characteristics of single-electron transistor. In 2018 IEEE international conference on electron devices and solid state circuits (EDSSC).* IEEE. doi:10.1109/EDSSC.2018.8487153

Jacob, A.P., Xie, R., Sung, M.G., Liebmann, L., Lee, R.T., & Taylor, B. (2017). Scaling challenges for advanced CMOS devices. *Int J High Speed Electron Syst, 26*(1-2).

Jain, A., Ghosh, A., Singh, N. B., & Sarkar, S. K. (2015). A new SPICE macro model of single electron transistor for efðcient simulation of single-electronics circuits. *Analog Integr Circ Sig Process, 82*(3), 653–662. doi:10.100710470-015-0491-5

Jia, C., Chaohong, H., Cotofana, S. D., & Jianfei, J. (2004). SPICE implementation of a compact single electron tunneling transistor model. *4th IEEE conference on nanotechnology,* 392–395.

Joyez, P., & Esteve, D. (1997). Single-electron tunneling at high temperature. *Physical Review, 56.*

Kang. (2003). *CMOS Digital Integrated Circuits: Analysis and Design* (3rd ed.). McGraw Hill Pub.

Karbasian, G., Orlov, A. O., & Snider, G. L. (2015). *Nanodamascene metal insulator-metal single electron transistor prepared by atomic layer deposition of tunnel barrier and subsequent reduction of metal surface oxide. In 2015 silicon nanoelectronics workshop (SNW)*. IEEE.

Kastner, M. A., & Goldhaber-Gordon, D. (2001). Kondo physics with single electron transistors. *Solid State Communications*, *119*(4-5), 245–252. doi:10.1016/S0038-1098(01)00106-5

Keyser, U., Schumacher, H. W., Zeitler, U., Haug, R. J., & Eberl, K. (2000). Fabrication of a single-electron transistor by current-controlled local oxidation of a two-dimensional electron system. *Applied Physics Letters*, *76*(4), 457–459. doi:10.1063/1.125786

Khadem Hosseini, V., Dideban, D., Ahmadi, M. T., & Ismail, R. (2018). An analytical approach to model capacitance and resistance of capped carbon nanotube single electron transistor. *AEÜ. International Journal of Electronics and Communications*, *90*, 97–102. doi:10.1016/j.aeue.2018.04.015

Khan, F. H. (2014). Chemical hazards of nanoparticles to human and environment (a review). *Oriental Journal of Chemistry*, *29*(4), 1399–1408. doi:10.13005/ojc/290415

Kim, N., Hansen, K., Paraoanu, S., & Pekola, J. (2003). Fabrication of Nb based superconducting single electron transistor. *Phys B*, *329*, 1519–1520. doi:10.1016/S0921-4526(02)02419-5

Koppinen, P. J., Stewart, M. D. Jr, & Neil, M. (2013). Fabrication and Electrical Characterization of Fully CMOS-Compatible Si Single-Electron Devices. *IEEE Transactions on Electron Devices*, *60*(1), 60. doi:10.1109/TED.2012.2227322

Korotkov, A. N. (1994). Intrinsic noise of the single-electron transistor. *Physical Review*, *49*(15), 10381–10392. doi:10.1103/PhysRevB.49.10381 PMID:10009861

Krautschneider, W., Kohlhase, A., & Terletzki, H. (1997). Scaling down and reliability problems of gigabit CMOS circuits. *Microelectronics and Reliability*, *37*(1), 19–37. doi:10.1016/0026-2714(96)00236-3

Kuhn, K., Kenyon, C., Kornfeld, A., Liu, M., Maheshwari, A., Shih, W., Sivakumar, S., Taylor, G., VanDerVoorn, P., & Zawadzki, K. (2008). Managing process variation in Intel's 45 nm CMOS technology. *Information Technology Journal*, *12*, 2.

Kumar, O., & Kaur, M. (2010). Single Electron Transistor: Applications & Problems. *International Journal of VLSI Design & Communication Systems, 1*(4).

Lageweg, C., Cotofana, S., & Vassiliadis, S. (2002). Static buffered SET based logic gates. In *Proceedings of the 2nd IEEE conference on nanotechnology*. IEEE. 10.1109/NANO.2002.1032295

Lee, Joshi, Orlov, & Snider. (2010). Si single electron transistor fabricated by chemical mechanical polishing. *Journal of Vacuum Science and Technology*, 28.

Li-Na, S., Li, L., Xin-Xing, L., Hua, Q., & Xiao-Feng, G. (2015). Fabrication and characterization of a single electron transistor based on a silicon-on-insulator. *Chinese Physics Letters*, 32(4), 047301. doi:10.1088/0256-307X/32/4/047301

Maeda, K., Okabayashi, N., Kano, S., Takeshita, S., Tanaka, D., Sakamoto, M., Teranishi, T., & Majima, Y. (2012). Logic operations of chemically assembled single-electron transistor. *ACS Nano, 6*(3), 2798–2803. 10.1021/nn3003086

Mahapatra, S., & Ionescu, A. M. (2005). Realization of multiple valued logic and memory by hybrid SETMOS architecture. *IEEE Transactions on Nanotechnology*, 4(6), 705–714. doi:10.1109/TNANO.2005.858602

Matsumoto, K. (1998). Application of Scanning Tunneling/Atomic Force Microscope Nanooxidation process to room temperature operated Single Electron Transistor and other devices. *Scanning Microscopy, 12*, 6169.

Matsumoto, K. (2000). Room-Temperature Single Electron Devices by Scanning Probe Process. *International Journal of High Speed Electronics and Systems, 10*(01), 83–91. doi:10.1142/S0129156400000118

Matsumoto, K., Ishii, M., Segawa, K., Oka, Y., Vartanian, B.J., & Harris, J.S. (1996). Room temperature operation of a single electron transistor made by the scanning tuneling microscope nanooxidation process for the TiOx/Ti system. *American Institute of Physics, 68.*

Miralaie, M., Leilaeioun, M., Abbasian, K., & Hasani, M. (2014). Modeling and analysis of room-temperature silicon quantum dot-based single-electron transistor logic gates. *Journal of Computational and Theoretical Nanoscience, 11*(1), 15–24. doi:10.1166/jctn.2014.3311

Miralaie, M., & Mir, A. (2016). Performance analysis of single-electron transistor at room-temperature for periodic symmetric functions operation. *Journal of Engineering (Stevenage, England), 10*(10), 352–356. doi:10.1049/joe.2016.0139

Mishra, B., Kushwah, V. S., & Sharma, R. (2021). Power consumption analysis of MOSFET and Single electron transistor for inverter circuit. *Materials Today: Proceedings, 47*(Part 19), 6600–6604. doi:10.1016/j.matpr.2021.05.094

Neelakandan, S., Rene Beulah, J., Prathiba, L., Murthy, G. L. N., Irudaya Raj, E. F., & Arulkumar, N. (2022). Blockchain with deep learning-enabled secure healthcare data transmission and diagnostic model. *International Journal of Modeling, Simulation, and Scientific Computing, 2241006.*

Ono, Y., Inokawa, H., Takahashi, Y., Nishiguchi, K., & Fujiwara, A. (2010). Single-electron transistor and its logic application. *Nanotechnol Online, 20,* 45–68.

Parekh, R. (2019). Design and simulation of single electron transistor-based SRAM and its memory controller at room temperature. *Int J Integrat Eng, 11*(6), 186–195.

Patel, Agrawal, & Parekh. (2020). Single-electron transistor: review in perspective of theory, modelling, design and fabrication. *Microsystem Technologies-micro-and Nanosystems-information Storage and Processing Systems,* 1-13.

Patel, R., Agrawal, Y., & Parekh, R. (2018). *A vector ðle generation program for simulating single electron transistor based computing system. In 2018 IEEE electron devices Kolkata conference (EDKCON).* IEEE. doi:10.1109/EDKCON.2018.8770464

Patel, R., Agrawal, Y., & Parekh, R. (2019). Design of prominent set-based high performance computing system. *IET Circuits, Devices & Systems, 14*(2), 159–167. doi:10.1049/iet-cds.2019.0166

Pushparaj, T. L., Raj, E., Rani, E., Darwin, S., & Appadurai, M. (2022). Employing Novel Si-Over-Si Technology to Optimize PV Effect in Solar Array. *Silicon,* 1-13.

Rai, C., Khursheed, A., & Haque, F. Z. (2019). Review on Single Electron Transistor (SET): Emerging Device in Nanotechnology, Austin. *Journal of Nanomedicine & Nanotechnology, 7*(1).

Raut, V., & Dakhole, P. (2016). *Design and implementation of quaternary summation circuit with single electron transistor and MOSFET. In 2016 international conference on electrical, electronics, and optimization techniques (ICEEOT).* IEEE. doi:10.1109/ICEEOT.2016.7755088

Sahað, A., Moaiyeri, M. H., Navi, K., & Hashemipour, O. (2013). Efðcient single-electron transistor inverter-based logic circuits and memory elements. *Journal of Computational and Theoretical Nanoscience, 10*(5), 1171–1178. doi:10.1166/jctn.2013.2824

Sahatiya. (2013). Single Electron Transistor: A Review. *International Journal of Scientific & Engineering Research, 4.*

Singh, V.P., Agrawal, A., & Singh, S.B. (2012). Analytical Discussion of Single Electron Transistor (SET). *International Journal of Soft Computing and Engineering, 2.*

Sui, B., Fang, L., & Zhang, C. (2011). Reconðgurable logic based on tunable periodic characteristics of single-electron transistor. In *2011 24th Canadian conference on electrical and computer engineering (CCECE)*. IEEE. 10.1109/CCECE.2011.6030497

Tannu, S., & Sharma, A. (2012). *Low power random number generator using single electron transistor. In 2012 international conference on communication, information and computing technology (ICCICT)*. IEEE. doi:10.1109/ICCICT.2012.6398099

Tsiolakis, T., Alexiou, G. P., & Konofaos, N. (2010). Low power single electron OR/NOR gate operating at 10GHz. *2010 IEEE computer society annual symposium on VLSI*, 273–276.

Tucker, J. (1992). Complementary digital logic based on the Coulomb blockade. *Journal of Applied Physics*, 72(9), 4399–4413. doi:10.1063/1.352206

Ubbelohde, N., Fricke, C., Flindt, C., Hohls, F., & Haug, R. J. (2012). Measurement of finite-frequency current statistics in a single-electron transistor. *Nature Communications*, 3(1), 612. doi:10.1038/ncomms1620 PMID:22215087

Venkataratnam, A., & Goel, A. K. (2008). Design and simulation of logic circuits with hybrid architectures of single-electron transistors and conventional MOS devices at room temperature. *Microelectronics Journal*, 39(12), 1461–1468. doi:10.1016/j.mejo.2008.08.002

Wasshuber, C. (2001). *Computational single-electronics*. Springer.

Wen, Y., Ares, N., Schupp, F., Pei, T., Briggs, G., & Laird, E. (2020). A coherent nano mechanical oscillator driven by single-electron tunnelling. *Nature Physics*, 16(1), 75–82. doi:10.103841567-019-0683-5 PMID:31915459

Yu, Y., Oh, J. H., & Hwang, S. (2000) Implementation of single electron circuit simulation by SPICE: KOSEC-SPICE. *Proc Asia Pac Workshop Fundam Appl Adv Semicond Device, 100*(150), 85–90.

Zhang, J., Liu, S., Kong, L., Nshimiyimana, J. P., Hu, X., Chi, X., Wu, P., Liu, J., Chu, W., & Sun, L. (2018, May). Room Temperature Carbon Nanotube Single-Electron Transistors with Mechanical Buckling–Defined Quantum Dots. *Advanced Electronic Materials*, 4(5), 1700628. doi:10.1002/aelm.201700628

Zhu, Gu, Dick, Shang, & Knobel. (2009). Characterization of Single-Electron Tunneling Transistors for Designing Low-Power Embedded Systems. *IEEE Transactions on Very Large Scale Integration Systems, 17.*

Chapter 6
Electronic Cooling

Shankara Murthy H. M.
(iD) https://orcid.org/0000-0003-3491-4356
Sahyadri College of Engineering and Management, India

Niranjana Rai
Canara Engineering College, India

Ramakrishna N. Hegde
Srinivas Institute of Technology, India

ABSTRACT

Through the ongoing downsizing and fast growth of heat flow of electronic components, cooling concerns are confronting severe tasks. This chapter examines the recent advancements and modernization in the cooling of electronics. The most popular electronic cooling technologies, which are classed as direct and indirect cooling, are examined and described in depth. The best prevalent methods of indirect cooling by employing heat pipes, microchannels, PCM are discussed. The efficiency of cooling strategies for various levels of electronic cooling requirements, as well as approaches to increase heat transfer capabilities, are also discussed in depth. Meanwhile, by considering the intrinsic thermal characteristics, optimization approaches, and pertinent uses, the advantages and disadvantages of various thermal management systems are examined. Furthermore, the present issues of electronic cooling and thermal management technologies are discussed as well as the prospects for future advancements.

INTRODUCTION

Electronic equipment has penetrated almost all facets of new life, from toys and

DOI: 10.4018/978-1-6684-4974-5.ch006

Figure 1. Major Applications of Integrated Circuits

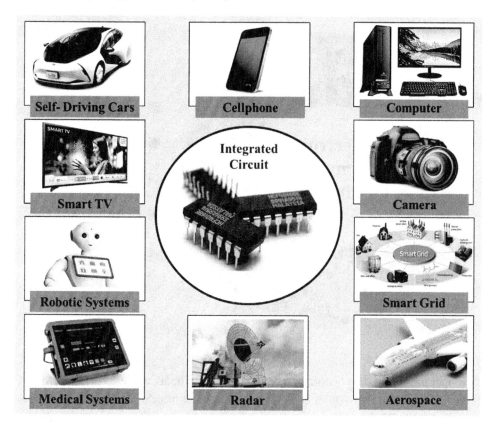

appliances to powerful processors. The worthiness of an electronic system is a key aspect in the total reliability of the system. Guarnieri M (2016) specified in his article that, integrated circuits (IC) have progressed significantly later in 1949 Werner Jacobi published the first conception of IC. An IC is a tiny chip constructed of the semiconductor material silicon that may hold lots of microelements viz., capacitors, transistors and resistors. It is often created using the several-nanometer method. As seen in Figure 1, integrated circuits are now employed in practically all electronic equipment, and modern life is closely entwined with numerous electronic items. These apps have greatly increased the efficiency and quality of labor, production, and living for modern people.

Ho-Ming Tong et al. (2013) depicted the failure reasons of electronic equipment in percentages as shown in Figure 2. Temperature, vibration, humidity, and dust are the most common reasons for electronic component failure. The greatest risk of failure is owing to heat production, which causes component temperatures to rise (Upto 55%). Electronic components rely on the flow of electricity to accomplish

Figure 2. Dissemination of failure causes of electronic equipment.

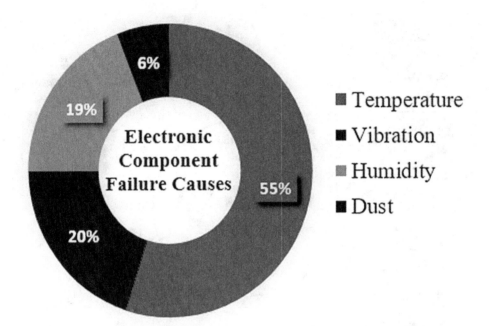

their operations and thus become incredibly powerful heat sources when the current passes through the resistance, causing continual heat buildup. In his book, Yunus Cengel A (2003) noted that the constant downsizing of electronic structures has resulted in a significant increase in the rate of heat generation by the volume of each unit. The failure of electronic equipment increases dramatically as the temperature rises. Furthermore, temperature changes create a rise in heat at electrical device joints on PCBs, which is a major cause of failure.

Murshed S S et al. (2017) calculated the maximum energy required by the chip and the maximum heat flux created by the chip during the previous two decades as shown in Figure 3. As demonstrated in figure 3, the power needs and heat flux created have grown significantly between 2001 and 2018. More heat created causes high operating temperatures in electronic systems, endangering their protection and consistency if not adequately planned and controlled. As a result, heat management in the design and operation of electronic equipment has become important. Maintaining the operating temperature through adequate cooling is the only way to overcome these challenges in electronic systems. Heat generation rate and cooling mechanism are carefully selected based on the various electronic applications. This chapter discusses the cooling of electronic devices using different heat transfer augmentation methods.

Figure 3. Maximum power consumption and heat flux density in the last 2 decades

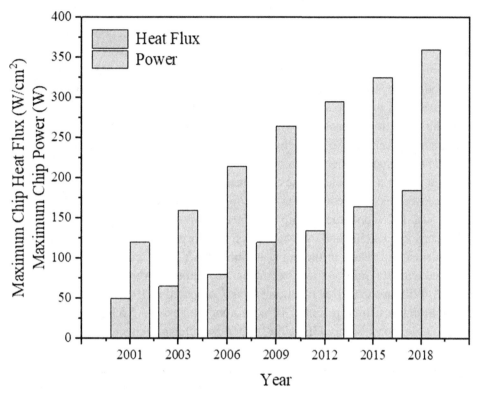

The main objective of this chapter is to introduce the reader regarding the heat generation in electronic devices and their cooling and strategies/methods used for the management of heat in electronic components. Further, the chapter depicts the current status of research and development in "Electronic Cooling" and suggests the best cooling techniques.

This chapter explains the necessity of cooling for electronic devices, develops a history of cooling techniques used for electronic equipment, techniques/methods of thermal management in electronic devices, solutions and recommendations for effective cooling strategy, upcoming research methods and conclusion gives the summary of the electronic cooling.

BACKGROUND

Latest advancements in semiconductor and electronic equipment leads to a significant improvement in power density, especially for high-end devices. Even with significant advances over the last few decades, heat control of electronics devices like microprocessors remains a major technical difficulty. The elimination of heat generation and uneven power indulgence are two major cooling difficulties. The highest dissipation in power and heat from microprocessors were predicted to be roughly 360 W and 190 W/cm² by 2020 as shown in figure 3. In reality, the heat produced by many high-performance electronic gadgets is presently far higher than the iNEMI roadmap's forecasts. Several electronic component manufacturers confronting the tough problem of eliminating the generated heat and keeping the temperature lower than 85°C, according to research published by Agostini B et al. (2007). Furthermore, when electronic systems become more integrated, dissipation of power on the IC's or electronic components becomes more complicated.

Moreover, electronic component shrinking in 1960 led to medium-scale integration having 50 to 1000 constituents/chip, in 1960 extensive incorporation having 1×10^3 to 1×10^5 constituents/chip and in 1980 extensive incorporation having 1×10^5 to 1×10^7 constituents/chip stated in the book authored by Yunus Cengel A (2003). Chips equipped with 100 million transistors/square centimeters were already being made in 2006. Williams C (2015) mentioned that Compared to 1990s 32-bit processors (3.1 million transistors) potential 10-core Xeon processors (5.5 billion transistors) are available today. Moore GE (1998) stated that electronic technologies continue to obey Moore's law, which calls for lesser features, higher transistor incorporation, enhanced quickness of the circuit, and higher chip efficacy. This indicates a steady drop in transistor dimension to 6 nm by 2020, as well as an increase in the number of transistors to almost 20 billion transistors/cm². Correspondingly, the chip dimension was expected to be roughly 100 mm². As a result, these conditions provide significant difficulties to thermal management solutions in terms of keeping activities within acceptable operating limits. Tong H.M et al. (2013) mentioned as temperature issues are the reason for 55% of electronic component letdowns, as illustrated in figure 2. A similar reason was stated by Black J.R (1969) for the failure of the electronic components. As a result, dealing with electronic device temperature issues is becoming progressively essential.

The current electronics design style is founded on the idea that the smaller and speedier device. However, this tendency is contributing to increased power densities and operating temperatures, as well as poor performance and device longevity. If heat is not evacuated at an amount equivalent to or higher than its generated amount, component and device temperatures continue to rise, affecting device reliability and performance and potentially leading to device failure. In reality, when the operational temperature rises, the probability of failure of electrical equipment rises virtually exponentially. According to a report set by the United States Department of Defense on electronic equipment reliability prediction, the failure factor, which is the comparative chance of failure at all temperatures above rate of failure at 75

Figure 4. Failure factor of electronic devises vs Temperature

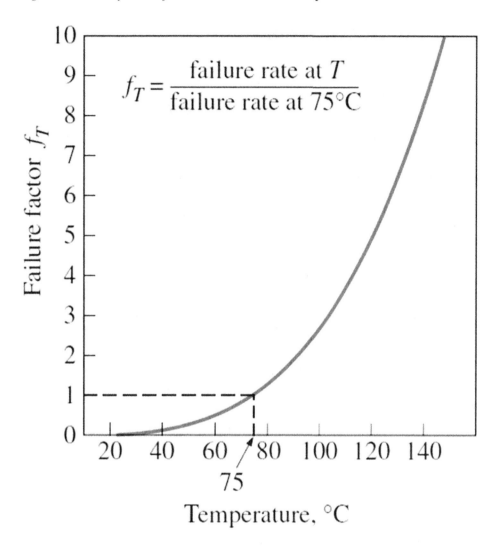

°C, tends to increase exponentially with the growing temperature of electronic equipment, as shown in figure 4 presented by Yunus Cengel A (2003).

As we know, there is no availability of devices with 100% efficiency. Energy can't be generated or destroyed; it can only be transformed into different forms. All of the energy that enters the gadget must depart as the same or a different kind of energy. Electrical energy "consumed" by electronic equipment is turned to light, motion (including sound,) radio waves, or heat in most cases. The method of heat conversion is through resistive "losses." Because of the inherent resistance of the metallic wire, some are unavoidable. Others are built within the gadgets themselves.

Figure 5. Classification of most prevalent thermal management techniques.

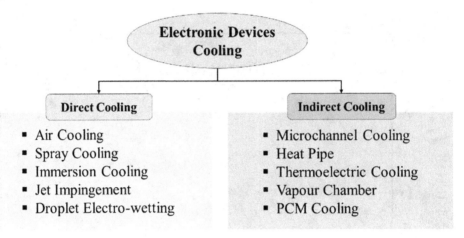

Transistors aren't infallible switches. Biasing networks, which "waste" power in resistance, are required for analog circuits. Filtering undesired radio emissions absorbs (converts to heat) the RF energy and prevents it from radiating. Because the electricity-conducting parts inside modern electronics are inefficient, they heat up. This is because all of the metals being used daily in electronics have some resistance under normal settings. That implies part of the energy stored in the wire will be converted to heat. That heat builds up over time, causing that particular electronic equipment to become hot.

As previously said, heat generation is rising in tandem with the advancement of the integrated circuit fabrication process. As a result, the integrated circuit industry needs effective thermal management solutions that will continue to improve the stability and efficiency of electronic items. However conventional air cooling may handle heat removal difficulties by improving the heat sink strategy for conventional electronic equipment, modern high-performance electronic devices require more innovative cooling techniques suggested by Wang C C (2017).

Figure 5 illustrates the different methods of electronic cooling to solve the heat generation problem and maintain the stability and efficiency of electronic devices. Active and passive cooling strategies are usually categorized into three thermal management approaches as addressed by Murshed. S.M.S (2016). The basic difference between these is passive methods depend on free convection, whereas the active method relies on additional energy to boost the heat sink's heat dissipation capabilities. Because active cooling systems have a superior ability to transfer heat, most cooling methods rely on the active method and are extensively employed for thermal control in electronic equipment. Heat transfer management is also separated into direct or indirect contact cooling, as indicated in figure 5, depending on whether the cooling medium comes into direct touch with the targets.

The environment wherein the electronic devices will function is a crucial factor to consider when choosing a cooling solution. Simple vent holes on the casing may be all that's needed to keep low-power-density devices like a TV or VCR cool in space, and a fan enough to keep a home computer cool. Thermal control of an aircraft's electronics, on the other hand, will present a challenge to thermal designers because the climatic factors in this scenario will twist from one point of absurdity to the other in a short amount of time. In the design phase, the estimated period of functioning in a harsh atmosphere is also taken into account. The thermal environment of electronics for an airplane flies for hours, it would take off and be altered compared to a missile with a few minutes of operation time. Because the main heat sink in maritime applications is water (temperature range of 0°C to 30°C), the thermal environment is reasonably steady. The ultimate heat sink for ground-based applications is atmospheric air, which has a temperature range of 50°C in arctic areas and a pressure range of 0.70 bar at 3000 m altitude to 1.07 bar at 500 m beneath sea level. The heat transfer coefficient for coupled convection and radiation was found to be 10 W/m^2 °C and 80 W/m^2 °C in clear weather and 100 km/h winds respectively. Further, the device exteriors facing the sun unswervingly can be exposed to a sun heat flux of 1 KW/m^2. The thermal environment in airborne applications can shift in minutes from 1 atm and 35°C on the earth to 19 kPa and 60°C at an altitude of 12 km. At supersonic velocity, some parts of aircraft approach the average temperature of 200°C above the ambient temperature.

Because of the large range of environmental factors, electronic equipment is rarely directly subjected to uncontrolled ambient circumstances. As an alternative, a conditioned fluid, including air, water, or dielectric fluid, is utilized as a medium for heat dissipation and an interface among electronic devices and the environment. Because it is safe, widely accessible, and not subject to leakage, treated air is the chosen cooling medium. Due to the poor thermal conductivity of air, its usage is restricted to small electronic devices. In this chapter, the issues in cooling technology are discussed and comprehensive analysis of several standards and new technologies for the effective thermal management of electronic components are reviewed.

ELECTRONIC COOLING TECHNIQUES

Technology has grown at an extraordinary rate in the last century, transforming everything from industrial processes to how people behave and interact with one another. The fast advancement of electronics and computer technologies has altered how technicians and engineers approach temperature management. Electronics that are speedier, compact, and more powerful generate more waste heat, which must be carefully handled. Heat exchangers have emerged to provide the superior temperature management systems required to effectively cool contemporary devices to meet these problems.

Contemporary thermal management had to develop when technology began to shrink. Large air conditioning systems would no longer be sufficient to keep smaller machinery and devices cool. Many electronic components today, for example, are measured in the nm range, and technology is ready to advance much farther to function on even smaller planes. Despite their little size, tiny electronics are more efficient than their predecessors, generating far more waste heat throughout their systems. The best part is now that heat exchangers may be built to fit into small places while still delivering efficient, high-performance thermal control. Heat exchangers, rather than cooling electronic components, continually move heat away from them, frequently in a closed-loop system. They're also excellent at transporting heat uniformly across a wide range of surfaces, making them more adaptable to the needs and problems of contemporary electronics.

Electronic cooling techniques may be classified into two parts: passive cooling and active cooling. Passive cooling cools a component by using natural conduction, convection, and radiation. Active cooling necessitates the use of energy dedicated solely to the cooling of the component. Heat sink and heat spreader design are two current examples of these two cooling categories, both of which use basic heat transfer principles. For example, the NVIDIA Jetson Nano comes pre-assembled with a heat sink connected to the development board. This fanned-fin heat sink will disperse enough energy away from the GPU to keep the development board running smoothly. Because the fan requires electricity, this thermal management system has now evolved into an active cooling system. Devices that generate substantial amounts of thermal energy, such as computers, gaming devices, televisions, and automobiles, use active cooling technology.

CONVENTIONAL COOLING TECHNIQUES

Despite significant development that has been achieved in electronics cooling technology in the modern era, removing excessive heat from powerful electronic equipment is still a difficult and insufficient task. Traditional cooling systems are summarized below, along with their classes depending on heat transfer processes and coolants employed, along with their cooling efficacy. Cooling mechanisms may be categorized into four main groups related to heat transfer efficacy:

- Natural convection

- Forced convection
- Air and liquid cooling
- Liquid evaporation.

For reliable operation, electronic device makers normally define the level of heat disposal and the maximum permitted component temperature. These two values assist us in determining which cooling strategies are appropriate for the gadget in question. Figure 6 depicts the heat fluxes achievable at given temperature changes for various common heat transfer methods. When a device power rating is specified, the heat flux is computed by dividing the rated power by the device or component's exposed surface area. Then, based on the specifications that the temperature differential between the device's surface and atmosphere medium does not exceed the permissible limit, appropriate heat transfer mechanisms may be found in figure 6 presented by Yunus Cengel A. (2003).

In addition to the main categories listed above, cooling methods may be categorized into numerous varieties. Based on the coolants used, cooling methods are divided into air, liquid and refrigeration cooling.

Kim YJ et al. (2008) stated that direct liquid immersion (DLI) cooling has a greater capacity for heat dissipation, which means the chip surface temperature rises less than the liquid coolant temperature. In addition, when compared to air-cooling, this cooling provides improved chip temperature consistency. Some DLI cooling techniques are recognized and frequently employed in a variety of temperature control systems.

Irrespective of the method employed for electronic device cooling, conveying heat to a fluid necessitates heat removal from the environment. This is usually achieved by forced convection, which is insufficient for high heat transfer settings. As a result, it's critical to effectively disperse the heat generated by the coolants. To overcome these problems conventional cooling and emerging cooling techniques were proposed by the researchers for effective heat disposal.

EMERGING COOLING TECHNIQUES

Emerging cooling technologies may be grouped into many sorts based on their efficacy and materials or processes used in cooling electronics, including the following examples:

- Free cooling
- Heat pipes
- Microchannels

Figure 6. Attainment of Heat fluxes vs temperature with different cooling methods (Yunus Cengel A. (2003)).

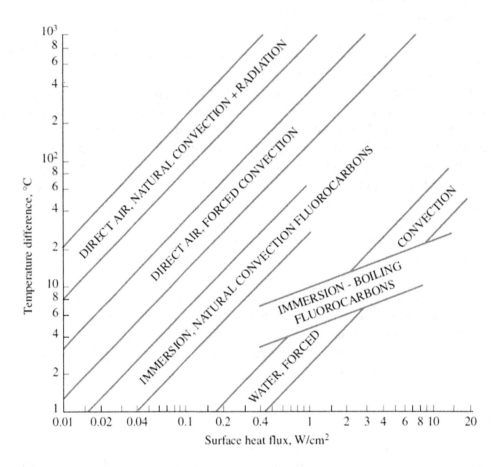

- Heat pumps
- PCM based cooling
- Thermoelectric cooling
- Spray cooling

As previously stated, traditional cooling approaches are insufficient to adequately cool recent electronic components with greater power capacity, such electronics require novel methods and procedures to increase the rate of heat dissipation to decrease working temperature and increase durability.

Though, utmost development systems for cooling high-tech devices, as listed above showed significant promise. As a result, the Electronic industry has been paying close attention to these developing approaches in recent years. In reality, just a few of these new approaches have been employed commercially to cool electronic equipment. A few of the most popular new cooling approaches have been described below.

Free Cooling

Free cooling is a cost-effective method that employs low exterior air temperatures to cool servers or data centers stated by Capozzoli A et al. (2015). To cool down most electronic units (such as a data center), it mostly employs natural cold and humid air, this cooling also employs the utilization of water. Heat can naturally move from the data center to the environment due to lower ambient air temperature. Direct and indirect cooling are the two types of free cooling. There are three types of free cooling, based on flow direction and method used free cooling is divided into airside, waterside and heat pipe. Cooling by heat pipe is a novel form of data center cooling technology that is rapidly gaining popularity. It functions in the same way as traditional heat pipe technology.

Heat Pipes

Heat pipes, which rely on working medium phase change in pipes to cool electronic equipment including computers, laptops, telecommunications, and satellite modules, are a viable method of cooling electronic devices presented by Faghri A (2012) and Chang YW et al (2008). Heat pipes are among the best feasible option for effective thermal management in heat-producing electronic systems, like CPUs owing to their relatively greater thermal conductivity and quite limited effective thermal resistance. Heat pipes have various commercial cooling applications nowadays, and due to their excellent heat removal capability, this cooling approach has piqued the interest of the electronic industries.

The massive amount of industrial heat pipe manufacturing attests to this. Heat pipes are produced in bulk quantities every month for heat management in compact electronic devices as mentioned by Faghri A (2012). Cooling Pentium CPUs in laptops is one of the most common uses of heat pipes. Chen X et al. (2016) have published a complete assessment of tiny heat pipe design, manufacturing, and performance analysis for electronics cooling. They also demonstrated how tiny heat pipes may be used to cool small electrical gadgets and components. The intriguing aspect about heat pipes is they don't have any moving components, making them very dependable and low-maintenance. Heat pipes also have several other benefits, including no required external power, noiseless operation, augmented durability, thus lower process and total costs, superior functional adjustability, small size and

Figure 7. Working of heat pipe

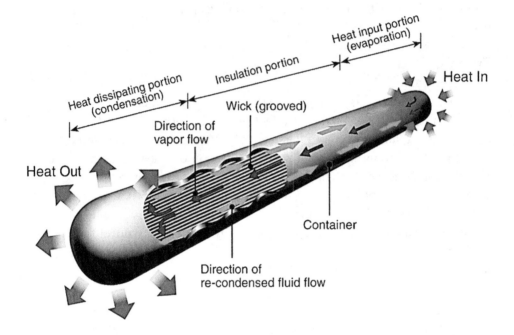

weight, operation in varied configuration, and secured body of heat pipe cause no consequence on the surroundings or electronic components.

Metal envelope, wick, and cooling fluid make up a heat pipe. The envelope is separated into three sections: heat input portion, insulation portion, and heat dissipation portion. The most significant component is the wick that is generally attached to the pipe's inner wall. It works as a capillary pump, pushing the fluid against gravity from the condenser part to the evaporator. This enables the heat pipe to work in either direction. The most common wick kinds are fluted wick, fused wick, and screen mesh wick.

Figure 7 depicts the heat pipe's simple operating concept, which is constructed by a metal envelope and filled with heat transfer fluid. The working fluid receives heat from the electronic components to be cooled at the evaporator section and liquid converting into vapour with higher pressure resulting in pressure difference in the heat pipe causing movement of vapours from the evaporator to condenser section. At the condenser part, latent heat of the fluid dissipated to the heat sink and condensed to liquid.

In the meantime, as the liquid is depleted by vaporization in the evaporator portion, the working fluid enters the wick surface, creating a tube pressure. Then condensed liquid is reversed to the evaporator by capillary pressure for continuous operation.

This way working fluid experienced simultaneous evaporation and condensation for absorption of heat and dissipation. Heat pipes are two-phase passive heat transfer devices that may transfer a considerable quantity of heat while maintaining a very low-temperature decrease. A heat pipe's excellent heat transmission capability is owing to the working fluid's high latent heat.

Microchannels

Microscale cooling systems can cool electrical equipment and appliances that generate a lot of heat. Figure 8 shows the schematic of the microchannel cooling plate used for the electronic cooling of Integrated Circuits and electronic components. A microchannel heat sink's heat transfer capability is far superior to that of any typical heat exchanger. For high-heat-generating electronic equipment, cooling by microchannel is the most capable and effective heat management technique. Wei Y & Joshi YK (2004) stated that, developing a cooling method is also adaptable to on-chip integration, in addition to drastically reducing package size.

After achieving a maximum heat dissipation of 790 W/cm^2 using an integrated microchannel with a temperature increase of 71°C in 1981, Tuckerman D & Pease R (1981) presented microchannel as an effective cooling solution. Microchannel-based cooling solutions have propelled electronics cooling to new heights. Since Tuckerman D & Pease R (1981) pioneering work shows that microchannels become a crucial technique for heat management in various electronic components, notably VLSI and ULSI presented by Kandlikar SG (2013). Here is a few typical research on microchannel-based cooling.

Colgan et al (2007) designed and constructed a microchannel heat sink using silicone for cooling very high power electronics like microprocessors. Such microchannel coolers have been shown to cool chips at average output densities of 400 W/cm^2. With a flow rate of 1.5 lpm, Kandilikar S et al (2005) reported that increased microchannel cooling employing offset fins and a separated flow configuration, reaching almost heat dissipation of 300 W/cm^2 at 0.24 bar. Dix et al (2008) conducted investigations on a microchannel with water as a working fluid using experimental and computational methods. Results show that performance was improved by altering channel geometry.

In addition, two-phase cooling in microchannels has appeared as a viable cooling approach for electronics that generate a lot of heat. Qu W et al (2004) conducted an investigation on a microchannel with 21 equivalent waterways and observed a fast rise in heat flux with minor temperature increases till a wall heat flux reaches critical heat flux (CHF). In research by Yao S et al (2012), considerable improvements in pool boiling heat flow were recorded using silicon microchannels with nanostructured walls.

Figure 8. Microchannel cooling system

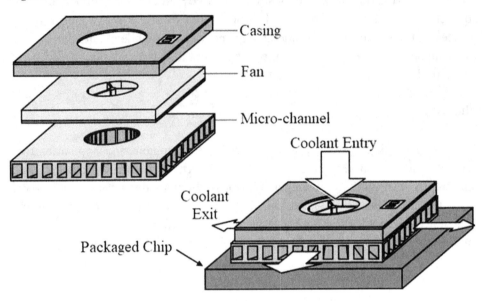

Kandlikar SG (2013) provided an in-depth analysis of the evolution of microchannel-based high heat removal methods. A road map describing the difficulties and prospects of this new cooling system for maximum heat-producing electronics was created. At a temperature differential (junction to ambient) of 50 ^{0}C, liquid coolant microchannels were expected to attain a rate of heat removal up to 1 kW/cm^2. They conducted an acute evaluation of the present status of exploration of heat transport in microchannels, obstacles and requirements of upcoming development.

Khaleduzzaman et al (2018) compared the thermal capabilities of various nanoparticles in the cooling of electronic equipment. In a microchannel fluid block, they tested the thermal performance of nanofluids. The researchers employed three distinct nanofluids that contained SiC, CuO, and alumina nanoparticles suspended in water. The greatest enhancement values were 11.36% and 11.98% for the CuO-water and alumina-water nanofluids, respectively. Furthermore, using nanofluids as the working fluid instead of water resulted in the highest heat transfer increase of 8.51% for the CuO nanofluid, 6.44% for the alumina nanofluid, and 5.60% for the SiC nanofluid, respectively. Finally, managing the size and shape of nanoparticles is a significant difficulty in nanoparticle production, especially in practical applications such as using nanofluids in liquid blocks, which can have a significant impact on thermal performance.

Phase Change Material (PCM)

PCM is a substance that, through a phase change process may absorb or release heat across a narrow temperature range. PCMs are now widely employed in a variety of industries and they also have benefits in electronic thermal control stated by Zhang S et al (2021). For an electronic gadget with intermittent fluctuations in heat generation, for example, PCM can collect heat when power usage is maximum and discharge heat whenever the power usage drops, assuring the electronic equipment's smooth operation. Typically, the PCM is not directly used for cooling IC's and electronic devices, but rather in conjunction with heat sinks, such as fins to maximize PCM properties and improve thermal management performance.

The characteristics of the PCM have a considerable impact on the cooling capacity as a carrier for absorption, accumulation and discharge of heat. As a result, several studies have focused on improving the heat transfer ability of PCM. Yang et al (2018) introduced the low melting point metal (LMPM), a novel kind of PCM with strong thermal conductivity and latent heat properties. Al_2O_3 nanoparticles were introduced to the Tricosane by Krishna et al (2017). According to the findings, this type of enriched nano PCM might increase heat conductivity by 32%. Farzanehnia et al (2019) investigated for fully mixed MWCNTs to the Paraffin wax, this drops the time of cooling by 6% and reduced the integrated circuit's peak temperature when compared to PCM without MWCNTs.

Further, the PCM is used in conjunction with heat management techniques like a heat pipe, a vapour chamber, or a TE module to provide a benefit in heat removal. After combining heat pipe evaporator temperature with PCM, Krishna et al (2017) discovered that the surface temperature of heat pipe was reduced by 25.75% while saving 53% of fan power. In their investigation, Behi et al (2017) discovered that PCM possibly delivers 86.7% of thermal management. Simultaneously, PCM-heat pipe heat sink structure optimization will be explored. Ho et al (2020) used a suspension of nano-condensed PCM and water to apply to the microchannel. The heat transfer performance of this type of PCM-microchannel heat sink might be improved by 70%, according to their findings. Meanwhile, they discovered that at high Re, its performance index reduced owing to improved viscosity and a decrease in perceptible heat.

Finally, despite its widespread application in electronic device thermal management, PCM still has several difficulties that need to be thoroughly investigated. Although the introduction of nanoparticles may enhance thermal management, the influence of concentration on heat transfer properties such as latent heat is yet unknown. For a complete evaluation of PCM operating functioning and performance under various situations, more precise models and analytical methodologies are also required. With the increased use of PCM in numerous industries, not only are additional chemical or physical measures required to improve PCM's thermal performance but the environmental friendliness of PCM should also be considered.

SOLUTIONS AND RECOMMENDATIONS

Despite significant advances in recent decades, the electronic and semiconductor industries continue to face significant technological hurdles in managing the heat of high-end electronic systems. This is due to traditional techniques and working fluids failing to fulfill the growing thermal management requirements of more heat-producing electronic gadgets. The majority of electrical equipment, on the other hand, uses traditional cooling methods. As a result, high-performance electronics goods (such as chips, devices, and so on) require novel mechanisms, procedures, and working fluids with greater capability to eliminate produced heat and maintain the desired efficiency. The most promising technologies, according to studies, are heat pipes and microchannels.

Nanofluids, on the other hand, as a new family of coolants, have much greater thermal properties than their basic conventional fluids with higher conductivity and convective heat transfer. These findings show that nanofluids are superior to their traditional counterparts when it comes to cooling. Furthermore, results from limited investigations on nanofluid applications revealed that this developing fluid worked superior as working fluids for electronics than traditional fluids.

Emerging cooling approaches, such as micro-channel systems, combined with these advanced fluids can significantly enhance heat removal rate and fulfill the cooling requirements of heat-producing electronic components. As a result, nanofluid-based micro-channels might be a next-generation cooling technique for electronics. However, additional research is needed before they can be realized and used on a commercial basis in the electronics industry.

FUTURE RESEARCH DIRECTIONS

The following might be considered based on the findings of the literature to stimulate the future development of sophisticated thermal management technologies:

- In electronics heat control, improvements in structural designs will continue to be critical. The system cooling capabilities may be enhanced effectively by enhancing the main configuration of augmentation methods like nozzle in spray cooling. In addition, electronic equipment is becoming increasingly miniaturized, portable, and flexible, posing new difficulties to heat management systems. To adapt to the evolution of electronic gadgets, investigating and inventing innovative techniques is essential.

- In electronic cooling, a nanofluid with a greater thermal conductivity was utilized. However, it still confronts challenges, such as the simple production of obstructions in flow-path and nozzles, which can harm thermal management systems and even electrical components.

- The surface qualities of the cooling fluid, such as wettability and roughness, have a substantial impact on the flow and thermal management of the working fluid. Clarifying the influence of different exterior qualities on their cooling performance is therefore critical for the design of various cooling techniques.

- The thermal conductivity of conventional materials is crucial to heat control. To enhance heat conductivity, appropriate alternative materials or manufacturing processes must be found. To identify suitable materials for effective cooling, more breakthroughs are required. Furthermore, the impact of heat stress and rust on materials must be taken into account.

- The thermal management system is frequently made up of several components, and optimization and innovative approaches to lowering total thermal resistance and energy consumption are still needed. The problem of being eco-friendly and recyclable must also be taken into account when designing a thermal management system.

CONCLUSION

The most recent advancements of frequently used approaches and technologies utilized for electronic equipment cooling are systematically examined in this work. This chapter addresses the major direct and indirect cooling methods to offer a broad summary of the cooling methods available and maybe inspire novel ones. Their heat transmission efficiency and use are emphasized in detail. It may be established that liquid cooling is more effective than air cooling and adaptive to the necessities of effective heat control as the heat flow of electronic devices increases. Because the cooling fluid contact with the heated exterior electronic equipment, direct contact cooling such as spray, jet impingement, and immersion cooling poses little heat conflict. Because it can regulate the directional travel of the liquid droplet to the target location, droplet electro-wetting can improve the cooling capabilities of hot areas. In terms of indirect cooling technologies, microchannel cooling uses two-phase heat transfer to enable effective heat dissipation and higher temperature homogeneity. Its tiny and lightweight design also allows it to be used with a variety of electrical equipment and even in other sectors. Heat pipes and vapour chambers, as types of heat spreaders, have a great capacity to convey heat to an external heat sink, providing more architectural concepts for electronic device thermal management systems. Finally, because of its unique heat management capabilities, PCM may be used in conjunction with various categories of heat distributors.

ACKNOWLEDGMENT

This research received no specific grant from any funding agency in the public, commercial, or not-for-profit sectors.

REFERENCES

Agostini, B., Fabbri, M., Park, J. E., Wojtan, L., Thome, J. R., & Michel, B. (2007). State of the art of high heat flux cooling technologies. *Heat Transfer Engineering*, *28*(4), 258–281. doi:10.1080/01457630601117799

Bahiraei, M., & Heshmatian, S. (2018). Electronics cooling with nanofluids: A critical review. *Energy Conversion and Management*, *172*, 438–456.

Behi, H., Ghanbarpour, M., & Behi, M. (2017). Investigation of PCM-assisted heat pipe for electronic cooling. *Applied Thermal Engineering*, *127*, 1132–1142.

Black, J. R. (1969). Electro migration - a brief survey and some recent results. *IEEE Transactions on Electron Devices*, *16*(4), 338–347. doi:10.1109/T-ED.1969.16754

Capozzoli, A., & Primiceri, G. (2015). Cooling systems in data centers: State of art and emerging technologies. *Energy Procedia*, *83*, 484–493. doi:10.1016/j.egypro.2015.12.168

Chang, Y. W., Cheng, C. H., Wang, J. C., & Chen, S. L. (2008). Heat pipe for cooling of electronic equipment. *Energy Conversion and Management*, *49*(11), 3398–3404. doi:10.1016/j.enconman.2008.05.002

Chen, X., Ye, H., Fan, X., Ren, T., & Zhang, G. (2016). A review of small heat pipes for electronics. *Applied Thermal Engineering*, *96*, 1–17. doi:10.1016/j.applthermaleng.2015.11.048

Colgan. (2007). A practical implementation of silicon microchannel coolers for high power chips. *IEEE Transactions on Components and Packaging Technologies*, *30*(2), 218–225.

Dix, J., Jokar, A., & Martinsen, R. (2008). A microchannel heat exchanger for electronics cooling applications. *Proceedings of 6th International Conference on Nanochannels, Microchannels and Minichannels.*

Faghri, A. (2012). Review and advances in heat pipe science and technology. *Journal of Heat Transfer, 134*(12), 123001. doi:10.1115/1.4007407

Farzanehnia, A., Khatibi, M., Sardarabadi, M., & Passandideh-Fard, M. (2019). Experimental investigation of multiwall carbon nanotube/paraffin based heat sink for electronic device thermal management. *Energy Conversion and Management, 179*, 314–325.

Guarnieri, M. (2016). The Unreasonable Accuracy of Moore's Law [Historical]. *IEEE Industrial Electronics Magazine, 10*(1), 40–43. doi:10.1109/MIE.2016.2515045

Ho, C., Liu, Y. C., Ghalambaz, M., & Yan, W. M. (2020). Forced convection heat transfer of nano-encapsulated phase change material (NEPCM) suspension in a mini-channel heatsink. *International Journal of Heat and Mass Transfer, 155*, 119858.

Kandilikar, S., & Upadhye, H. (2005). Extending the heat flux limit with enhanced micro-channels in direct single-phase cooling of computer chips. *Proceedings of 21st Semi Therm Symposium,* 8-15.

Kandlikar, S. G., Colin, S., Peles, Y., Garimella, S., Pease, R. F., Brandner, J. J., & Tuckerman, D. B. (2013). Heat transfer in micro-channels, *status and research needs. Journal of Heat Transfer, 135*, 091001.

Kim, Y. J., Joshi, Y. K., & Fedorov, A. G. (2008). An absorption miniature heat pump system for electronics cooling. *International Journal of Refrigeration, 31*(1), 23–33. doi:10.1016/j.ijrefrig.2007.07.003

Krishna, J., Kishore, P., & Solomon, A. B. (2017). Heat pipe with nano enhanced-PCM for electronic cooling application. *Experimental Thermal and Fluid Science, 81*, 84–92.

Moore, G. E. (1998). Cramming more components onto integrated circuits. *Proceedings of the IEEE, 86*(1), 82–85. doi:10.1109/JPROC.1998.658762

Murshed, S. S., & De Castro, C. N. (2017). A critical review of traditional and emerging techniques and fluids for electronics cooling. *Renewable & Sustainable Energy Reviews, 78*, 821–833. doi:10.1016/j.rser.2017.04.112

Murshed, S.M.S. (2016). *Electronics Cooling.* Intech Open.

Qu, W., & Mudawar, I. (2004). Measurement and correlation of critical heat flux in two-phase microchannel heat sinks. *International Journal of Heat and Mass Transfer, 47*, 2045–2059.

Tong, H.-M., Lai, Y.-S., & Wong, C. (2013). *Advanced Flip Chip Packaging*. Springer. doi:10.1007/978-1-4419-5768-9

Tong, H.-M., Lai, Y.-S., & Wong, C. P. (2013). *Advanced Flip Chip Packaging*. Springer. doi:10.1007/978-1-4419-5768-9

Tuckerman, D., & Pease, R. (1981). High-performance heat sinking for VLSI. *IEEE Electron Device Letters, 2*(5), 126–129. doi:10.1109/EDL.1981.25367

Wang, C.-C. (2017). A quick overview of compact air-cooled heat sinks applicable for electronic cooling - recent progress. *Inventions (Basel, Switzerland), 2*(1), 5. doi:10.3390/inventions2010005

Wei, Y., & Joshi, Y.K. (2004). Stacked microchannel heat sinks for liquid cooling of microelectronic components. *J Elect Pack, 126*(6).

Williams, C. (2015). *MGMT9*. 4LTR Press.

Yan, W. M., Ho, C., Tseng, Y. T., Qin, C., & Rashidi, S. (2020). Numerical study on convective heat transfer of nanofluid in a minichannel heat sink with micro-encapsulated PCM-cooled ceiling. *International Journal of Heat and Mass Transfer, 153*, 119589.

Yang, X. H., Tan, S. C., He, Z. Z., & Liu, J. (2018). Finned heat pipe assisted low melting point metal PCM heat sink against extremely high power thermal shock. *Energy Conversion and Management, 160*, 467–476.

Yao, S., Lu, Y. W., & Kandlikar, S. G. (2012). Pool boiling heat transfer enhancement through nanostructures on silicon microchannels. *ASME J Nanotechnol Eng Med, 3*, 031002.

Yunus Cengel, A. (2003). *Heat Transfer-A Practical Approach*. McGraw-Hill.

Zhang, S., Feng, D., Shi, L., Wang, L., Jin, Y., Tian, L., Li, Z., Wang, G., Zhao, L., & Yan, Y. (2021). A review of phase change heat transfer in shape-stabilized phase change materials (ss-PCMs) based on porous supports for thermal energy storage. *Renewable & Sustainable Energy Reviews, 135*, 110127.

ADDITIONAL READING

Anandan, S. S., & Ramalingam, V. (2008). Thermal management of electronics: A review of literature. *Thermal Science*, *12*(2), 5–26. doi:10.2298/TSCI0802005A

Azar, K. (1997). *Thermal Managements in Electronic Cooling*. CRC Press.

Bang, I. C., & Chang, S. H. (2005). Boiling heat transfer performance and phenomena of Al_2O_3-water nanofluids from a plain surface in a pool. *International Journal of Heat and Mass Transfer*, *48*(12), 2407–2419. doi:10.1016/j.ijheatmasstransfer.2004.12.047

Escher, W., Brunschwiler, T., Shalkevich, N., Shalkevich, A., Burgi, T., Michel, B., & Poulikakos, D. (2011). On the cooling of electronics with nanofluids. *Journal of Heat Transfer*, *133*(5), 051401. doi:10.1115/1.4003283

Ijam, A., & Saidur, R. (2012). Nanofluid as a coolant for electronic devices (cooling of electronic devices). *Applied Thermal Engineering*, *32*, 76–82. doi:10.1016/j.applthermaleng.2011.08.032

Liu, Z. H., & Li, Y. Y. (2012). A new frontier of nanofluid research – application of nanofluids in heat pipes. *International Journal of Heat and Mass Transfer*, *55*(23-24), 6786–6797. doi:10.1016/j.ijheatmasstransfer.2012.06.086

Roberts, N. A., & Walker, D. G. (2010). Convective performance of nanofluids in commercial electronics cooling systems. *Applied Thermal Engineering*, *30*(16), 2499–2504. doi:10.1016/j.applthermaleng.2010.06.023

Scott, W. A. (1974). *Cooling of Electronic Equipment*. John Wiley and Sons.

KEY TERMS AND DEFINITIONS

Conduction: It is a mode of heat transfer that happens without molecular motion, caused due to temperature differences.

Convection: Convection is stated as the heat transfer due to the bulk motion of a liquid caused by density changes in the liquid.

Heat Pipe: A heat pipe is a heat-transfer equipment that transfers heat between two solid contacts via phase transition.

Heat Pump: A heat pump transfers heat energy from one location to another. They are frequently used to transmit thermal energy by collecting heat from a cold environment and transferring it to a warmer environment.

Heat Sink: A heat sink is a device that removes heat from electronic components or chips.

Microchannel: In microtechnology, a microchannel is a channel having a hydraulic diameter of less than 1 mm. Fluid control and heat transmission are two applications for microchannels.

Nanofluid: A nanofluid is a fluid that contains nanoparticles, which are nanometer-sized particles. These fluids are colloidal nanoparticle suspensions in a base fluid that has been prepared.

Radiation: It is the mode of heat transfer and flows in the form of waves without media.

Chapter 7
AIoT and Deep Neural Network–Based Accelerators for Healthcare and Biomedical Applications

Jothimani K.

iD https://orcid.org/0000-0002-4179-3348
Graphic Era University, India

Bhagya Jyothi K. L.
KVG College of Engineering, India

ABSTRACT

Convolutional neural network (CNN) systems have an increasing number of applications in healthcare and biomedical edge applications due to the advent of deep learning accelerators and neuromorphic workstations. AIoT and sense of care (SOC) medical technology development may benefit from this. In this chapter, the authors show how to develop deep learning accelerators to address healthcare analytics, pattern classification, and signal processing problems using emerging restrictive gadgets, field programmable gate arrays (FPGAs), and metal oxide semiconductors (CMOS). Neuromorphic processors are compared with DL counterparts when it comes to processing biological signals. In this study, the authors focus on a range of hardware systems that incorporate data from electromyography (EMG) and computer vision. Inferences are compared using neuromorphic processors as well as integrated AI accelerators. In the discussion, the authors examined the issues and benefits, downsides, difficulties, and possibilities that various acceleration and neuromorphic processors bring to medicine and biomedicine.

DOI: 10.4018/978-1-6684-4974-5.ch007

INTRODUCTION

Artificial intelligence can best manage the expanding needs of the universal healthcare system (Goyal, 2019). By 2022, it is anticipated that the healthcare sector will generate over $10 trillion in revenue, and the related burden for medical professionals would increase at a similar rate (Ditterich Thomas G, 2000) To monitor the patient for abnormalities and/or to anticipate illnesses, a smart DL system may be used to process data obtained from the persistent, which may be a mix of biomodels, medical imaging, disease, progress, etc. Deep Learning (DL), whose accuracy is rising, has infiltrated many facets of healthcare, including monitoring, prediction, diagnosis, therapy, and prognosis (Frenkel, 2019). In a closed-loop scenario, DL systems may be utilized to provide prognosis and treatment options, which influence monitoring and prediction.

AIoT is a helpful tool for infrastructure development, traffic monitoring, and other areas of transportation. AIoT connects with the signal in the contemporary healthcare system because of this benefit (Y. Geng, 2018).

The healthcare system offers the initial link (T. Hirtzlin, 2019) and a higher standard of living (Frenkel, 2019). Machine learning for prediction is the primary goal of this effort. Connecting sensors with AIoT enables the collection of uninterrupted time series information and cloud-based data processing. Heart illness is challenging to diagnose (O. Krestinskaya, 2018), and many times individuals are not even conscious that they are ill until they have cardio problems like arrhythmia or even stroke. A trained physician must examine the individual to identify the usual symptoms of a cardiac illness to make the diagnosis of heart disease. Currently, there are insufficient doctors, and most countries do not believe computers can accurately detect cardiac problems with clarity and precision (X. Zhang, 2017). To ensure that customers obtain the results before the deadline, existing healthcare systems connect before devices for patient information processing. These systems are constructed using cloud or fog computing platforms that are AIoT driven. Many earlier studies tried to predict heart-related health conditions using AIoT, but they were able to do so with the degree of accuracy necessary by the strict standards of medical standardizing bodies. Given the prominence of deep learning in recent years, more modern technology can even identify cardiac problems with more precision than doctors can (A. R.Aslam, 2019).

To address the doctor shortage issue, this effort attempts to combine deep learning with AIoT in the healthcare sector. It is hoped that this would encourage medical standards organizations to embrace a model that offers low latency and high accuracy. Few researchers, such as (C. Lammie, 2020), have attempted to combine these two paradigms, but none have taken use of edge computing's dispersed nature to boost accuracy using ensemble deep learning models. Data from AIoT systems smart devices is collected, aggregated, and then processed and stored on network edge or public cloud.

An integrated Edge- DoctorAI based computing paradigm is needed to provide healthcare and other delay discoveries with quick responses, low energy consumption, and high accuracy. This is crucial to provide effective computing services for the benefit of cardiac patients and other clients for the need of actual outcomes. We are inspired by the dearth of simulations which is fusing the strength of edge computing nodes with low latency with high accuracy deep learning methods.

In this work, we introduce **DoctorAI**, a smart health service for heart disease diagnostics. It is based on deep learning and the Artificial Internet of Things (AIoT) and completely handles patient information through cloud facilities. Using the **DoctorAI** framework (X. Zhang, 2017), **DoctorAI** provides this service and demonstrates engineering ease and application enabling for doing the same.

Contributions are as follows:

- The utilization of three distinct developing and proven hardware technologies for DL acceleration is the subject of our study, which is focused on biomedical products.
- For the first time, we describe how DL accelerators may be used in conjunction with event-based neuromorphic processors for biological signal processing.
- The created DoctorAI, a portable solution for automatically diagnosing heart patient data using ensemble deep learning.

The relevant study of current healthcare systems is presented in Section 2. Another Section provides information on the backgrounds of DoctorAI and Aneka. DL accelerators and system design are described in Section 4 and Section 5. The experimental setup is described in Section 6 along with the performance assessment findings and results are discussed in Section 7. Section 8 concludes with suggestions for more research.

An overview of the (a). the use of DL in an intelligent healthcare environment, which usually entails checking, forecasting, diagnosing, treating, and predicting outcomes. The many components of the healthcare system based on DL can run on (b). the edge devices, edge nodes, and cloud—the three AIoT tiers. However, edge learning and inference are chosen for AIoT and PoC processing in healthcare.

The potential of cloud computing has been proved by specialized embedded DL accelerators such as the Movidius Brain Compute Stick (S. Sagar, 2018), the Nvidia Jetson & Xavier series (M. Payyand, 2020), and Nvidia Jetson and Xavier series (A. R. Aslam, 2019). A more modern edge accelerator made specifically for the healthcare industry is the Nvidia Clara Embedded. This is an edge-enabled AI computer system for the Internet of Health Things (AIoT). Many cutting-edge algorithms greatly

Figure 1. An overview of the (a). the use of DL in an intelligent healthcare environment.

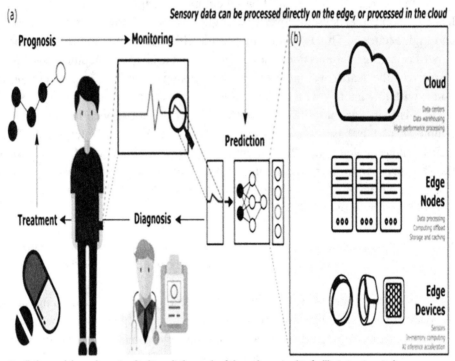

exceed the memory capacity of systems with constrained resources, yet embedded systems continue to be comparatively costly and power-hungry. Systems that assist personalized medicine in the environment, are not proven for the greatest hardware technology. To make the availability of these products and edge gadgets across the world, we must be able to satisfy their exacting requirements for implementation.

In this study, we discuss how three different hardware technologies can be used to create specialized deep network accelerators, with a focus on biomedical and healthcare applications. These three technologies—memristors, field-programmable gate arrays, and CMOS—are what this essay examines (FPGAs). Though the technique and hardware benefits are aimed at edge inference engines in the biomedical field, these techniques will likely be beneficial for efficient offline deep CNN learning or live on-chip learning as well. A device that can do DL inference and possibly training is alluded to as a "DL accelerator" in this context.

RELATED WORK

The fog data center is a new paradigm for effectively handling healthcare data from various AIoT devices. Fog computing may manage the heart patients' information through different types of network borders or nodes with appropriate managing capability to decrease inactivity and reaction time. AIoT handles this using cloud servers, and it avoids latency.

A Low Price Health Observing (LPHO) methodology was put up by (Gia et al.,2019) to collect data on the health of various cardiac patients. Additionally, sensor nodes continuously track and evaluate the Electro Cardio Graph (ECG) to process patient data effectively, although LCHM's slower reaction time lowers performance. Additionally, sensor nodes collect data on body temperature, respiration rate, and ECG and communicate it wirelessly to a smart gateway so that it may make an immediate decision to assist the patient. The performance of the LCHM model is tested on an Orange Pi One-based tiny testbed in terms of execution time, however, LCHM uses more energy during data collecting and transmission.

On a Software Defined Network (SDN) application, (Ghazal & Ali, 2019) presented an Internet of Things (AIoT) e-health services company that gathers data via smartphone utilizing speech recognition for evaluating the patient's health progress. Another AIoT e-health service uses a conceptual system based on a phone app to identify the kind of heart attack, even if the efficiency of the suggested application is not assessed in cloud settings. To diagnose cardiac irregularity using an ECG, (Krestinskaya et al, 2018) proposed an ECG-based Healthcare Insurance (ECGH) system; however, the system has a poor sensitivity and a high quick response when attempting to identify anomalous events because it uses retrieving the info directly without ever using analytics or other methodologics for extracting features.

(Choi et al., 2017) created visualization-built focused method in healthcare transfer learning which is supplementing digital health records with data structures from medical ontologies. Additionally, GRAM's training accuracy performance has improved. When compared to RNN using a moderately limited dataset, GRAM outperforms RNN with the accuracy level. The probability of a heart attack is calculated by GRAM using predictive analysis. Large datasets could make GRAM perform worse. By highlighting the function of the Smart Fog Gateway (SFG) in managing the data collecting conditioning, clever filtering, intelligent analysis, and selected transmission to the cloud, (Nicholas et al., 2017) presented the DoctorAI Portal (SFG) architecture data processing in wearable AIoT devices.

In addition to exploring the use of Cloud Fog abilities in interoperable Healthcare solutions that go beyond conventional Cloud-based structure, (Sagar et. al,2018) provided an AIoT-Healthcare appropriate design. Additionally, the operation of the FIH solution is evaluated only in terms of power and latency consumption using the iFogSim simulator. The accuracy and performance time of the FIH approach may be used to gauge its effectiveness. A fog-based technique for computer vision for smart networks big data analytics, known as DoctorAI, was presented by (Rabindra and Rojalina 2021) for application in Ganga Basin Management using real feature data for identifying diabetics with diabetes.

(Greco L, Percannella, et al, 2020) gave a small review of the use of IoT methods and its solutions for health care, which starts from the early stages using wearable device sensors with various modern tendencies around fog environment for smart health-based systems. They have given a general overview of multilevel architecture which contains three levels namely, Edge level, Fog level, and Cloud level.

(Manjula K, Bhavana VN, et al, 2022) gave a deep network accelerator method that contains the CNN method for health care and biomedical applications. They have given brief introductions about various technologies which are used in healthcare industries.

(Mostafa Rahimi Azghadi et al, 2020) provided a tutorial giving various technologies and methods including the latest emerging tools and devices and Field Programmable Gate Arrays (FPGAs), with (CMOS) Complementary Metal Oxide Semiconductors which are applied to design robust deep learning accelerators to give a solution to different health-related applications with the help of advanced methodologies. They also explored how neuromorphic processors will complement their DL technologies for processing various biomedical signals. They also presented various literature surveys on neural networks and neuromorphic hardware used in the healthcare domains. Computer vision is combined with electromyography (EMG) signals in different sectors. Inference latency and energy are compared between neuromorphic processors and embedded AI accelerators. Finally, they analyzed the advantages, limitations, challenges, and possible various opportunities in health care systems.

BACKGROUND TECHNOLOGIES

The structure for creating and deploying integrated, cross-platform Cloud cover settings with English proficiency is called DoctorAI (Ditterich Thomas G, 2000). DoctorAI connects different AIoT sensors, including those utilized in the healthcare sector, in connection to gateway devices, for giving information to fog employee nodes. On fog broker nodes, tasks are started, and resources are managed. DoctorAI combines blockchain, identity, and encryption approaches that improve the robustness and dependability of the cloud environment to guarantee data integrity, secrecy, and security. Using HTTP Rest API for communication, DoctorAI can quickly connect a fog installation to the Cloud thanks to the Aneka technology platform (A. Valentian, 2019).

Distributed applications can be more easily contained and deployed onto clouds thanks to a software system and design called Aneka (C. Barbieri, 2019). The APIs provided by Aneka allow developers to access virtual resources hosted in the cloud. As part of the Aneka architecture, several foundational components are created and applied in a provider-based manner. Dynamic provisioning, as the name implies, refers to the ability to deal with changing resources as they become available and incorporate them into existing systems and infrastructure as needed. The famous resource purchased from a cloud provider that offers infrastructure as a service (IaaS) is virtual machines (VMs). By providing services for creating effective connections from public cloud providers to supplement regional supplies, Aneka's Network Tools enable production migration. The interplay of two systems, the Scheduler Service, and the Provisioning Service, largely makes this feasible. Bag of Tasks, Shared Threads, Hadoop, and Parameter Sweeping are the four programming paradigms that Aneka currently supports (M. Payyand, 2020).

DEEP NEURAL ACCELERATORS FOR HEALTHCARE & BIOMEDICAL

We proceed using CMOS and different DL acceleration members. We go over the many methods that employ to accomplish MAC parallelism and decreased memory access, two of the main DNN acceleration objectives. This article also explores and reviews FPGAs as a unique programmable Deep Neural accelerator tech that has a lot of potential in the biological and medical fields.

DNN Based CMOS Accelerators

Algorithms of medical AIoT and PoC systems can leverage common edge-AI CMOS acceleration processors. Because of this, we first evaluate a few of these technologies in this section and give illustrations of potential smart healthcare advancements they may facilitate. We will also go through a variety of popular methods for CMOS-driven AI algorithm acceleration that take advantage of edge-AI devices as well as offline data centre scale speed. Massive MAC concurrency and constrained memory access are used in these methods. The need for ASICs, which concentrate on a more current generation of microprocessors designed with computer vision and DNNs in mind, has grown dramatically in recent years, as has research in this field. Several significant AI acceleration chips that are created primarily for pattern classification (DL) employing DNNs, RNNs, or both are chosen in this article. There are several

Figure 2. Different operational models are adopted by DNNs and SNN neuromorphic processors.

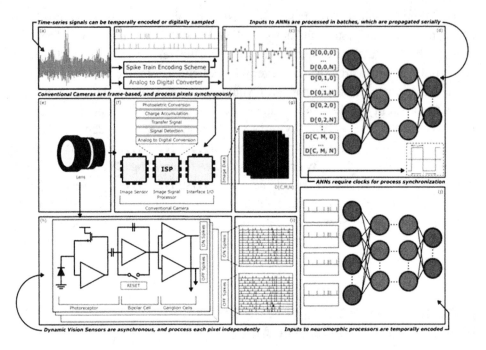

such examples of AI acceleration devices (for a comprehensive analysis, (Barbieri, 2019)). Additionally, there are a few only those AI accelerators from Huawei and Google (X. Zhang, 2017).

Most of these devices generally only consume a few hundred microwatts (mW) of power per chip (Frenkel, 2019). To avoid using bulky heat sinks and to meet the requirements of portable batteries, this is necessary. Additionally, it provides computing power per second in the "Computational Power (GOP/s)".

For example, VGG CNN, which has been shown to work with Cambricon-x (A. Valentian, 2019), has been successfully used in (T. Hirtzlin, 2019) to analyze ECog signals. A portable automatic Eco analyzer again for PoC diagnosis of several cardiovascular disorders may be constructed utilizing Cambricon-high x because of its power efficiency. Everiss can carry it out similarly to VGG-16, which has been used well to identify thyroid cancer. Eyeriss may also utilize AlexNet for several purposes, including diagnostic imaging. While describes a CNN-based ECG analysis for heart monitoring or offers a two-stage final CNN for person action detection for aging and rehab monitoring, origami may be used to develop a smart health AIoT edge device. This shows that AlexNet, which might be utilized in a proof-of-concept sonar image processing system, can run on the CNN processor recommended (T.

Hirtzlin, 2019). Another accelerate that can execute lengthy CNNs is Envision. To extract EEG/ECog properties for epilepsy diagnosis, it may also be employed as an edge inference for an inter CNN (A. Valentian, 2019), Another CNN accelerator that has been demonstrated to be capable of running Fully Convolutional CNN and being utilized for edge skin cancer detection is the neural processor (Frenkel, 2019). The only CNN accelerator that can train and infer a convolutional model, like AlexNet and VGG16, is LNPU (Y. Geng, 2018). This contrasts with the other CNN accelerators.

SYSTEM DESIGN

The DoctorAI model is a public cloud for health AIoT-based fog-enabled framework that can handle the data of heart patients and evaluate their health condition to determine the severity of heart disease. By combining a range of hardware instruments with software components, DoctorAI provides organized and seamless end-to-end connectivity of Edge-Fog Cloud for speedy and accurate results delivery. The hardware and software components that make up DoctorAI's architecture, as seen in Figure 1, will be covered in greater detail below.

DoctorAI Hardware Components

The following hardware elements are included in the DoctorAI model (Shreshth Tuli, 2020):

1.) **Network of Body Area Sensors:** The activity, environment, and health sensors make up most of this component. Examples of medical sensors include electrocardiogram (ECG), electroencephalogram (EEG), electromyography (EMG), oxygen level, temp, respiration rate, and glucose level sensors. It collects data from heart disease patients and sends it to neighboring gateway devices.

2) **Access Point:** Mobile phones, laptops, and tablets are three distinct types of gateway devices that act as fog devices to gather sensed data from various sensors and transmit it to Broker / Worker nodes for additional dispensation.

3) **DoctorAI Broker Node:** This part accepts information entry requests and/ or task requests from Gateway devices. Before data transfer, the demand input module receives work orders from Gate devices. To increase system dependability and data integrity, the System Security module provides

Figure 3. DoctorAI System Architecture

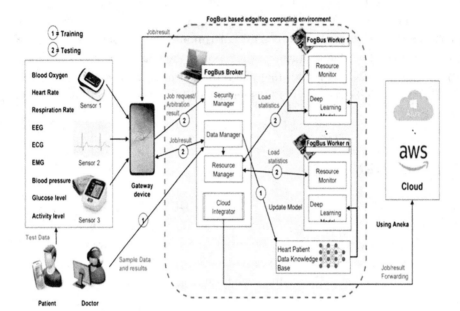

secure communication between various elements along with safeguards, where the data is accumulated from unauthorized or damaging data alteration. The Scheduler in the broker base class Arbitration module uses the data from each working node as an input to decide the node which is assigned to perform the actual tasks.

4) **DoctorAI Worker Node:** The Resource Manager of the Broker node delegated these duties to this component. Worker nodes can be created using embedded hardware or Single Board Computers (SBCs) like Raspberry Pis. Complex deep learning algorithms may be used by DoctorAI's worker nodes to process, analyze, and provide results from input data. Additional components for information processing, data gathering, information extraction, big data analytics, or storage may also be included in the Worker node. The Worker nodes create results after getting input directly from the Gate devices and disseminate them to the same devices. In the DoctorAI architecture, the Broker node can serve as a Worker node.

5) **Virtual Datacenter:** When the cloud network is overloaded, services are delayed, or the input data size is much larger than usual, DoctorAI uses cloud data center resources (CDC). Because of this, it is more resilient,

quick to manage heavy workloads, and location independent while processing data.

Figure 4. Resource Scheduling in DoctorAI

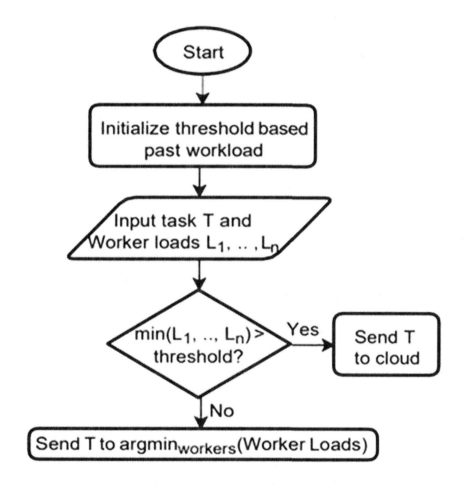

IMPLEMENTATION

The aspects described in Section 5 were accomplished in a variety of ways using programming languages. The ensemble deep learning and pre-processing components were implemented using Python. Using the distribution of the dataset's highest and lowest field parameter values, the pre-processing module generalizes the data.

SciKit Learn Library was utilized by the ensemble deep learning application (X. Zhang, 2017). To create our voting system, we utilized BaggingClassifier from the SciKit Learn Library. The model accepts as inputs the kind of base classifier—in our example, a deep neural network—and the total number of classifiers. To train the classifiers, the model now randomly divides the data among them. It uses all anticipated classes as input at a diagnostic time and delivers the majority prediction. The criteria for the primary structure are as follows:

- input layer size: Thirteen
- The output layer's size is 2, and it contains binary classification information on the patient's cardiac condition.
- 3 hidden layers total
- The layers are described as connected levels having 20 nodes, FC level with the number of nodes, and FC level with 10 nodes.
- Adam Optimizer
- ReLU Stimulation Function

- **LR Ratio 0.0001**

MIT's App Inventor1 was used to create the Android application connected to the DoctorAI Broker node. The data properties are saved by the Android application in a Comma Separated Value (.csv) file, which is then uploaded via Hyper Text Transfer Protocol to the Data Collection Module on the broker node.

The Negotiation Module on the broker node chooses which worker node will carry out a job. According to the standard DoctorAI policy of choosing workers with the least amount of CPU burden, this worker selection procedure is carried out. The CSV file is forwarded to the chosen worker for analysis. Each worker's Execution Interface Module gets the data and creates an instance of the Ensemble Deep Learning code to analyze it. The designated worker/broker node received the originally submitted relevant files and produced the outcomes. The outcome is combined using the trapping technique and sent to the entryway device.

RESULTS

The impact of training accuracy is illustrated in Figure 6. It shows how the edge node affects the system. As can be seen, training accuracy steadily rises as the number of slave node grows. This is because each device receives fewer cases as the number of nodes grows since every single node builds a model for the data it gets. At the

Figure 5. Different modules in doctorai

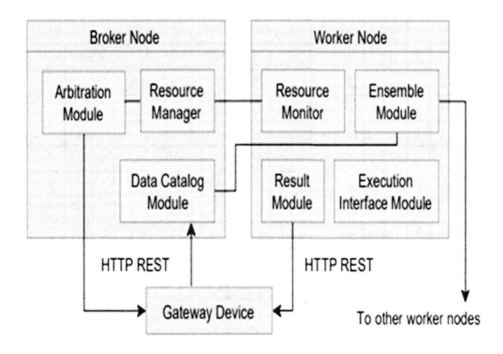

outset, overfitting the samples occurs during model training over numerous epochs, increasing training accuracy.

The change in test data accuracy as the number of Edge nodes rises is seen in Figure 7. The test accuracy declines with increasing node count due to the fact that each node receives a smaller portion of training data, preventing the model from being generalized. Furthermore, ensemble learning almost always results in substantially higher accuracy than it alone.

The likelihood p0 (chance of no disease) and p1 are generated each time the deep learning system analyses and produce the outcome of the patient according to his health, with the outcome that $p_0 + p_1 = 1$. The assurance interval for a prediction is [0,100], and it is computed as 100 (2 max (p_0, p_1) 1). As a result, if the forecast probabilities are (0.5, 0.5) and (0.9, 0.1), respectively, the forecast class is 0% and the trust is 80%.

Figure 8 displays the variance in the binary classifier's level of confidence for the entire test dataset, the subsets for accurate predictions, as well as the subsets for which it made inaccurate predictions. When comparing data points where the forecast was accurate to those where it was inaccurate, we can observe that the confidence

Table 1. Sample data

age	sex	cp	trestbps	chol	fbs	restecg	thalach	exang	oldpeak	slope	ca	thal	target
63	1	3	145	233	1	0	150	0	2.3	0	0	1	1
37	1	2	130	250	0	1	187	0	3.5	0	0	2	1
41	0	1	130	204	0	0	172	0	1.4	2	0	2	1
56	1	1	120	236	0	1	178	0	0.8	2	0	2	1
57	0	0	120	354	0	1	163	1	0.6	2	0	2	1
62	0	0	140	268	0	0	160	0	3.6	0	2	2	0
63	1	0	130	254	0	0	147	0	1.4	1	1	3	0
53	1	0	140	203	1	0	155	1	3.1	0	0	3	0
56	1	2	130	256	1	0	142	1	0.6	1	1	1	0
48	1	1	110	229	0	1	168	0	1	0	0	3	0

is higher for the right data points. The model's highest error rate is **60.77%,** thus if the confidence level is lower than 50%, our model advises the patient to visit a physician as the forecast could not be accurate.

CONCLUSION

Providing healthcare is a huge undertaking. We exclusively discuss the medical needs of people with heart illnesses. In this paper, we recommend DoctorAI, a cutting-edge smart medical system for the automatic identification of diseases based on DoctorAI. DoctorAI offers healthcare as a DoctorAI service and effectively handles patient data received from various AIoT devices for heart disease patients. For the examination of a's understanding of cardiac disease, DoctorAI employed deep learning on Edge desktop PCs. Because they did not use deep learning, earlier studies for this kind of study on heart patients had very poor prediction accuracy and were therefore useless in practical uses. During both training and during prediction, high-accuracy deep learning algorithms require a lot of processing resources (CPU and GPU). This study made it possible to incorporate complex deep learning networks into edge paradigms, producing excellent accuracy and incredibly low latencies. This was done by using cutting-edge modeling distribution and communications

Figure 6. Training accuracy

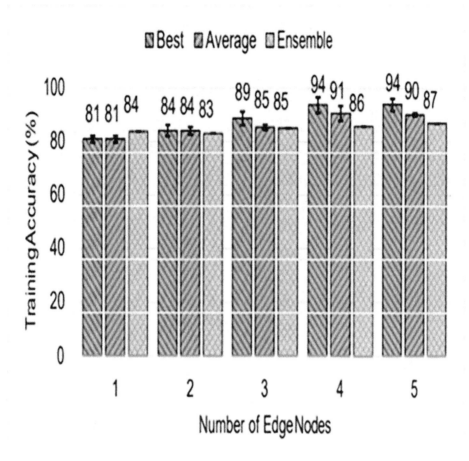

methodologies, such as assembly. This was also confirmed for the analysis of real-life cardiac patient information by constructing in real-time and training neural networks on well-known datasets. To validate DoctorAI in a cloud computing environment, we evaluated the efficacy of the proposed system in terms of energy consumption, bandwidth, speed, jitter, training efficiency, testing accuracy, and execution time that use the DoctorAI framework.

Figure 7. Test accuracy

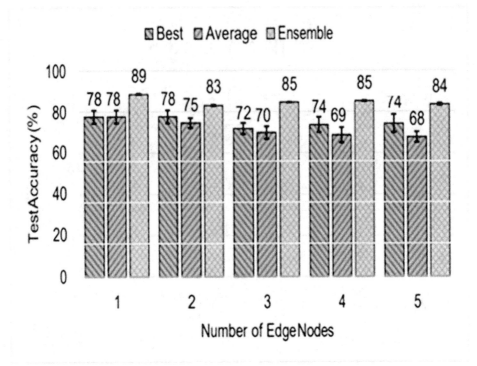

Figure 8. Model confidence for various Cleveland Data subsets

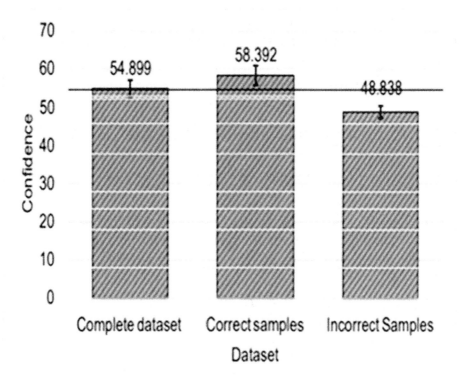

REFERENCES

Aslam, A. R., & Altaf, M. A. B. (2019). An 8 Channel patient specific neuromorphic processor for the early screening of autistic children through emotion detection. *Proc. IEEE Int. Symp. Circuits Syst. (ISCAS)*, 1–5.

Azghadi, M. R., Eshraghian, J. K., & Linares-Barranco, B. (2020). Hardware Implementation of Deep Network Accelerators Towards Healthcare and Biomedical Applications. *IEEE Transactions on Biomedical Circuits and Systems*.

Barbieri, C. (2019). Development of an artificial intelligence model to guide the management of blood pressure, fluid volume, and dialysis dose in end-stage kidney disease patients: Proof of concept and first clinical assessment. *International Journal of Medical Informatics*, 5(1), 28–33.

Choi, E., Bahadori, M. T., Song, L., Stewart, W. F., & Sun, J. G. R. A. M. (2017). Graph-based Attention Model for Healthcare Representation Learning. *KDD: Proceedings / International Conference on Knowledge Discovery & Data Mining. International Conference on Knowledge Discovery & Data Mining*, 787–795. doi:10.1145/3097983.3098126

Dietterich, T. G. (2000). *Ensemble methods in machine learning*. In *International workshop on multiple classifier systems* (pp. 1-15). Springer.

Frenkel, J.-D. L., & Bol, D. (2019). MorphIC: A 65-nm 738kSynapse/mm 2 Quad-Core Binary-Weight Digital Neuromorphic Processor With Stochastic Spike-Driven Online Learning. *IEEE Transactions on Biomedical Circuits and Systems*, *13*(5), 999–1010.

Geng, Y., Mingzhe, J., Wei, O., & Guangchao, J. (2018). AIoT-based remote pain monitoring system: From device to cloud platform. *IEEE Journal of Biomedical and Health Informatics*, *22*(6), 1711–1719. doi:10.1109/JBHI.2017.2776351 PMID:29990259

Goyal, A., Narang, K., & Ahluwalia, G. (2019). Seasonal variation in 24 h blood pressure profile in healthy adults-A prospective observational study. *Journal of Human Hypertension*.

Greco, L., Percannella, G., Ritrovato, P., Tortorella, F., & Vento, M. (2020). Trends in IoT-based health care solutions: Moving AI to the edge. *Pattern Recognition Letters*, 346–353. doi:10.1016/j.patrec.2020.05.016

Guan, Q., Wang, Y., Ping, B., Li, D., Du, J., Qin, Y., Lu, H., Wan, X., & Xiang, J. (2019). Deep Convolutional Neural Network VGG-16 Model for Differential Diagnosing of Papillary Thyroid Carcinomas in Cytological Images: A Pilot Study. *Journal of Cancer*, *10*(20), 4876.

Guang, N. L. L., Logenthiran, T., & Abidi, K. (2017). Application of Internet of Things (IoT) for home energy management. *IEEE PES Asia-Pacific Power and Energy Engineering Conference (APPEEC)*, 1-6. doi: 10.1109/APPEEC.2017.8308962

Hirtzlin, T., Bocquet, M., Penkovsky, B., Klein, J.-O., Nowak, E., Vianello, E., Portal, J.-M., & Querlioz, D. (2019). Digital Biologically Plausible Implementation of Binarized Neural Networks With Differential Hafnium Oxide Resistive Memory Arrays. *Frontiers in Neuroscience*, *13*. PMID:31998059

Hsieh, J. H., Lee, R. C., Hung, K. C., & Shih, M. J. (2018). Rapid and coding efficient SPIHT algorithm for wavelet-based ECG data compression. *Integration, the VLSI Journal*, *60*, 248-256.

Krestinskaya, O., Salama, K. N., & James, A. P. (2018). Learning in Memristive Neural Network Architectures Using Analog Backpropagation Circuits. *IEEE Transactions on Circuits and Systems. I, Regular Papers, 66*(2), 719–732.

Lammie, C., Xiang, W., Linares-Barranco, B., & Azghadi, M. R. (2020). *MemTorch: An Open-source Simulation Framework for Memristive Deep Learning Systems.* arXiv preprint arXiv:2004.10971.

Manjula, K., & Bhavana, V.N. (2022). Deep Network Accelerators Towards Healthcare Edge Applications And Systems. *International Research Journal of Modernization in Engineering Technology and Science, 4*(7).

Payvand, M., Demirag, Y., Dalgaty, T., Vianello, E., & Indiveri, G. (2020). Analog weight updates with compliance current modulation of binary ReRams for on-chip learning. In *Proceedings of the IEEE International Symposium on Circuits and Systems (ISCAS).* IEEE.

Priyadarshini, R., Barik, R. K., Dubey, H. C., & Mishra, B. K. (2021). A Survey of Fog Computing Based Healthcare Big Data Analytics and Its Security. *International Journal of Ambient Computing and Intelligence, 12*(2), 53–72.

Sagar, S., Keke, C., & Amit, S. (2018). Toward practical privacy-preserving analytics for AIoT and cloud-based healthcare systems. *IEEE Internet Computing, 22*(2), 42–51. doi:10.1109/MIC.2018.112102519

Tuli, S., Basumatary, N., Gill, S. S., Kahani, M., Arya, R. C., Wander, G. S., & Buyya, R. (2020). HealthFog: An Ensemble Deep Learning based Smart Healthcare System for Automatic Diagnosis of Heart Diseases in Integrated IoT and Fog Computing Environments. *Future Generation Computer Systems, Elsevier, 104*, 187–200.

Valentian, A., Rummens, F., & Vianello, E. (2019). Fully Integrated Spiking Neural Network with Analog Neurons and RRAM Synapses. *Proceedings of the IEEE International Electron Devices Meeting (IEDM)*, 14.13.1–14.13.4.

Zhang, X., & Wang, D. L. (2017). Deep learning based binaural speech separation in reverberant environments. *IEEE/ACM Transactions on Audio, Speech, and Language Processing, 25*(5), 1075–1084.

Section 2

Low–Power Technologies in IoT, WSNs, and Embedded Systems

Chapter 8

Review of Applications of Energy Harvesting for Autonomous Wireless Sensor Nodes

Wilma Pavitra Puthran
Microsoft, India

Sahana Prasad
T-Systems International GmbH, Germany

Rathishchandra Ramachandra Gatti
iD https://orcid.org/0000-0002-4086-2778
Sahyadri College of Engineering and Management, India

ABSTRACT

Energy harvesting has been the empowering innovation in the internet of things to power the wireless sensors envisioned to be deployed ubiquitously. In recent decades, there has been an increasing drift towards remote sensor systems from wired networks in commercial and industrial applications due to expensive cabling and their non-feasibility in remote locations. The challenge is to convert these remote sensor systems into self-powered wireless sensor networks using energy harvesters. A brief review of current trends in the applications of energy harvesting in remote sensor systems is discussed in this chapter. A generic architecture of the energy harvesters and their transduction mechanisms and the design methodology of energy harvesters is introduced. The existing business products and the potential prototypes of the energy harvesters with their application domains are reported.

DOI: 10.4018/978-1-6684-4974-5.ch008

INTRODUCTION

The emergence of global computing or Internet of things (IoT) is feasible by massive deployment of inexpensive sensors(Miller 2015) (Weber et al. 2010) in the things that need to be controlled using internet protocols for remote sensor networks. The dependence of the energy consumption of wireless sensor nodes (WSNs) on their batteries is a significant barrier to such a widespread deployment (STMicroelectronics 2013). Batteries are unreliable and cumbersome to be replaced if the WSNs are located at hard to reach locations or hazardous areas such as radiation dose monitoring in nuclear plants(Liu et al. 2017). Another drawback of the batteries is their size when estimated to the shrinking size of the wireless sensor motes. Energy harvesting has thus emerged as a replacement to these batteries for powering the sensors using the ambient energy sources present in their vicinity. The objective of this chapter is to identify the state-of-art as well as the potential applications of energy harvesters. A brief review of the energy harvester concept, how energy harvesters need to be designed and the current energy transducing technologies is done in the introductory part followed by detailed review of each application.

ENERGY HARVESTER AND ITS NEED IN WSNS

Energy harvesters are energy transducers that generate negligible amount of electrical energy from the ambient energy sources available around the wireless sensor nodes and power them (STMicroelectronics 2013). Over the years, the cost of sensors has drastically decreased from an average of $1.30 to 60 cents per sensor as reported by Goldman Sachs (Jankowski et al. 2014) and is assumed to continuously decrease in the future thus fueling the deployment of smart sensors everywhere as envisioned in the concept of internet of things (IoT).

A typical smart WSN is as shown in the Fig.1 (a). It comprises of a microcontroller for processing sensor data, wireless antenna, a single sensor or range of sensors and battery. The new concept for IoT is to replace the battery with an energy harvester and rechargeable battery as shown in Fig.1 (b). Also, the microcontroller is replaced by the ultra-low powered microcontrollers such as MSP430 (Borgeson 2012) or the STM8L/32L series (STMicroelectronics 2017), thus reducing the power consumption. The power required for such sensor is less and intermittent.

Figure 1. Schematic of smart wireless sensor nodes (a) traditional smart wireless sensor node (b) modern IoT ready self-powered smart WSN.

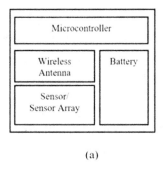
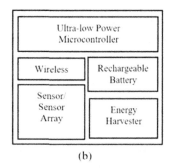

(a)　　　　　　　　　　　　　　　　(b)

3. ROLE OF ENERGY HARVESTERS IN LOW-POWER COMPUTING

There are two perspectives on achieving energy efficiency in computing devices. One is to reduce the energy consumption using LP and ULP circuits and devices. Another strategy is to integrate auxiliary energy providing devices such as energy harvesters to the existing computing circuits. Often, both the energy efficiency must work in tandem to achieve the power optimization goals and deliver the required power that will suffice for both normal and ad-hoc computing requirements of the deployed computing devices.

The role of energy harvesters is very crucial in Low power-edge computing devices. In such devices, energy harvesters can power devices such as sensors, microcontrollers, and radios and extend their battery life. Energy harvesters can also be used in stand-alone devices such as mobile phones, data loggers, low-power SoC and FPGA controllers.

One predominant application of energy harvesters in LPC is Intermittent Computing Systems (ICS)(Lucia et al. 2017). Intermittent computing systems. ICSs are used where monitoring and control actions are necessary only at intermittent times, such as animal wildlife monitoring, detecting occupancy in infrastructure monitoring, and pedestrian monitoring. Another example of application of energy harvesters is in the Programmable Array Accelerators that are used to accelerate deep vision applications of edge nodes (Das et al. 2017).

ENERGY HARVESTER DESIGN PROTOCOL

A generic energy harvesting design protocol is as shown in the Fig.2. Initially, the WSN application where energy harvesting is feasible is chosen and the energy harvester is designed for it. The feasibility of the energy harvester depends not only on the economics but also on the availability of ambient power sources surrounding the WSN application. The ambient energy source power characteristics are then analyzed. For example, Time domain accelerometer signals can be analyzed from the random vibrations of the engine by converting it into frequency domain using spectral methods like Fast-Fourier transforms (FFTs) and understand the prominent frequencies. A suitable energy transduction technology is then chosen. To cite another example, electromagnetic vibration energy harvesters are chosen for low frequency and high amplitude vibrations while piezoelectric energy harvesters are chosen for high frequency, low amplitude vibrations. The energy harvester is then designed keeping into account the power density and optimal performance during the period of operation. This is followed by embedding the energy harvester with power management system of the wireless sensor node as shown in Fig.1 (b).

TYPES OF ENERGY HARVESTERS

Energy harvesters (EHs) are usually classified based the power source they are harvesting (e.g., thermal, vibration) and are sub-classified by the physics involved (e.g., Electromagnetic EHs, Piezoelectric vibration EHs) in energy transduction. The primary sources of ambient energy include motion, vibration, temperature difference, light, RF waves and chemical reactions.

Vibration Energy Harvesters

Vibration energy harvesters (VEHs) convert kinetic energy, especially ambient vibrations into electrical power (Gatti, Shetty, and Rao 2022) which is either stored in storage device or directly supplied to load. This has been demonstrated by Perpettuum (Tiemann 1996) where in vibration EHs were used to power the condition monitoring sensors of rotary machines. The transduction mechanisms currently employed to convert vibrations into electrical energy which includes piezoelectricity (Dutoit, Wardle, and Kim 2005), magnetostriction (Wang and Yuan 2008), electromagnetism (Gatti 2013) and electrostatics(Tao et al. 2017). Out of these, piezoelectric vibration energy harvesters (VEH) and electromagnetic VEHs are commonly used.

Figure 2. Energy harvester design protocol

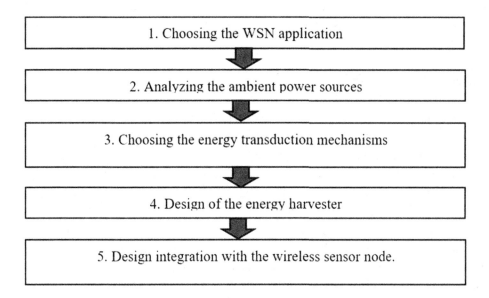

Piezoelectric VEHs

Piezoelectric material produces electrical charge when mechanical tension is applied (Caliò et al. 2014). Inverse piezoelectric effect is said to occur if stress is produced when electric field is applied. Piezoelectricity occurs in polycrystalline materials such as quartz, lead-zirconate-titanate (PZT), lead-titanate (PbTiO2) (Chawanda and Luhanga 2012). If the piezoelectric crystal gets deformed even by 0.01% of the total magnitude of the device, then electrical power is attained. In piezoelectric energy harvester, a mass is suspended by a beam where piezoelectric layer is coated at the tip of it. When pressure is applied on the mass, then the piezoelectric layer gets deformed mechanically due to tension and as a result, voltage is produced.

Electromagnetic VEHs

Electromagnetic VEHs are based on Faraday's law of electromagnetic induction, in which change in magnetic field produces electricity. Magnetic field is produced by movement of electric charge. Linear electromagnetic generators can be used to convert vibration energy into electrical energy (von Büren and Tröster 2007). Electromagnetic harvesters are known for their high output power and low output voltage (Ali, Leong, and Mustapha 2016). There is emerging need for this type of harvesters in terms of miniaturization, high efficiency, and better output performance (Tibu et al. 2015). The electric energy obtained from these micro generators depends on various factors such as flux linkage gradient, number of turns in a coil and impedance of a coil.

Magnetostrictive VEHs

These transducers mainly deal with the conversion from magnetically induced strain energy to mechanical energy and vice versa (Dapino et al. 1999). When magnetostrictive elements are subjected to external pressure, strain is produced, which alters the magnetic shape of the material. These elements have high energy density and large coupling factor (Dai et al. 2009), as a result their bidirectional coupling provides good transduction capability to actuators and sensors. The transduction capability of magnetostrictive elements makes it capable for underwater sea operations (Meeks and Timme 1977). Terfenol-D has magnetostriction nearly 1600 ppm (parts per million) and has expansion and contraction characteristics. Galfenol is recently classified magnetostrictive material which has magnetostriction of 400 ppm. It has less expansion and contraction characteristic in magnetic field (Deng, Asnani, and Dapino 2015). A magnetostrictive effect that is caused in these elements is called Wiedemann effect (Tzannes 1966). The rods are twisted through which electric current flows in a helical magnetic field. Matteuci effect (Garshelis 1974) is magneto-mechanical effect and is converse of Wiedemann effect. This effect takes place in torque transducers and is useful for energy harvesting.

Electrostatic VEHs

Electrostatic VEHs are high voltage and low output capacitance devices. In an Electrostatic VEH, the vibration induces relative motion between the parallel plate capacitors and this generates voltage. Electrostatic VEHs are active energy harvesters since a negligible amount of power is consumed by them to maintain voltage between two conducting plates (Galayko et al. 2011).

Light Energy Harvesters

Two principles for conversion of light energy into electrical energy are photoelectric effect and electro-photosynthesis (Fujishima and Honda 1971). Photovoltaic cell directly converts light energy to electrical energy. There are various types of photovoltaics which are broadly classified as outdoor photovoltaics and indoor photovoltaics. The indoor photovoltaic cells are for critically low power consumption devices. The efficiency of photovoltaic converters for indoor sources ranges from 46%

to 67% for artificial lights (Freunek, Freunek, and Reindl 2012). One such technique used in indoor photovoltaics is maximum power point tracking that increases the efficiency of photovoltaic cells by 30% (Wang et al. 2008).

Electro Photosynthesis converts light energy into chemical energy. Another technique is Microalgae Energy Harvester (Gajda et al. 2015). There cannot be single best method of harvesting microalgae. Algae species, algae production, growth medium, product, and production cost benefit are the choices made in the harvesting technology. Algae size is a vital factor since minimal filtration techniques are relevant just to harvest genuinely large microalgae. It has two fundamental processes bulk processing and thickening. Bulk processing usually begins with flocculation followed by filtration, flotation, and gravity sedimentation. Thickening mainly concentrates on biomass after it is separated. After algae harvesting is done, it is followed by dewatering process. Efficient dewatering procedures are important for minimizing energy and expenses.

Bio-flocculation is a conceivably ease, low-energy harvesting strategy. Utilization of Tetrahymena to algal cells will initiate rapid bio-flocculation. Harvesting and dewatering often go hand in hand. We consider a Microbial Fuel Cell (MFC) (Shelef, Sukenik, and Green 1984) used for energy harvesting. The anaerobic and phototrophic biofilm forms the anodic and cathodic half-cell. Anodic half-cell produces current, whereas the cathode provides the oxygen and form biomass. The cathodic portion of the MFC simultaneously generates electricity and regenerates biomass, subject to the nutrient content of the anodic feedstock.

Thermal Energy Harvesters

Thermal Energy Harvesters(TEHs) are preferred because of their high efficiency, less weight and good durability with less maintenance (Stark 2006). The harvested thermal energy can be utilized in domestic and other industrial applications. It involves conversion of heat into electrical or mechanical energy, which is done by using thermoelectric generators (TEGs). Thermoelectric materials are utilized in the transportation of heat from one trace to another. These harvesters are typically used in Wireless Body Area Networks (WBANs), where they can harvest the heat energy generated by the human body (Hoang et al. 2009). In order to improve the efficiency of WSNs, low temperature systems can be used for heat energy harvesting (Lu and Yang 2010).

Electrochemical Energy Harvesters

As a gastric battery in electrochemical energy harvesting, a microfabricated electrochemical cell on flexible parylene film is proposed (Mostafalu and Sonkusale

2014). Zinc and palladium are used as anode and cathode, respectively. Gastric juice is used as an unlimited source of electrolytes. The two electrodes are biocompatible, and parylene gives the system adaptability. Generated power of 1.25 mW for a surface area of 15 mm is sufficient for most implantable endoscopy applications (Beckett 2015). The output voltage of this battery in an open circuit is 0.75V. Its low intrinsic weight, flexibility, and biocompatibility make it appropriate for use as a power harvesting source for biomedical implants.

Hybrid Energy Harvesters

In Hybrid energy harvesters (HEHs), two or more energy sources can be used in one platform. Consequently, the output power can be increased compared to each individual harvester. Additionally, by harvesting two or more energy sources in a single device, near-continuous output is possible even when only one energy source is available. Compared to a device that harvests only one type of energy, the hybrid device can enable efficient energy collection in particular environments. HEHs are categorized into two types. The first type of HEHs uses different transduction technologies in order to harvest energy from the same source. An example is piezoelectric-electromagnetic HEHs that use piezoelectric effect and electromagnetic induction to harvest vibration power. The second set of hybrid EHs harvest energy from different source with different transduction mechanisms, which are discussed below.

Solar-Acoustic HEHs

In Solar-Acoustic HEH, both solar energy and acoustic energy can be converted simultaneously to electric energy. For direct solar energy conversion, silicon nanopillar solar cells (Lee et al. 2013) are utilized in its design. Due to their low reflection, high absorption, and potential for low-cost mass production, they are photovoltaic devices. Utilizing a technique known as plasma etching and annealing processes, the solar cells are manufactured. The entire solar-acoustic HEH is only a few hundred nanometers tall. The harvester could generate electricity from solar cells with a conversion efficiency of 3.29 percent. In addition, this Solar-Acoustic HEH could generate 0.8V of output voltage when exposed to 100 dB of sound pressure.

Thermal, Solar, and Mechanical HEHs

Thermal energy, Mechanical energy and solar energy comprise some of the major sources of energy harvesters. In HEHs, mechanical, thermal, and solar energy are harvested at the same time or exclusively to power electronic devices (Yang et al.

2013). To harvest pyroelectric and piezoelectric thermal and mechanical energy, a polarized Poly (VinyliDene Fluoride) (PVDF) film-based nanogenerator (NG) was utilized. Using aligned ZnO, nanowire arrays were fabricated on a flexible polyester (PET) substrate, and a ZnO–poly(3-hexylthiophene) (P3HT) heterojunction solar cell was designed to harvest solar energy. It contains both natural gas and solar cells. The Li-ion battery is used to store the harvested energy in order to power four light-emitting diodes that emit red light.

Thermal, Solar and Vibration HEHs

Thermal energy resulting from temperature gradients, ambient vibrations, and solar energy are some of the primary sources of energy that can be harvested. It is a power cell and acts as a storage device with the goal to enhance the system-level robustness and structural multi-functionality in energy harvesting (Tan and Panda 2010). It is found that 1 mAh of a thin battery could be charged in 20 minutes using solar energy, in 40 minutes using thermal energy and in eight hour using vibrational energy.

Comparison of the Energy Harvesting Transduction Mechanisms

The different types of energy harvesters discussed above are constructed based on the physics phenomena discovered till date that have their final ouput as electrical energy. The energy harvester type is predominantly selected based on the available ambient energy sources that can be tapped. The Table 1 below summarises the various types of energy harvesters based on their energy conversions, including their advantages and disadvantages.

APPLICATIONS OF ENERGY HARVESTERS

EHs are typically used to power wireless sensor networks where replacing batteries is difficult or impossible. The major industry that is anticipated to employ energy harvesters include the biomedical devices followed by consumer electronics, infrastructure, defense, and environment.

Biomedical Applications

Energy harvesting devices can harness power sources for various in-vivo (in-body) and ex-body (in-vitro) medical implants and accessories. The longevity of EH embedded implant will eliminate the need for repeated surgical procedures. In

Table 1. Types of energy harvesters

Sl. no	Energy harvester type	Ambient energy source	Primary features	Reported harvester energy/energy density	Advantages	Disadvantages
1	Vibration energy harvesters					
1a	Piezoelectric VEHs	High frequency, low amplitude vibrations	Based on piezoelectric effect. Requires special materials such as PZT.	16.4 mW cm^{-2} (Yuan et al. 2020)	High energy density, low volume, Ideal for high frequency vibration energy harvesting, Easily scalable to MEMS	high output impedance, depolarization, brittle, Expensive for large scale.
1b	Electromagnetic VEHs	Low frequency high amplitude vibrations	Based on Faraday's Law of induction. Intended for capturing energy from high amplitude vibrations.	0.454 mW cm^{-3} (Cai and Liao 2020)	High energy density, Cost effective, low impedance	Bulky. MEMS scalability is difficult. inefficient coupling, low resonant frequency
1c	Magnetostrictive VEHs	High frequency, low amplitude vibrations	Based on Wiedmann effect/ Mattuci effect.	87.12 mW cm^{-3} (Deng and Dapino 2017)	Highly resilient to strong vibrations, high energy conversion efficiency, no depolarization.	Non-linear, bulky bias magnets
1d	Electrostatic VEHs	High frequency, low amplitude vibrations	Based on electrostatic effect. Passive EH as it requires small potential difference between the capacitance plates	0.218 mW cm^{-2}(Boisseau, Despesse, and Sylvestre 2010)	Passive, Easily scalable to MEMS	Need external voltage, leaking of charge.
2	Light energy harvesters					
2a	Photovoltaic LEHs	Visible light between XX to XX nm	Based on photovoltaic effect with impinging photons on photoelectric surface generates electrons (electric current)	8.242 mW cm^{-2}(Saha et al. 2020)	Most viable as light source is ubiquitously available in major applications, photoelectric LEHs	Low energy conversion efficiency
2b	Electro-photosynthesis LEHs	Normal visible light range	Uses energy of photons to create complex molecules (chemical energy)as energy storage.	No sufficient data (Fromme et al. 2014)	Useful for storing energy as hydrocarbons	Need of electric charge or another energy harvester to create electric discharge
3	Thermal Energy Harvesting	Heat and/or temperature difference	Uses thermoelectric effect to convert thermal energy to electric energy.	0.00725 mW cm^{-2} (Suarez et al. 2017)	High scalability, useful for waste heat recovery	Low efficiency, affected by uneven temperature gradients
4	Electrochemical Energy harvesters	Electrolyte	Use electrolysis principle. Requires electrolyte and sacrificial electrodes.	8.33 mW cm^{-2}(Mostafalu and Sonkusale 2014)	Useful for prolonged invivo energy harvesting applications	Requires sacrificial electrode and/or electrolyte.
5	Hybrid energy harvesters	Solar-thermal-vibration or any other two or more different ambient energy sources	Harvests multiple sources of energy. As of now, only solar-thermal-vibration based HEHs are reported.	0.128 mW/ cm^{2} (Lee et al. 2013)	Robust and highly reliable due to capabilities of multi-source energy scavenging	Possibly high space consumption, complicated construction

addition, they enable the miniaturization of existing implants such as pacemakers and bio-signal recording equipment. Researchers have developed a device that converts the energy harvested from the resonance of heartbeats (Zurbuchen et al. 2013) through the chest into electricity to power a pacemaker. Diverse methods for harvesting body heat (Gyselinckx et al. 2005), motion, and vibration to power other implantable devices are under investigation. Pacemaker batteries are recharged using radio frequency (Wei and Liu 2008). Researchers have developed a chip that can be implanted in the inner ear with relative ease. The necessary power is generated by harvesting the energy in sound waves. This chip monitors biological activities such as hearing or balance impairments in the ears. Fitness buffs (Zareei and Deng 2016) are used by fitness enthusiasts to recover some of the energy expended at the gym. Here, knee mounted piezoelectric EHs devices are attached so that while walking or running on a treadmill, these devices or gadgets generate power. Another potential source of direct energy harvesting is the human digestive system, which could power a future wireless endoscopy capsule (Nadeau et al. 2017). The output voltage for an open circuit is 0.75 V. It offers a promising solution for implantable applications requiring power.

Agricultural Applications

Modern agriculture adopts techniques such as precision farming to optimize the workflow. To support this, agricultural machines are provided with varieties of wired sensors and actuators. Advancing the existing wired communication networks by wireless sensors results in new possibilities. This could be collecting and transmitting measurement data. To achieve real-time capability, low latency, and deterministic behavior for agricultural machinery, a protocol with timeslot architecture for wireless sensor network was used (Müller, Rittenschober, and Springer 2010). Piezoelectric material is used in the energy harvesting unit. This unit generates an average power of 200 μW to the sensor nodes.

The conversion from the wind energy to electrical energy is obtained by the bio inspired piezo-leaf architecture. This architecture is known for its wind-induced fluttering motion. Conventional fluttering devices are organized in parallel with the flow direction. Peak output power obtained from a single leaf is approximately 600 μW (Ramasur and Hancke 2012). Precision viticulture (Morais et al. 2008) was introduced in the demarcated regions of Doura and the major platform created for this cause was called MPWiNodeZ. The energy is harvested from the surrounding environment from up to three sources and this energy is used recharge batteries (power management). It acts as a wireless acquisition device for remote sensing in large areas.

Environmental Applications

The most prevalent sources of ambient energy are light, heat, wind, and vibration. Consumer and industrial applications frequently employ small solar cells. Examples include toys, watches, calculators, controls for street lighting, portable power supplies, and satellites. The wind is another renewable energy resource that is present abundantly on Earth. Future applications can benefit from a self-sufficient flow-sensing microsystem. This system can be employed for smart home applications, remote sensing, and environmental monitoring. Thin piezoelectric films are commonly used harvesters for the conversion of mechanical strain into an electrical charge (Liu et al. 2014).

Infrastructure

In 2009, Jindo Bridge (Jang et al. 2010) in South Korea was installed with wireless bridge monitoring system. Seventy-one WSS nodes were installed which contained 427 sensing channels. Structural health monitoring (SHM) is a greatest problem faced by civil infrastructures due to its cost. In this low-cost wireless smart sensor

(WSS) are used for continuous, reliable SHM. For health monitoring of wind turbine blades piezo-ceramic-based wireless sensor network (Chae et al. 2012) were used with active sensing approach. There are various access points that coordinate the network and PC is used to control the wireless nodes.

Defense and Spacecraft

Electro-dynamic tether (EDTs) (Bilen et al. 2012) could be used by space crafts for energy harvesting where a long conducting wire can be spread out from a tether satellite. This is based on the principle of electromagnetism where the electromotive force in wire is generated from Lorentz force. In addition, EDTs are used provide power to spacecraft's by harvesting energy from system's orbit. Large satellites can harvest as much as kilowatts at load. Harvesting of energy and power using spacecraft system reduces consumption of fuel and provides amplified capability and better lifetime (Xia et al. 2016). Defense clothing for the soldiers can be done using energy harvesting textiles (Gould 2003).

Consumer Electronics

In consumer electronics sector, there are promising developments of using energy harvested sensors in wearable electronics, intelligent clothing, mobile phones and laptops, appliances, and home automation. In self-powering of wearables and mobile devices, high reliance is on the ambient energy vibrations available in the human body such as body heat, kinetic motion of the body parts. Seiko Corporation developed wrist watches that are powered by hand movements. Thermal EHs are reported that convert the body heat into electrical energy for powering the wearables and mobile phones.

Intelligent clothing with self-powered capabilities has been reported for military purposes (Gould 2003) (Jao, Chang, and Lin 2017). In the home automation sector, smart switches capable of self-powering themselves under normal ON operation and using that stored energy for both sensing and actuation purpose are now commercially available.

Tyndall institute (Mathews et al. 2014) has developed indoor photovoltaic modules that can power the indoor wireless sensors required for home automation. Enocean (Bissengaliyeva 2020) has envisaged home automation can be controlled by self-powered sensors in number of home automation applications such as occupancy sensing to detect the presence of humans in the room to switch on/off the lights, room temperature sensor for efficient use of air-conditioning, maintenance of house concrete structures.

Industrial Electronics

In the industrial electronics sector, energy harvesting could be mainly used in structural health monitoring of machine structures or conditional monitoring of rotating machinery. One good example is the General Electric's electromagnetic energy harvester that is used to power the GE Insight mesh WSNs (Gatti n.d.). Another major aspect of this application is in increasing the update rates of the measurement sensors.

Perpetua Power packs (Kafka 2015), a commercially available thermal energy harvester has been successfully used to increase the update rate of sensors in various applications, which could be powering the yield quality sensors in oil wells, powering wireless monitoring and maintaining of the equipment of a remote oil & gas plant in Alaska's harsh environmental conditions. Traditionally, before the deployment of energy harvesters, the update rates of the measurement sensors were very slow, thus decreasing the sample rate and eventually increasing the standard deviation in detection of faults.

Transportation

Recently, RF energy harvesting circuit has been proposed to replace the 3V lithium coin cell battery in the car key fob for remote locking operations (Gambhir, Yadav, and Pawar 2017). Significant work is being sustained in harvesting energy from road vibrations and shock loads using regenerative suspensions in cars (Demetgul and Guney 2017) and trains (Mi et al. 2017). TEGs using liquid cooling approach were reported by Ataur (Rahman et al. 2015) that can be used for engine sensors.

Significant research interest is shown for carbon reinforced composites for automotive chassis and other automotive parts due to its incredible strength and light weight. Zinc oxide nanowires are reported to have significant energy harvesting capabilities and due to its piezoelectric properties, it can be utilized to create carbon reinforced composites (Masghouni et al. 2015).

Summary of Applications of Energy Harvesting

As discussed in the major applications discussed in the above sections 5.1 to 5.8, one can observe that the trend towards ubiquitous sensory data acquisition, ubiquitous computing requires WSNs to be self-powered by embedding energy harvesters to them. The above discussion can be summarized in the Table 2 below as a reference for selection of the appropriate energy transduction mechanism. Most of these energy harvesters have energy scavenging in the order of mW to 1W and sometimes even

Table 2. Summary of energy harvesting applications in major industrial sectors

Sl.no	Industry	Applications	Feasible ambient energy	Feasible energy harvesters	Examples
1	Bio-medical	Powering implanted sensors - in vivo (endoscopy capsules) and invitro sensors of implants, powering RF tracking of Animals.	For in-vivo applications, chemical energy and kinetic energy of fluid can be harnessed. For in-vitro applications, kinetic energy of animal or human motion can be harnessed	For in-vivo applications, electro-chemical EHs can be used. For in-vitro applications, kinetic energy can be used.	Energy harvesters for pacemakers(Zurbuchen et al. 2013) and capsule endoscopy(Nadeau et al. 2017)
2	Agricultural	Powering agricultural sensors for enabling self-powered wireless precision agriculture. Powering soil moisture sensors, PH sensors and water valve position sensors	Solar energy, vibrational energy of the pipelines, wind energy	Solar EHs, Wind powered EHs, vibration energy harvesters	Smart agriculture monitoring(Sharma, Haque, and Jaffery 2019)
3	Environmental	Powering of Environmental monitoring sensors for measuring air pollution, landslides, forest fires etc. Powering gas sensors, PH sensors, load cells, smoke detectors, humidity sensors.	Solar energy, kinetic energy of wind	Solar EHs, Wind powered Piezoelectric VEHs	Aquatic environmental monitoring(Alippi et al. 2010)
4	Infrastructure	Structural health monitoring, building information monitoring, pipe condition monitoring. Powering load cells and strain gauges,	Solar energy, kinetic energy of wind	Solar EHs, Wind powered Piezoelectric and/ or electromagnetic VEHs	Structural health monitoring of reinforced concrete (Cahill, Mathewson, and Pakrashi 2018)
5	Defence	SHM of defence aircraft and naval structures and rotating machinery, energy scavenging defence textiles.	Vibration and thermal energy of machines or bodies of soldiers (for wearables)	Vibration EHs, Thermal EHs	Defence textiles (Kandasubramanian and Ramdayal 2013) and EnHANT Tags (Gorlatova et al. 2010)
6	Consumer Electronics	Low power consumer electronics such as pedometers,watches and other wearable electronics, appliance monitoring for smart homes, mobile devices.	RF energy, Thermal and kinetic energy from the body.	RF energy harvesters, Thermal energy harvesters, Piezoelectric energy harvesters	Power generating shoes(Zhu et al. 2013) and mobile devices (Jabbar, Song, and Jeong 2010)
7	Industrial Electronics	Industrial WSN networks, Routers,Modems, Industrial portable sensors and low power electronics, Condition monitoring sensors of rotating machinery	Vibration and thermal energies of machinery, Solar energy (for external applications)	Solar EHs, Vibration EHs, Thermal EHs, Hybrid EHs	Solar EHs(Ibrahim et al. 2017), thermal EHs(Hou et al. 2018) and vibration EHs(Gatti n.d.) for Industrial WSNs
8	Transportation	Tire pressure sensors, Engine management,	Vibration and thermal energies of automotive components	Vibration EHs, Thermal EHs, Hybrid EHs	Piezoelectric VEHs in engines(Kim 2015)and for tyre pressure sensors(Maurya et al. 2018)

as little as μW range that is sufficient to power the sensing and data transmission for the intended WSN application.

CONCLUSION

Significant developments have been made in self-powering of portable electronics and wireless sensor networks by embedding energy harvesters into their sensor design. This area of research is still ongoing and in the infant stage with only few commercial products being reported. The highest requirement of this type of harvesting is in healthcare sector, followed by industrial sector particularly monitoring of rotating machinery.

Likewise, it can also be concluded that the design of the EHs need to be more application specific and not stand-alone generic energy harvester as reported in the existing reported research.

REFERENCES

Ali, N., Leong, K., & Mustapha, A. (2016). Experimental Investigation on Piezoelectric and Electromagnetic Hybrid Micro-Power Generator. *Journal of Engineering and Applied Sciences (Asian Research Publishing Network)*, *11*(10).

Alippi, C., Camplani, R., Galperti, C., & Roveri, M. (2010). A Robust, Adaptive, Solar-Powered WSN Framework for Aquatic Environmental Monitoring. *IEEE Sensors Journal*, *11*(1), 45–55. doi:10.1109/JSEN.2010.2051539

Beckett, R. G. (2015). Application and Limitations of Endoscopy in Anthropological and Archaeological Research. *The Anatomical Record*, *298*(6), 1125–1134. doi:10.1002/ar.23145 PMID:25998646

Bilen, S. G., McTernan, J. K., Gilchrist, B. E., Bell, I. C., Liaw, D., Voronka, N. R., & Hoyt, R. P. (2012). *Energy Harvesting on Spacecraft Using Electrodynamic Tethers*. Pennsylvania State Univ State College.

Bissengaliyeva, A. (2020). *A Home Automation System Using EnOcean Wireless Technology and Beckhoff Automation*. Academic Press.

Boisseau, S., Despesse, G., & Sylvestre, A. (2010). Optimization of an Electret-Based Energy Harvester. *Smart Materials and Structures*, *19*(7), 75015. doi:10.1088/0964-1726/19/7/075015

Borgeson, J. (2012). *Ultra-Low-Power Pioneers: TI Slashes Total MCU Power by 50 Percent with New 'Wolverine' MCU Platform.* Texas Instruments White Paper.

Cahill, P., Mathewson, A., & Pakrashi, V. (2018). Experimental Validation of Piezoelectric Energy-Harvesting Device for Built Infrastructure Applications. *Journal of Bridge Engineering*, 23(8), 4018056. doi:10.1061/(ASCE)BE.1943-5592.0001262

Cai, M., & Liao, W.-H. (2020). High-Power Density Inertial Energy Harvester without Additional Proof Mass for Wearables. *IEEE Internet of Things Journal*, 8(1), 297–308. doi:10.1109/JIOT.2020.3003262

Caliò, R., Rongala, U. B., Camboni, D., Milazzo, M., Stefanini, C., De Petris, G., & Oddo, C. M. (2014). Piezoelectric Energy Harvesting Solutions. *Sensors (Basel)*, 14(3), 4755–4790. doi:10.3390140304755 PMID:24618725

Chae, M. J., Yoo, H. S., Kim, J. Y., & Cho, M. Y. (2012). Development of a Wireless Sensor Network System for Suspension Bridge Health Monitoring. *Automation in Construction*, 21, 237–252. doi:10.1016/j.autcon.2011.06.008

Chawanda, A., & Luhanga, P. (2012). Piezoelectric Energy Harvesting Devices: An Alternative Energy Source for Wireless Sensors. *Smart Materials Research*, 2012.

Dai, X., Wen, Y., Li, P., Yang, J., & Jiang, X. (2009). A Vibration Energy Harvester Using Magnetostrictive/Piezoelectric Composite Transducer. Sensors. doi:10.1109/ICSENS.2009.5398445

Dapino, M. J., Deng, Z., Calkins, F. T., & Flatau, A. B. (1999). Magnetostrictive Devices. Wiley Encyclopedia of Electrical and Electronics Engineering, 1–35. doi:10.1002/047134608X.W4549

Das, S., Rossi, D., Martin, K. J. M., Coussy, P., & Benini, L. (2017). A 142mops/Mw Integrated Programmable Array Accelerator for Smart Visual Processing. In *2017 IEEE International Symposium on Circuits and Systems (ISCAS)*. IEEE. 10.1109/ISCAS.2017.8050238

Demetgul, M., & Guney, I. (2017). Design of the Hybrid Regenerative Shock Absorber and Energy Harvesting from Linear Movement. *Journal of Clean Energy Technologies*, 5(1), 81–84. doi:10.18178/JOCET.2017.5.1.349

Deng, Z., Asnani, V. M., & Dapino, M. J. (2015). Magnetostrictive Vibration Damper and Energy Harvester for Rotating Machinery. In Industrial and Commercial Applications of Smart Structures Technologies 2015 (Vol. 9433). SPIE.

Deng, Z., & Dapino, M. J. (2017). Review of Magnetostrictive Vibration Energy Harvesters. *Smart Materials and Structures, 26*(10), 103001. doi:10.1088/1361-665X/aa8347

Dutoit, N. E., Wardle, B. L., & Kim, S.-G. (2005). Design Considerations for MEMS-Scale Piezoelectric Mechanical Vibration Energy Harvesters. *Integrated Ferroelectrics, 71*(1), 121–160. doi:10.1080/10584580590964574

Freunek, M., Freunek, M., & Reindl, L. M. (2012). Maximum Efficiencies of Indoor Photovoltaic Devices. *IEEE Journal of Photovoltaics, 3*(1), 59–64. doi:10.1109/JPHOTOV.2012.2225023

Fromme, Moore, Moore, Rittmann, Torres, & Vermaas. (2014). *Microbial Electrophotosynthesis*. Academic Press.

Fujishima, A., & Honda, K. (1971). Electrochemical Evidence for the Mechanism of the Primary Stage of Photosynthesis. *Bulletin of the Chemical Society of Japan, 44*(4), 1148–1150. doi:10.1246/bcsj.44.1148

Gajda, I., Greenman, J., Melhuish, C., & Ieropoulos, I. (2015). Self-Sustainable Electricity Production from Algae Grown in a Microbial Fuel Cell System. *Biomass and Bioenergy, 82*, 87–93. doi:10.1016/j.biombioe.2015.05.017

Galayko, D., Guillemet, R., Dudka, A., & Basset, P. (2011). Comprehensive Dynamic and Stability Analysis of Electrostatic Vibration Energy Harvester (E-VEH). In *2011 16th International Solid-State Sensors, Actuators and Microsystems Conference*. IEEE. 10.1109/TRANSDUCERS.2011.5969592

Gambhir, A., Yadav, D., & Pawar, G. (2017). *Energy Harvesting in Automotive Key Fob Application*. SAE Technical Paper.

Garshelis, I. (1974). A Study of the Inverse Wiedemann Effect on Circular Remanence. *IEEE Transactions on Magnetics, 10*(2), 344–358. doi:10.1109/TMAG.1974.1058325

Gatti, R. R. (2013). *Spatially-Varying Multi-Degree-of-Freedom Electromagnetic Energy Harvesting*. Academic Press.

Gatti, R. R. (n.d.). *Novel Multi-Beam Spring Design for Vibration Energy Harvesters*. Academic Press.

Gatti, R. R., Shetty, S. H., & Rao, A. (2022). Building Autonomous IIoT Networks Using Energy Harvesters. In Enterprise Digital Transformation. Auerbach Publications. doi:10.1201/9781003119784-11

Gorlatova, M., Kinget, P., Kymissis, I., Rubenstein, D., Wang, X., & Zussman, G. (2010). Energy Harvesting Active Networked Tags (EnHANTs) for Ubiquitous Object Networking. *IEEE Wireless Communications*, *17*(6), 18–25. doi:10.1109/MWC.2010.5675774

Gould, P. (2003). Textiles Gain Intelligence. *Materials Today*, *6*(10), 38–43. doi:10.1016/S1369-7021(03)01028-9

Gyselinckx, B., Van Hoof, C., Ryckaert, J., Yazicioglu, R. F., Fiorini, P., & Leonov, V. (2005). Human++: Autonomous Wireless Sensors for Body Area Networks. In *Proceedings of the IEEE 2005 Custom Integrated Circuits Conference, 2005*. IEEE. 10.1109/CICC.2005.1568597

Hoang, D. C., Tan, Y. K., Chng, H. B., & Panda, S. K. (2009). Thermal Energy Harvesting from Human Warmth for Wireless Body Area Network in Medical Healthcare System. *2009 International Conference on Power Electronics and Drive Systems (PEDS)*. 10.1109/PEDS.2009.5385814

Hou, L., Tan, S., Zhang, Z., & Bergmann, N. W. (2018). Thermal Energy Harvesting WSNs Node for Temperature Monitoring in IIoT. *IEEE Access: Practical Innovations, Open Solutions*, *6*, 35243–35249. doi:10.1109/ACCESS.2018.2851203

Ibrahim, R., Chung, T. D., Hassan, S. M., Bingi, K., & Salahuddin, S. K. (2017). Solar Energy Harvester for Industrial Wireless Sensor Nodes. *Procedia Computer Science*, *105*, 111–118. doi:10.1016/j.procs.2017.01.184

Jabbar, H., Song, Y. S., & Jeong, T. T. (2010). RF Energy Harvesting System and Circuits for Charging of Mobile Devices. *IEEE Transactions on Consumer Electronics*, *56*(1), 247–253. doi:10.1109/TCE.2010.5439152

Jang, S., Jo, H., Cho, S., Mechitov, K., Rice, J. A., Sim, S.-H., Jung, H.-J., Yun, C.-B., Spencer, B. F. Jr, & Agha, G. (2010). Structural Health Monitoring of a Cable-Stayed Bridge Using Smart Sensor Technology: Deployment and Evaluation. *Smart Structures and Systems*, *6*(5–6), 439–459. doi:10.12989ss.2010.6.5_6.439

Jankowski, S., Covello, J., Bellini, H., Ritchie, J., & Costa, D. (2014). *The Internet of Things: Making Sense of the next Mega-Trend*. Goldman Sachs.

Jao, Y.-T., Chang, T.-W., & Lin, Z.-H. (2017). Multifunctional Textile for Energy Harvesting and Self-Powered Sensing Applications. *ECS Transactions*, *77*(7), 47–50. doi:10.1149/07707.0047ecst

Kafka, T. (2015). Industrial Application of Thermal Energy Harvesting. *IDTex.* http://perpetuapower.com/wp-content/uploads/2015/12/GE_IDTec hEx_Presentation_Berlin_20140402.pdf

Kandasubramanian, B., & Ramdayal, M. (2013). Advancement in Textile Technology for Defence Application. *Defence Science Journal*, *63*(3), 331–339. doi:10.14429/dsj.63.2756

Kim, G. W. (2015). Piezoelectric Energy Harvesting from Torsional Vibration in Internal Combustion Engines. *International Journal of Automotive Technology*, *16*(4), 645–651. doi:10.100712239-015-0066-6

Lee, D.-Y., Kim, H., Li, H.-M., Jang, A. R., Lim, Y.-D., Cha, S. N., Park, Y. J., Kang, D. J., & Yoo, W. J. (2013). Hybrid Energy Harvester Based on Nanopillar Solar Cells and PVDF Nanogenerator. *Nanotechnology*, *24*(17), 175402. doi:10.1088/0957-4484/24/17/175402 PMID:23558434

Liu, H., Zhang, S., Kobayashi, T., Chen, T., & Lee, C. (2014). Flow Sensing and Energy Harvesting Characteristics of a Wind-driven Piezoelectric Pb (Zr0. 52, Ti0. 48) O3 Microcantilever. *Micro & Nano Letters*, *9*(4), 286–289. doi:10.1049/mnl.2013.0750

Liu, Y., Nishimura, M., Li, L., & Colins, K. (2017). Study on a Low-Cost and Large-Scale Environmentally Adaptive Protocol Stack of Nuclear and Space Wireless Sensor Network Applications under Gamma Radiation. *Nuclear Technology*, *197*(1), 75–87. doi:10.13182/NT16-97

Lu, X., & Yang, S.-H. (2010). Thermal Energy Harvesting for WSNs. In *2010 IEEE International Conference on Systems, Man and Cybernetics*. IEEE.

Lucia, Balaji, Colin, Maeng, & Ruppel. (2017). Intermittent Computing: Challenges and Opportunities. *2nd Summit on Advances in Programming Languages (SNAPL 2017)*.

Masghouni, N., Burton, J., Philen, M. K., & Al-Haik, M. (2015). Investigating the Energy Harvesting Capabilities of a Hybrid ZnO Nanowires/Carbon Fiber Polymer Composite Beam. *Nanotechnology*, *26*(9), 95401. doi:10.1088/0957-4484/26/9/095401 PMID:25670370

Mathews, I., Kelly, G., King, P. J., & Frizzell, R. (2014). GaAs Solar Cells for Indoor Light Harvesting. In *2014 IEEE 40th Photovoltaic Specialist Conference (PVSC)*. IEEE. 10.1109/PVSC.2014.6924971

Maurya, D., Kumar, P., Khaleghian, S., Sriramdas, R., Kang, M. G., Kishore, R. A., Kumar, V., Song, H.-C., Park, J.-M. J., Taheri, S., & Priya, S. (2018). Energy Harvesting and Strain Sensing in Smart Tire for next Generation Autonomous Vehicles. *Applied Energy*, *232*, 312–322. doi:10.1016/j.apenergy.2018.09.183

Meeks, S. W., & Timme, R. W. (1977). Rare Earth Iron Magnetostrictive Underwater Sound Transducer. *The Journal of the Acoustical Society of America*, *62*(5), 1158–1164. doi:10.1121/1.381650

Mi, J., Xu, L., Guo, S., Abdelkareem, M. A. A., & Meng, L. (2017). *Suspension Performance and Energy Harvesting Property Study of a Novel Railway Vehicle Bogie with the Hydraulic-Electromagnetic Energy-Regenerative Shock Absorber*. SAE Technical Paper.

Miller, R. (2015). Cheaper Sensors Will Fuel The Age Of Smart Everything. *Tech Crunch*. Retrieved January 31, 2023 (https:// techcrunch.com/2015/03/10/cheaper-sensors-will-fuel-the-age-of-smart-everything/)

Morais, R., Fernandes, M. A., Matos, S. G., Serôdio, C., Ferreira, P. J. S. G., & Reis, M. J. C. S. (2008). A ZigBee Multi-Powered Wireless Acquisition Device for Remote Sensing Applications in Precision Viticulture. *Computers and Electronics in Agriculture*, *62*(2), 94–106. doi:10.1016/j.compag.2007.12.004

Mostafalu, P., & Sonkusale, S. (2014). Flexible and Transparent Gastric Battery: Energy Harvesting from Gastric Acid for Endoscopy Application. *Biosensors & Bioelectronics*, *54*, 292–296. doi:10.1016/j.bios.2013.10.040 PMID:24287419

Müller, Rittenschober, & Springer. (2010). A Wireless Sensor Network Using Energy Harvesting for Agricultural Machinery. *E & I Elektrotechnik Und Informationstechnik*, *3*(127), 39–46.

Nadeau, P., El-Damak, D., Glettig, D., Yong, L. K., Mo, S., Cleveland, C., Booth, L., Roxhed, N., Langer, R., & Chandrakasan, A. P. (2017). Prolonged Energy Harvesting for Ingestible Devices. *Nature Biomedical Engineering*, *1*(3), 22. doi:10.103841551-016-0022 PMID:28458955

Rahman, A., Razzak, F., Afroz, R., Akm, M., & Hawlader, M. N. A. (2015). Power Generation from Waste of IC Engines. *Renewable & Sustainable Energy Reviews*, *51*, 382–395. doi:10.1016/j.rser.2015.05.077

Ramasur, D., & Hancke, G. P. (2012). A Wind Energy Harvester for Low Power Wireless Sensor Networks. In *2012 IEEE International Instrumentation and Measurement Technology Conference Proceedings*. IEEE. 10.1109/I2MTC.2012.6229698

Saha, C. R., Huda, M. N., Mumtaz, A., Debnath, A., Thomas, S., & Jinks, R. (2020). Photovoltaic (PV) and Thermo-Electric Energy Harvesters for Charging Applications. *Microelectronics Journal*, *96*, 104685. doi:10.1016/j.mejo.2019.104685

Sharma, H., Haque, A., & Jaffery, Z. A. (2019). Maximization of Wireless Sensor Network Lifetime Using Solar Energy Harvesting for Smart Agriculture Monitoring. *Ad Hoc Networks*, *94*, 101966. doi:10.1016/j.adhoc.2019.101966

Shelef, Sukenik, & Green. (1984). *Microalgae Harvesting and Processing: A Literature Review*. Academic Press.

Stark, I. (2006). Invited Talk: Thermal Energy Harvesting with Thermo Life. In *International Workshop on Wearable and Implantable Body Sensor Networks (BSN'06)*. IEEE. 10.1109/BSN.2006.37

STMicroelectronics. (2013). *Accurate Power Consumption Estimation For STM32L1 Series of Ultra-Low-Power Microcontrollers*. Retrieved January 31, 2023 https://www.st.com/resource/en/technical_article/dm00024152.pdf

STMicroelectronics. (2017). *STM8L151x6/8 STM8L152x6/8 Datasheet*.

Suarez, F., Parekh, D. P., Ladd, C., Vashaee, D., Dickey, M. D., & Öztürk, M. C. (2017). Flexible Thermoelectric Generator Using Bulk Legs and Liquid Metal Interconnects for Wearable Electronics. *Applied Energy*, *202*, 736–745. doi:10.1016/j.apenergy.2017.05.181

Tan, Y. K., & Panda, S. K. (2010). Energy Harvesting from Hybrid Indoor Ambient Light and Thermal Energy Sources for Enhanced Performance of Wireless Sensor Nodes. *IEEE Transactions on Industrial Electronics*, *58*(9), 4424–4435. doi:10.1109/TIE.2010.2102321

Tao, K., Wu, J., Tang, L., Hu, L., Lye, S. W., & Miao, J. (2017). Enhanced Electrostatic Vibrational Energy Harvesting Using Integrated Opposite-Charged Electrets. *Journal of Micromechanics and Microengineering*, *27*(4), 44002. doi:10.1088/1361-6439/aa5e73

Tibu, M., Chiriac, H., Ovari, T., & Lupu, N. (2015). Efficient Electromagnetic Energy Harvesting Devices. In *2015 IEEE International Magnetics Conference (INTERMAG)*. IEEE.

Tiemann, J. J. (1996). *Apparatus for Converting Vibratory Motion to Electrical Energy*. Academic Press.

Tzannes, N. S. (1966). Joule and Wiedemann Effects-The Simultaneous Generation of Longitudinal and Torsional Stress Pulses in Magnetostrictive Materials. *IEEE Transactions on Sonics and Ultrasonics, 13*(2), 33–40. doi:10.1109/T-SU.1966.29373

von Büren, T., & Tröster, G. (2007). Design and Optimization of a Linear Vibration-Driven Electromagnetic Micro-Power Generator. *Sensors and Actuators. A, Physical, 135*(2), 765–775. doi:10.1016/j.sna.2006.08.009

Wang, L., & Yuan, F. G. (2008). Vibration Energy Harvesting by Magnetostrictive Material. *Smart Materials and Structures, 17*(4), 45009. doi:10.1088/0964-1726/17/4/045009

Wang, W. S., O'Donnell, T., Wang, N., Hayes, M., O'Flynn, B., & O'Mathuna, C. (2008). Design Considerations of Sub-MW Indoor Light Energy Harvesting for Wireless Sensor Systems. *ACM Journal on Emerging Technologies in Computing Systems, 6*(2), 1–26. doi:10.1145/1773814.1773817

Weber, Weber, Weber, & Weber. (2010). Internet of Things as Tool of Global Welfare. *Internet of Things: Legal Perspectives,* 101–25.

Wei, X., & Liu, J. (2008). Power Sources and Electrical Recharging Strategies for Implantable Medical Devices. *Frontiers of Energy and Power Engineering in China, 2*(1), 1–13. doi:10.100711708-008-0016-3

Xia, Q., Xie, K., Liu, X., Wu, Z., & Wang, N. (2016). Influence and Efficiency of Energy Harvesting on the Process of De-Orbiting Using Bare Electrodynamic Tether System. *52nd AIAA/SAE/ASEE Joint Propulsion Conference*. 10.2514/6.2016-5110

Yang, Y., Zhang, H., Zhu, G., Lee, S., Lin, Z.-H., & Zhong, L. W. (2013). Flexible Hybrid Energy Cell for Simultaneously Harvesting Thermal, Mechanical, and Solar Energies. *ACS Nano, 7*(1), 785–790. doi:10.1021/nn305247x PMID:23199138

Yuan, X., Gao, X., Yang, J., Shen, X., Li, Z., You, S., Wang, Z., & Dong, S. (2020). The Large Piezoelectricity and High Power Density of a 3D-Printed Multilayer Copolymer in a Rugby Ball-Structured Mechanical Energy Harvester. *Energy & Environmental Science, 13*(1), 152–161. doi:10.1039/C9EE01785B

Zareei, S., & Deng, J. D. (2016). Energy Management Policy for Fitness Gadgets: A Case Study of Human Daily Routines. In *2016 26th International Telecommunication Networks and Applications Conference (ITNAC)*. IEEE. 10.1109/ATNAC.2016.7878774

Zhu, G., Bai, P., Chen, J., & Zhong, L. W. (2013). Power-Generating Shoe Insole Based on Triboelectric Nanogenerators for Self-Powered Consumer Electronics. *Nano Energy*, *2*(5), 688–692. doi:10.1016/j.nanoen.2013.08.002

Zurbuchen, A., Pfenniger, A., Stahel, A., Stoeck, C. T., Vandenberghe, S., Koch, V. M., & Vogel, R. (2013). Energy Harvesting from the Beating Heart by a Mass Imbalance Oscillation Generator. *Annals of Biomedical Engineering*, *41*(1), 131–141. doi:10.100710439-012-0623-3 PMID:22805983

Chapter 9

An Exhaustive Analysis of Energy Harvesting Absorbers and Battery Charging Systems for the Internet of Things

C. Padmavathy
Sri Ramakrishna Engineering College, India

Dankan Gowda V.
🆔 https://orcid.org/0000-0003-0724-0333
B.M.S. Institute of Technology and Management, India

Vaishali Narendra Agme
Bharati Vidyapeeth College of

Engineering, India

Algubelly Yashwanth Reddy
🆔 https://orcid.org/0000-0001-6422-1943
Sree Dattha Group of Institutions, India

D. Palanikkumar
Dr. N.G.P. Institute of Technology, India

ABSTRACT

Nearly all application fields are paying increased attention to the internet of things (IoT). Nearly 20 billion devices are now linked to the internet. With several applications ranging from smart buildings and smart cities to smart devices, IoT has progressed over the last few decades. As a result, the quantity of sensors, sensor nodes, and gateways has increased, making these battery-powered devices power-hungry. It will be a laborious operation to change the battery in remote monitoring applications for these smart sensors or nodes. By gathering RF energy from the environment and converting it to DC power, RF energy harvesting is a cost-effective method of extending the lifespan of wireless sensor networks (WSNs). A brand-new, IoT-based smart universal charger is suggested in this chapter for charging multichemistry

DOI: 10.4018/978-1-6684-4974-5.ch009

batteries. The suggested charger has an advantage over traditional chargers since it can charge both already installed batteries and any future batteries.

INTRODUCTION

In the last few years, the demand for the power has increased enormously. Researchers are looking for alternate energy sources such as solar, wind, thermal, vibration and radio frequency (RF) for energy harvesting. Inexhaustible RF energy sources with zero harmful emissions can pave way for harvesting energy unlimitedly for powering low power sensors and microcontrollers. Internet of Things (IoT) enables to connect enormous sensor nodes for remote monitoring applications to the internet. RF Energy harvesting reduces the complexity of IoT nodes by avoiding power

Figure 1. RF energy harvesting System (RFEH)

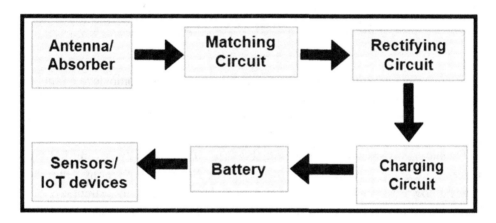

circuits, reduces the overall cost and improves the overall efficiency of the system. RF Energy harvesting or Green Energy harvesting is an excellent technique which allows size reduction in comparison with other harvesting systems such as photonic cells or wind turbines, where miniaturization really matters a lot in portable devices which uses battery (Almoneef, T. S. 2014). RF energy harvesting plays a key role in providing a sustainable energy source towards the future of wearable electronics too. Further, RF energy harvesting is reliable, portable environmental friendly and also cost effective. These advantages have attracted the researchers to unveil novel research in this field. A typical EHS consists of Antenna/absorber, matching circuit,

Figure 2. Material classification

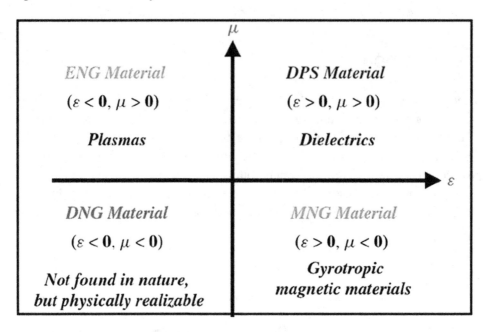

rectifier circuit, charging circuit, battery and low power electronic devices such as sensors, microcontrollers etc. as shown in Figure.1.

Generally materials are classified on the basis of permittivity (ε) and permeability (μ) in the four quadrants as shown in Figure.2.

Double positive materials (DPS) with ($\varepsilon > 0; \mu > 0$) i.e. naturally existing dielectric materials lie in quadrant-I. In such materials, the direction of poynting vector and wave number is same. These are known as right handed materials (Almutairi, A. F. 2019). The refractive index is positive in these materials. Epsilon negative materials (ENG) with ($\varepsilon < 0; \mu > 0$) i.e. plasmas and metals at optical frequencies belong to quadrant-II. The wave travels in evanescent mode in these materials. Double negative materials (DNG) with ($\varepsilon < 0; \mu < 0$) are referred as metamaterials and lie in quadrant-III. The refractive index is negative in these materials and are also known as Left handed materials (LHM). The effective (ε & μ) of a material can be tailored to design an artificial structure for any specific application. The real part of the permittivity $\varepsilon'(\omega)$ defines the dielectric constant of the material and the imaginary part $\varepsilon''(\omega)$ defines the attenuation provided by the material (Aminov, P. 2018). The real part of permeability $\mu'(\omega)$ defines the energy stored in the magnetic field and the imaginary part $\mu''(\omega)$ gives the measure of energy dissipation in the material for the incident magnetic field. by combining the structural elements of electric and magnetic response together in a unit cell a material with effective

Figure 3. Negative Index metamaterial (a) Design of an SRR array (b) Transmitted power vs Frequency

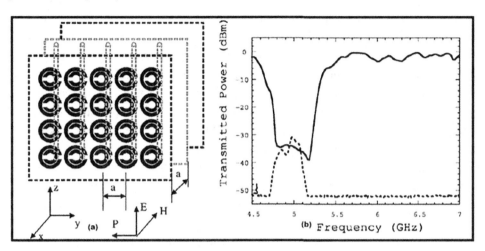

metamaterial properties can be designed. The electric and magnetic responses can be generated by coupling between metamaterial elements. Moreover, the flexibility of engineering 'μ' and 'Ɛ' values match the impedance of a designed metamaterial absorber with free space to reduce the reflection (Amiri, M. 2018). Subsequently, it leads to the perfect absorption at the desired frequency. To synthesize negative permeability was demonstrated as shown in Figure.3(a). The transmission response for SRR is represented by bold line and the response for the combined structure (thin wire and SRR) is shown by dashed curve in Figure.3(b). For the bulk structure, if SRR is alone present, there is no transmission around 5 GHz. But when SRR is combined with thin wire, energy is transmitted. The effect of the combined negative permeability and negative permittivity derives the energy transmission.

To understand the wave absorption theory, the interaction of the incident wave with the boundary surface is studied. The ways through which an incident wave can interact with a boundary surface includes: reflection, transmission, absorption, scattering and the excitation of surface waves. The surface electromagnetic waves (SEWs) (Ritchie, 1957; Watts, Liu and Padilla, 2012) are generated due to the incident wave (Ang, K. H., Chong, G. 2015). The effect of these SEWs can be studied based upon their propagation length (Lp). Though the propagation length (Lp) gives insufficient information about an incident wave coupled with the boundary surface, it defines the distance over which the intensity of SEWs decays exponentially (Arrawatia, M. 2016). In order to describe the coupling of the incident wave with the boundary surface, an equation for Figure of merit is described by Gil et al. (2007). The scattering of the waves at the interface is considered to be negligible if the operating wavelength is much larger than the surface roughness ($\lambda > Rs$) causing the surface waves to die out before the occurrence of re-scattering. Further, based on the Fresnel equation and transmission matrix approach, the basic absorption theory for a magneto dielectric medium backed by a conductive ground metal was discussed by Watts, Liu and Padilla (2012). In this, all possible ways an incident wave can interact with boundary surface is considered. The metamaterial absorber consists of a dielectric medium which is supported by an opaque metallic surface as the ground plane, therefore there is zero transmission (S21) and the maximum absorption solely depends upon the minimum reflection. Most of the researchers focussed on metamaterial absorber design with single or dual band absorption characteristics (Bagmanci, M. 2018). Very few works have been published on absorbers with wide band absorption characteristics. Hence, the present research work was envisioned to explore the metamaterial based RF energy harvesting for ultra-wide frequency band (UWB) with maximum absorptivity. The existing multichemistry batteries such as lead-acid, Li-ion, etc., requires different chargers for charging (Bakır, M., Karaaslan. 2017). However, universal chargers can afford solution to some extent. Moreover, the existing chargers will not support, when a new type of battery is invented. Hence, these sort of issues necessitate an alternative way to address these problems. As an alternative, a novel IoT based smart universal charger has been proposed in this chapter for charging multichemistry batteries.

LITERATURE SURVEY

WSNs are one of the prominent areas of research which provides a background framework for Internet of Things (IoT). The key components of WSNs are sensors, actuators, embedded controllers and connectivity devices through which data or information is communicated. One of the major challenge lies in the incessant power supply required for these enormous smart connected devices which are mainly battery operated. RF energy harvesting provides an encouraging and economically friendly solution to provide low power DC supply from the harvested RF energy sources such as Wi-Fi, GSM, Radio/TV broadcasting stations, radar, etc. Numerous research works are carried out in the development of RF Energy harvesting systems (RFEHs) with research challenges in the antenna design, bandwidth and DC conversion efficiency (Baqir, M. A., Ghasemi, M. 2019). An overview of the different RFEH systems along with the challenges in the antenna design, rectifier design, matching circuit design are reviewed to afford a general framework for designing RF energy harvesting systems. In addition, an IoT based universal battery charging system has been developed to replace the conventional batteries. Advancement in the technologies has brought about a demand for instant access to everything. Engineers are bound

Figure 4. Various Energy Harvesting sources in the wireless networks

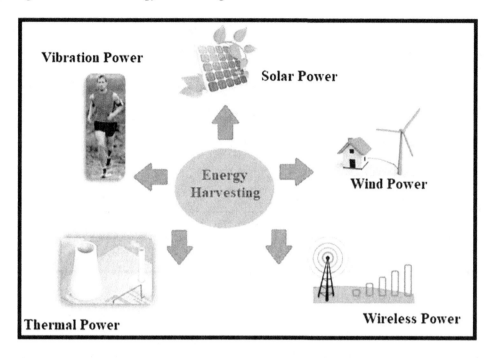

to revamp the existing technologies such as IoT (Benayad, A., & Tellache. 2020), 5G technologies, artificial intelligence (AI), (Bently, W. F., & Heacock. 1996) and wearable electronics (Brown, W. C. 1980). These technologies involve huge measure of sensors which are deployed in the remote places, but these sensors require batteries to satisfy their power requirement. But replacing these batteries periodically from these ubiquitous sensors is quite a herculean task, which makes the environment more and more hazardous. Energy Harvesting (EH) or Green Energy (GE) could be an alternate or a feasible solution to this environmental challenge which involves extracting energy from the readily available renewable energy sources. Figure.4. illustrates the various energy harvesting sources for WSNs. The ambient energy from the various sources such as solar power, wind power, thermal power, vibration power and dedicated RF energy sources such as Wi-Fi, WLAN, Bluetooth, AM/FM etc. are converted into electrical energy. The right choice of energy harvesting helps in reducing the complexity of end devices (WSN/IoT nodes). This is done by avoiding the power circuits which reduces the overall cost and improves the overall efficiency of the devices/system. Figure.5. represents the architecture of a wireless sensor node (WSN). A typical WSN consists of energy harvesting device, power management unit and storage device. Energy harvesting device converts the incident ambient energy from various energy sources such as solar, wind, thermal,

Figure 5. Architecture of a WSN node

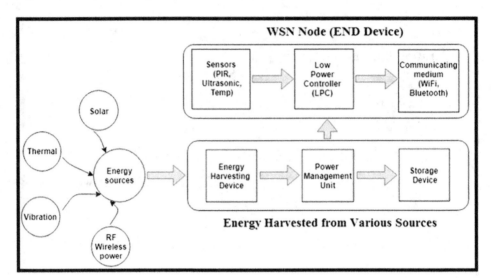

vibration, and RF wireless energy into DC voltage. The power management unit (PMU) is responsible for either storing the harvested energy or regulating the power to the WSN Node (End Device). The storage device conserves the harvested power for future use.

In the past, several reviews were reported by the researchers on different energy harvesting techniques. (Chen, T., Li, S.J. 2019) demonstrated energy harvesting from body heat (thermal energy) to power WSN node. (Chen, W. C. 2018) demonstrated piezoelectric techniques for energy harvesting. (Chen, X., Tomasz. 2017) focussed on small scale wind mills for powering low power WSN nodes. Though, battery has been the main source of power, replacing and discarding them remains a challenge for uninterrupted data and also hazardous to the environment (Dawar, P., Raghava. 2019). Therefore, GE or environment friendly technology is required to avoid the toxic chemicals and metals which pollute the environment. Energy harvesting was achieved through near field meth (Derbal, M. C., & MouradNedil. 2020). In the chapter, Far field method is being focussed more for RF energy harvesting.

NEED FOR IoT BASED UNIVERSAL CHARGERS

The rapid evolution of technology has yielded advances in introducing portable devices and gadgets with profound impacts on society. Promisingly, in every domain, new portable devices keep updating the market. Rechargeable batteries are one of the major sources to operate these portable devices. Hence, development of new rechargeable batteries has become a major research area in low power electronics. Since its invention, researchers have concentrated mainly on increasing charge density, decreasing charging time, reducing size and minimizing the cost of the rechargeable battery (Devi, K. K. A., Hau, N. C. 2018). This has resulted in the development of new rechargeable batteries with different chemical composition. Currently, development of such batteries are in a greater pace, as it replaces the conventional batteries. Although different kinds of batteries utilize different charging algorithm, the major problem was every battery should follows its charging algorithm to get maximum life time, reduced charging time and no hazards. Secondly, different batteries have to use different chargers (Devi, K. K. A., Sadasivam. 2019). The need of universal chargers are much important which supports charging many kinds of batteries instead of buying separate chargers, reduces the overall size and cost. IoT refers to the interconnection of embedded devices through internet. IoT enables the devices to share and interact each other without manual intervention. The application of IoT has expanded to every field to facilitate real time monitoring by accessing remote database. Moreover, incorporation of IoT platform with the universal charger seems to be a novel alternative to add new battery profile from the server to the client, which in turn paves the way to charge existing battery profile as well as upcoming batteries in the near future. number of cells for NiCd, NiMH, Li-ion and Li-Polymer batteries (Ding, F., Dai, J. 2018). It had the advantage of charging Constant Current (CC), Constant Voltage (CV) with a wide range of currents and voltages. But the CMOS processor used was more expensive and also adds more complexity to the design. The design experiment was carried out using DU2004S3 a Nickel/Li-ion development system. But the limitations were manual selection of batteries and fixed battery profiles. The PWM signal generated by the microcontroller was able to feed the buck convertor circuit that controls the voltage and current on the output using a PI controller (Divakaran, S. K., Krishna. 2018). The charging of only two kinds of batteries and the unavailability of addition of new kind of battery are considered as the major disadvantage of this system. A patented work by (Dolgov, A., Zane. 2019) demonstrated a universal charger in which RFID was used to select the battery profile for charging. Based on the RFID tag information, the current and voltage required to charge the device was selected by charging selector. But, the major disadvantage of the charger was in the selection of only predefined battery profiles. This survey has attempted to review and critically assess the reported design techniques for RFEH in the literature and battery charging techniques for various multichemistry batteries. The design techniques for antenna/absorber was classified into four categories – Miniaturization, polarization, reconfigurability and harmonic rejection. Also the design considerations for matching circuits are classified as diode and Mosfet based rectifier circuits. The survey has given an overview that, metamaterial based circular absorbers with diode based rectifier circuits can play a significant role in reducing the electrical size of the antenna and exhibits circular polarization with wide absorption characteristics and provide maximum efficiency. Similarly from the literatures it was observed that, the requirement of different chargers for charging multichemistry batteries such as Lead-acid, Lithium-ion etc. is a major concern. The need for universal chargers are very essential which can support charging many kinds of batteries instead of buying separate chargers. So the need of the hour is a novel IoT based technique for charging existing chargers and upcoming any upcoming batteries. The present work was aimed at designing wide band metamaterial absorbers. Under wideband configuration three absorber designs have been proposed. At first, a polarization insensitive dual band square based metamaterial was designed (Escala, O. A. 2019). The proposed ultrathin absorber size was 16mm x 16mm and and documented to provide high absorptivity of more than 90% in the desired bands. Though, the proposed absorber was found insensitive to various incident angles of electromagnetic wave, it affords only linear polarization with dual bands. Thereafter, a circular shaped metamaterial absorber with concentric circular rings was proposed with optimized design to make the unit cell size smaller, compared to the operating wavelength (Falade, O. P., Rehman. 2018). The proposed circular absorber with compact size of 10mm x 10mm resonated at three different frequency bands with narrow bandwidth. The proposed absorber was found to have a good broad band characteristic and provided FWHM (Full Width Half Maximum). The absorber was thin in thickness with an overall thickness of ~ $\lambda/25$ corresponding to the absorption frequency of 7.6 GHz. The simulations were also carried out to determine the absorption characteristics for different values of incident wave as shown in the Figure.4. The proposed absorber provided wide angle insensitivity i.e. 00, 300, 450, 900 for various oblique angles of incident wave.

TRIPLE BAND METAMETERIAL ABSORBER

Besides, a compact triple band metamaterial with narrow band absorption characteristics was proposed. The proposed metamaterial structure contained three concentric circular rings with notch structures at the top and bottom with thickness of $0.04\lambda0$ for the attained frequency. The size of the proposed circular absorber was reduced to 16% with resonant frequency at three different bands (triple band). The proposed metamaterial absorber was designed with patch (copper) at the top layer, having three asymmetrical concentric circular rings with notched structures etched at the top and bottom. Figure.6(a) represents the geometry of the proposed absorber (Fante, R. L., & Mccormack. 1988). Substrate size of 10 mm x10 mm, ground width size of 2 mm printed on the FR4 dielectric material with the thickness 1.6 mm, loss

Figure 6. (a) Unit cell absorber (b) Top View (c) Bottom View (d) Trimetric view

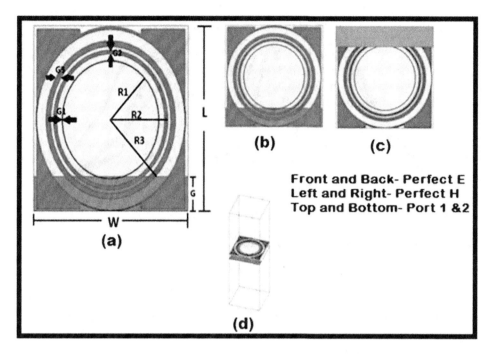

tangent 0.02 mm and permittivity 4.4 mm.The optimal radius of the three circular rings and the notched shape at the top and bottom was attained after the parametric analysis. Figure.6(a) illustrates the proposed absorber, Figure.6(b) and (c) represents the top view and the bottom view of the proposed absorber. Figure.6(d) represents the trimetric view of the proposed unit cell absorber.

Figure.7. represents the normalized absorption plot with various values for R1. When the radius R1 is 3.2 mm, two absorption peaks are obtained and their peak normalised values are 0.99 and 0.89. Similarly, when the value of R1 is 3.25 mm, similar two absorption peaks are obtained with normalised absorption values above 0.80. But when the value of R1 is 3.3mm, three absorption peaks with normalised absorption values of 0.96, 0.90 and 1 are obtained (Faruque, M. R. I. & Islam. 2018). The optimum R1 value is attained when the value of R1 is at 3.15 mm and hence the value of R1 is assigned at 3.15mm.

Figure.8. illustrates the normalized absorption plots for the various values of R2. From the figure, it is observed that when R2 value is 3.65 mm only two absorption peaks are attained with maximum absorption rates of more than 0.80. When the R2 value is 3.7 mm, three rates of 0.95, 0.90, and 1(unity) is observed. With 3.75 mm as the R2 value, the absorber exhibits the absorption rate of 1(unity), 0.90 and 0.96. But when R3 value is at 3.8 mm, the absorption peaks obtained are at 0.92, 1(unity)

Figure 7. Absorptivity plots optimizing R1

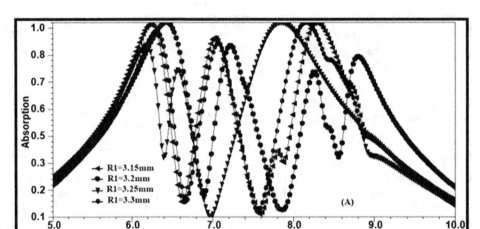

and 0.94. The absorption rates are almost similar to each other. From the parametric study, the value of R2 at 3.7 mm is assigned as the optimum value.

The parameters magnetic permeability (μ) and Electric permittivity (ε) are the parameters which determine the metamaterial behavior when exposed to a time varying electromagnetic field (Fowler, C., & Zhou, J. 2017). Based on the values of (μ) & (ε), metamaterials can be classified as DNG (both μ, ε <0), ENG (ε <0) and MNG (μ<0). To analyze the absorber's metamaterial characteristics, the values

Figure 8. Absorptivity plots optimizing R2

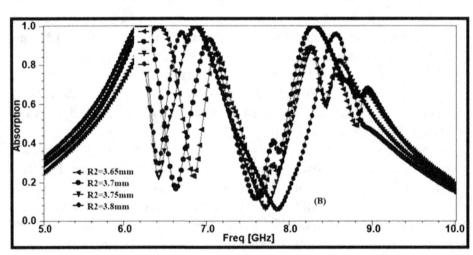

Figure 9. Simulated Real permittivity (ℰ) and permeability (μ) plots at 6.3GHz frequency

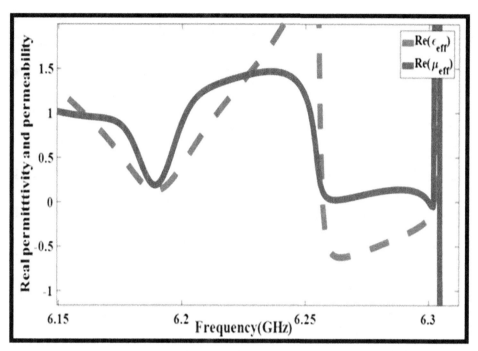

of μ and ε should be retrieved from the S parameters (S11 and S12). To retrieve the values of μ and ε the method proposed by Singh, Abegaonkar and Koul (2019).

It is obseved that at 6.3GHz both the permittivity and permeability values are negative with permeability value μ = -0.1 and the permittivity value ℰ = -1.2. At 7 GHz both the permittivity and permeability values are in the negative region with permittivity ℰ = -1.4 and the permeability μ = -0.4. At the peak frequency of 8.35 GHz the value of permittivity is around ℰ= -0.2 and permeability is μ = +0.4. So it is concluded that at 6.3 GHz and 7 GHz, the material behaves as DNG metamaterial or a perfect metamaterial absorber and at 8.35 GHz it behaves as ENG metamaterial absorber (Ghosh, S., Bhattacharyya. 2018). From the above figures, it can be observed that in all the three frequencies (6.3GHz, 7GHz and 8.35GHz), the values of negative permittivity and the negative permeability lies in the positive region in their respective graphs (Gil, I., Martin, F. 2017). This clearly indicates that the imaginary part of permeability or permittivity always remains positive, irrespective to the metamaterial type such as MNG or ENG or DNG.

When Z (ω) is matched with free space impedance (Z0) then the material is said be a perfect absorber. Figure.9. and Figure.10. shows the plot of the refractive index

Figure 10. Simulated Real permittivity (ε) and permeability. (μ) plots at 7 GHz frequency

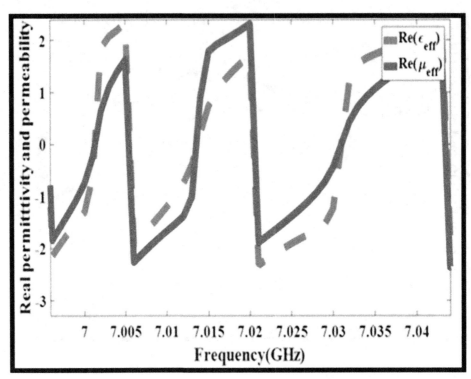

and impedance at frequencies 6.3GHz, 7GHz and 8.35GHz respectively. The values of the refractive index (n) is negative at the frequencies 6.3GHz (n = - 0.1) and & 7GHz (n = -1.1), proving it to be DNG absorber and the value of n is positive (n = 0.1) at 8.35GHz proving it be ENG absorber.

The parameters μ and ε constitutes the metamaterial characteristics. These values were retrieved and analysed to realise the type of metamaterial absorber (MTM) (Gozel, M. A., Kahriman. 2018). The results prove that at frequencies 6.3 GHz and 7 GHz, the proposed absorber behaves as a perfect absorber with more than 90% absorption(Guha, D., Biswas. 2019). To achieves desirable size reduction compared to the existing dimensional structures (Guo, L., & Lu, Q. 2018). Though the work done by Singh (Hashemi, S. S., Sawan. 2018) has exhibited the same dimensions, the maximum absorption rate was comparatively low compared to the proposed work.

SINGLE BAND ABSORBER AND RECTIFIER

This section discusses a compact single band circular metamaterial absorber to harvest DC energy from very low ambient microwave signals in particular for indoor environments such as office, buildings etc. The most widely used 2.45 GHz frequency band which is used for Wi-Fi, Bluetooth, Zigbee, and RFID is preferred for RF energy harvesting. The RF energy harvested from the designed absorber is converted to dc power using a tuned single stage and a three stage voltage rectifier circuit. The absorber and the rectifying circuit are analysed for maximum dc output voltage and maximum efficiency. The L network is a simple circuit with few components compared to T or II network. The advantage of using L network is using few components for matching but with narrow band characteristics. The simulation is carried out with advanced design system (ADS) software with smart Smith Chart features for tuning the L matching circuits. For matching the input impedance of rectifier circuit (Zin) with normalised impedance Zo (50Ω), the value of Zin at the desired frequency (2.45 GHz) was computed. From the results, it was observed that the designed rectifier circuit with L-matching circuit, attained the maximum efficiency of 87.2% at the input power of 10dBm reported in the Figure 5.20. The maximum output volatage attained at -10 dBm, 0 dbm and 10 dBm were 1.07V 0.45V and 1.78V, respectively. The rectifier voltage doubler circuit designed for single stage gave maxium voltage around 1.8V and an efficiency of 87% at an input power of 10dBm. Further, doubler was increased to three stages. In each stage, two schottky diodes HSMS2850, filter capacitor and a load resistor of 100kΩ were used. The schottky diode is preferred due to its small resistance (Rs = 25Ω) and small barrier capacitance (Cp = 0.18pF) with high cut off frequency and high conversion efficiency. The Smith Chart utility provides synthesis of matching networks and enables impedance matching. Using Smith chart utility, L matching network was tuned and the values of L1 & L2 were obtained. The desired frequency 2.45GHz was attained after Zo is matched with Zin. The 3 stage rectifier voltage doubler was analysed for output voltage and efficiency. The output voltage was analysed for various input power levels. A peak voltage of 7.9V was attained around 10dBm with a maximum efficiency of 65% with a 100K load. From the efficiency plots, it is observed that despite improvement in the output voltage, the efficiency is reduced due to the dissipation losses in the schottky diode (Hassan, N., Hisham. 2018). In continuation, simulation of rectifier voltage doubler circuit for single stage and three stage was analyzed. Schotty diode HSMS 2850 was selected due to its high cut off frequency and high conversion efficiency. The single stage RFEH system showed a conversion efficiency of 87% at the input power of 10 dBm with an output voltage of 1.78V. As the stages are increased, the efficiency is reduced to 65%. But the output voltage is increased to 7.9V at 10dBm. This may be due to the dissipation of power due to the non linear elements in the circuit. The attained results are compared with the previous works and are tabulated. The results concludes that the RFEH system is a suitable candidate to harvest energy for indoor applications at the 2.45GHz frequency band.

IoT BASED SMART CHARGING SYSTEM

The intelligent charger is user friendly, which allows the user to select any of the existing battery profiles and also facilitates to update new battey profiles from the server. The proposed design consists of two sections: IoT section and charging Section. The IoT section consists of Raspberry pi controller as client and PC as server, both are remotely connected via internet. Both server and client contains the databases of various battery profiles (Hong, H., Cai, X., Shi, X. 2018). The charging section consists of a microcontroller based buck converter which monitors the nominal battery voltage, current and temperature. It also supplies the required output voltage and current for charging the batteries using pulse width modulation (PWM) technique. The charging profile of NiCd and NiMH batteries are same. The battery follows a Constant Current (CC) method for charging and the charging voltage is in the range of 1.4V-1.6V per cell. From the cell voltage of 0.075V to 0.625V, the battery follows a slow charging at 0.1C. From 0.625V onwards, it follows fast charging at 1C. Vpeak is represented as the maximum voltage at which the cell attains 1.4V. The maximum safe voltage (Vmax) attained by cell is 1.2V. It is imperative to determine the detection of full charge in the cell as overcharging could damage the cell. The detection of full charge was determined by the dropping of voltage from Vmax in a duration time of 1 minute. The proposed design is the combination of both hardware and software and is divided into two sections: IoT section and the charging section. In the IoT section, personal computer (PC) is server and it is connected to the client (raspberry pi) through Wi-Fi module. In the charging section, the modules used are Atmega 32 microcontroller, MOSFET IRF 540, DC – DC Buck converter, current sensor, relay module and the battery which is to be charged. Figure.11. illustrates proposed system.

The proposed block diagram has two main sections as shown in the Figure.11. In the IoT section, PC is the server and the client is the raspberry pi. The raspberry pi is a mini computer running on Linux based platform which facilitates both hardware and software applications. Both the server and client have the databases of different battery profiles. The server and client are connected remotely to the internet. The python script helps to connect, raspberry pi(client) to the servers database. Therefore, whenever a new battery profile is available at the server, the client can download and update the new battery profile to its database. In the charger section, the main blocks used are Atmega32 microcontroller, buck converter and a current sensor.

Figure 11. proposed system

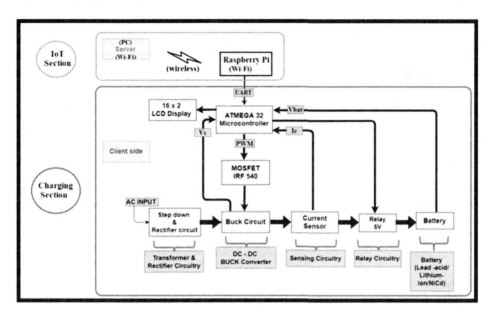

The microcontroller communicates with the Raspberry pi through a serial UART interface. Also it generates the desired PWM voltage and current pulses by varying the duty cycle for CV and CC charging stages. Then the PWM pulses generated are fed to the buck converter circuit. The Buck converter circuit contains MOSFET IRF540, inductor and a capacitor. In the converter design, the design of L and C is very crucial. The voltage and the current values of the buck circuit has to be monitored continuously. So these values are fed back to the microcontroller. The current sensor ACS712 generates the equivalent analog voltage with respect to the current that is sensed. The voltage and current fed back is retained constantly by the PID controller for charging the battery. The PID controller continuously maintains the voltage and current constant in accordance to the set points of voltage and current. The battery charging is done in two modes i.e constant current and constant voltage, referred as charging algorithm. For charging a particular type of battery, the user enters the CC and CV values. Before the start of battery charging, the microcontroller checks for the battery voltage. If the battery voltage is less than the full discharge voltage, then the battery is detected as faulty. If the voltage is greater than the full discharge voltage, then the controller follows the constant current algorithm and the current is set according to the CC value. The PID control is applied to the CC values to maintain the current constant throughout the CC mode. As the battery voltage reaches more than 80% of full charge or Nominal voltage, the microcontroller enters CV mode and follows a constant voltage algorithm and the voltage is set according to the CV value. The PID control is applied to the CV values to maintain constant voltage during the CV mode. The voltage (V) and current (I) are constantly updated in the LCD display. As the battery attains the maximum voltage value and if the current drops below 4% of the rated current, then the charge cycle is stopped and updated in the LCD display. In the proposed system, addition of upcoming new battery profiles is possible along with the existing battery profiles. This has an added davantage over the existing chargers reported in the literatures. In this proposed work, an intelligent charging system is implemented which can charge different batteries simultaneously. The novelty of the proposed system is the ability to update new charging profiles from the server. The proposed system is simulated and tested for various battery profiles and the results are found to be reliable. The results confirm that the designed charger can be implemented with differerent multichemistry batteries ranging from low capacity to high capacity. Therefore, it can be realized as a universal charging system for all kind of portable home appliances.

CONCLUSION

In this chapter, a wide band RF energy harvesting metamaterial absorber was proposed to harvest energy in the frequency range of 3.1 GHz to 10.6 GHz. Adding up, a single stage and three stage rectifier voltage doubler circuit were investigated for a single band circular absorber at 2.45 GHz frequency. Moreover, a novel IoT based battery charger has been designed as an alternative to the conventional chargers. From the attained results, the designed absorbers are proved to be a better candidate for energy harvesting applications. A novel IoT based intelligent universal charger was proposed in this study. The designed charger consisted of two sections such as IoT section and charging section. The IoT section consisted of Raspberry pi controller as client and PC as server. The server contains the databases of battery profiles and remotely connected to the controller through Internet. The charging section consists of microcontroller based buck converter which monitors the nominal battery voltage, current and temperature. The required output voltage was provided to charge the battery using Proportional Integral Derivative (PID) algorithm through PWM technique. Unlike the conventional chargers that could charge only the existing batteries, the proposed charger was able to charge diverse multichemistry batteries. The uniqueness of this charger is recognized based on the updation of existing battery profile as well as the upcoming new battey profile from the server. In the future, flexible electronics will play a major role in applications such as wearables, ingestible smart pills etc. The designs of various absorbers in this thesis considered is FR4 material with dielectric constant of 4.4. Hence flexible dielectric materials like polydimethylsiloxane (PDMS), paper and jeans cloth will be concentrated in the future work.

ACKNOWLEDGMENT

We would like to acknowledge the support given by BMS Institute of Technology and Management, Bangalore, India.

REFERENCES

Almoneef, T. S., & Ramahi, O. M. (2014). Can Split Ring Resonators be viable for electromagnetic energy harvesting? IEEE Antennas and Propagation Society, 424-425.

Almutairi, A. F., Islam, M. S., Samsuzzaman, M., Islam, M. T., Misran, N., & Islam, M. T. (2019). A Complementary Split Ring Resonator Based Metamaterial with Effective Medium Ratio for C-band Microwave Applications. *Results in Physics*, *15*, 102675. doi:10.1016/j.rinp.2019.102675

Aminov, P., Jai, P., & Agrawal. (2018). RF Energy Harvesting. *IEEE 64th Electronic Components and Technology Conference (ECTC)*, 1838-1841.

Amiri, M., Tofigh, F., Shariati, N., Lipman, J., & Abolhasan, M. (2020). Wide angle metamaterial absorber with highly insensitive absorption for TE & TM modes. *Scientific Reports*, *10*(1), 2. doi:10.103841598-020-70519-8 PMID:32788706

Ang, K. H., Chong, G., & Li, Y. (2015). PID control system analysis, design, and technology. *IEEE Transactions on Control Systems Technology*, *13*(4), 559–576.

Arrawatia, M., Baghini, M. S., & Kumar, G. (2016). RF energy harvesting system at 2.67 and 5.8GHz. *Asia-Pacific Microwave Conference*, 900-903.

Bagmanci, M., Karaaslan, M., Altintaş, O., Karadag, F., Tetik, E., & Bakir, M. (2018). Wideband metamaterial absorber based on CRRs with lumped elements for microwave energy harvesting. *The Journal of Microwave Power and Electromagnetic Energy*, *52*(1), 45–59. doi:10.1080/08327823.2017.1405471

Bakır, M., Karaaslan, M., Unal, E., Akgol, O., & Sabah, C. (2017). Microwave metamaterial absorber for sensing applications. *Opto-Electronics Review*, *25*(4), 318–325. doi:10.1016/j.opelre.2017.10.002

Baqir, M. A., Ghasemi, M., Choudhury, P. K., & Majlis, B. Y. (2019). Design and analysis of nanostructured subwavelength metamaterial absorber operating in the UV and visible spectral range. *Journal of Electromagnetic Waves and Applications*, *29*(18), 2408–2419. doi:10.1080/09205071.2015.1073124

Benayad, A., & Tellache, M. (2020). A compact energy harvesting multiband rectenna based on metamaterial complementary split ring resonator antenna and modified hybrid junction ring rectifier. *International Journal of RF and Microwave Computer-Aided Engineering*, *30*(2), e22031. doi:10.1002/mmce.22031

Bently, W. F., & Heacock, D. K. (1996). Battery management considerations for multichemistry systems. *IEEE Aerospace and Electronic Systems Magazine*, *11*(5), 23–26. doi:10.1109/62.494184

Brown, W. C. (1980). The history of the development of the rectenna. *Proc. SPS Microwave Systems Workshop at JSC-NASA*, 271-280.

Chen, T., Li, S. J., Cao, X. Y., Gao, J., & Guo, Z. X. (2019). Ultra-wideband and polarization-insensitive fractal perfect metamaterial absorber based on a three-dimensional fractal tree microstructure with multi-modes. *Applied Physics. A, Materials Science & Processing*, *125*(4), 232. doi:10.100700339-019-2536-6

Chen, W. C., Totachawattana, A., Fan, K., Ponsetto, J. L., Strikwerda, A. C., Zhang, X., Averitt, R. D., & Padilla, W. J. (2018). Single-layer terahertz metamaterials with bulk optical constants. *Physical Review. B*, *85*(3), 035112. doi:10.1103/PhysRevB.85.035112

Chen, X., & Tomasz, M. (2017). A Robust method to retrieve the constitutive effective parameters of metamaterials. *Physical Review. E*, *70*(1), 016608. doi:10.1103/PhysRevE.70.016608 PMID:15324190

Dawar, P., Raghava, N. S., & De, A. (2019). UWB Metamaterial-Loaded Antenna for C-Band Applications. *International Journal of Antennas and Propagation*, *2019*, 1–13. doi:10.1155/2019/6087039

Derbal, M. C., & Nedil, M. (2020). A High Gain Dual Band Rectenna for RF Energy Harvesting Applications. *Progress in Electromagnetics Research Letters*, *90*, 29–36. doi:10.2528/PIERL19122604

Devi, K. K. A., Hau, N. C., Chakrabarty, C. K., & Din, N. M. (2018). Design of Patch Antenna Using Metamaterial at GSM 1800 for RF Energy Scavenging. *IEEE Asia Pacific Conference on Wireless and Mobile*, 157-161.

Devi, K. K. A., Sadasivam, S., Din, N. M., & Chakrabarthy, C. K. (2019). Design of a 377 Ω Patch Antenna for Ambient RF Energy Harvesting at Downlink Frequency of GSM 900. *The 17th Asia Pacific Conference on Communications,* 492-495.

Ding, F., Dai, J., Chen, Y., Zhu, J., Jin, Y., & Bozhevolnyi, S. I. (2018). Broadband near-infrared metamaterial absorbers utilizing highly lossy metals. *Scientific Reports,* 6(1), 39445. doi:10.1038rep39445 PMID:28000718

Divakaran, S. K., & Krishna, D. D. & Nasimuddin. (2018). RF energy harvesting systems: An overview and design issues. *International Journal of RF and Microwave Computer-Aided Engineering, 29,* 1.

Dolgov, A., Zane, R., & Popovic, Z. (2019). Power Management System for Online Low Power RF Energy Harvesting Optimization. *IEEE Transactions on Circuits and Systems, 57*(7), 1802–1811. doi:10.1109/TCSI.2009.2034891

Escala, O. A. (2019). *Study of the Efficiency of Rectifying Antenna Systems for Energy Harvesting* [Thesis]. UPC-Barcelona, Spain.

Falade, O. P., Rehman, M. U., Gao, Y., Chen, X., & Parini, C. G. (2018). Single Feed Stacked Patch Circular Polarized Antenna for Triple Band GPS Receivers. *IEEE Transactions on Antennas and Propagation, 60*(10), 4479–4484. doi:10.1109/TAP.2012.2207354

Fante, R. L., & Mccormack, M. T. (1988). Reflection Properties of the Salisbury Screen. *IEEE Transactions on Antennas and Propagation, 36*(10), 1443–1454. doi:10.1109/8.8632

Faruque, M. R. I., & Islam, T. (2018). Novel triangular metamaterial design for electromagnetic absorption reduction in human head. *Progress in Electromagnetics Research, 141,* 463–478. doi:10.2528/PIER13050603

Fowler, C., & Zhou, J. (2017). *A Highly Efficient Polarization-Independent Metamaterial-Based RF Energy-Harvesting Rectenna for Low-Power Applications.* arXiv, 1705.07717.

Ghosh, S., Bhattacharyya, S., Chaurasiya, D., & Srivastava, K. V. (2018). An Ultra-wideband Ultrathin Metamaterial Absorber Based on Circular Split Rings. *IEEE Antennas and Wireless Propagation Letters, 14,* 1172–1175. doi:10.1109/LAWP.2015.2396302

Gil, I., Martin, F., Rottenberg, X., & De Raedt, W. (2017). Tunable stop-band filter at Q-band based on RF-MEMS metamaterials. *Electronics Letters, 43*(21), 1153–1154. doi:10.1049/el:20072164

Gozel, M. A., Kahriman, M., & Kasar, O. (2018). Design of an efficiency-enhanced Greinacher rectifier operating in the GSM 1800 band by using rat-race coupler for RF energy harvesting applications. *International Journal of RF and Microwave Computer-Aided Engineering*, *21*(1), e21621. doi:10.1002/mmce.21621

Guha, D., Biswas, S., & Antar, Y. M. M. (2019). *Microstrip and Printed Antennas: New Trends, Techniques and Applications*. John Wiley & Sons.

Guo, L., & Lu, Q. (2018). Potentials of piezoelectric and thermoelectric technologies for harvesting energy from pavements. *Renewable & Sustainable Energy Reviews*, *72*, 761–773. doi:10.1016/j.rser.2017.01.090

Hashemi, S. S., Sawan, M., & Savaria, Y. (2018). A high-efficiency low-voltage CMOS rectifier for harvesting energy in implantable devices. *IEEE Transactions on Biomedical Circuits and Systems*, *6*, 326335. PMID:23853177

Hassan, N., Hisham, A. B., Fareq, A. M. M., Abidin, A. M. Z., Bakar, H., Noor, A. M. S., & Khairy, I. M. (2018). Radio frequency (RF) energy harvesting using metamaterial structure for antenna/rectenna communication network: A review. *Journal of Theoretical and Applied Information Technology*, *96*(6), 1538–1550.

Hong, H., Cai, X., Shi, X., & Zhu, X. (2018). Demonstration of a highly efficient RF energy harvester for Wi-Fi signals. *International Conference on Microwave and Millimetre Wave Technology (ICMMT)*, 1–4.

Chapter 10
Wireless Sensor and Actuator Networks–Based Reliable Data Acquisition Mechanism

Anil Sharma
(iD) https://orcid.org/0000-0002-7115-6278
College of Computing Sciences and IT, Teerthanker Mahaveer University, India

Dankan Gowda V.
(iD) https://orcid.org/0000-0003-0724-0333
B.M.S. Institute of Technology and Management, India

A. Yasmine Begum
Sree Vidyanikethan Engineering College, Mohan Babu University, India

D. Nageswari
Nehru Institute of Engineering and Technology, India

S. Lokesh
PSG Institute of Technology and Applied Research, India

ABSTRACT

The objective of deploying a wireless sensor network is to collect data about the environment in which it will be utilised and then to transmit that data to a distant sink where it will be used to estimate or reconstruct the environment or event. In order for the centralised sink to be able to accurately reconstruct or estimate the event and take the necessary actions on time, the wireless sensor and actuator network must be able to guarantee delivery of a sufficient amount of the information gathered by the deployed sensor nodes in a time-bound and coherent manner. In addition to the above-mentioned fundamental problem, reliability also refers to the network's capacity to tolerate defects up to a certain point without compromising performance. This chapter introduces a brand-new, dependable data acquisition technique that makes use of wireless sensor and actuator networks.

DOI: 10.4018/978-1-6684-4974-5.ch010

INTRODUCTION

A Wireless Sensor Network (WSN) is deployed with the purpose of acquiring information and then passing on the acquired information to a remote sink where the information can be used to estimate or re-construct the environment or event. This requires the WSN to have the ability to sense the parameter or event under consideration, in the region of deployment, and then ensure reliable delivery of the information to a centralized sink where the information sensed by it can be used to re-construct the events (Abroshan, S., & Moghaddam, M. H. Y. 2014). However, in context of the Wireless Sensor & Actuator Networks the Network Latency time assumes much greater importance since the action taken by the Actuator nodes has to be time coherent with the event sensed by the deployed nodes failing which the control loop will become un-stable and erratic.

This chapter presents a novel Reliable Data Acquisition methodology using Wireless Sensor & Actuator Networks which meets the criterion of reliability as mentioned above. The proposed methodology, ensures, regarding a sensed event is reported to the centralized sink / Actuators for acceptable estimation or re-construction of the event detected, within the time constraint fixed by the control action to be taken by the Actuator nodes.

The objective in this approach is for the Sensor Network to continue its operation with as high reliability as is feasible under the given fault condition in the network (Adelstein, 2005). The fault conditions could vary from temporary loss of communication between a set of nodes, to permanent loss of a node because of damage or end of battery life of the node. The focus of design of algorithm tends to be on ensuring detection of event and then delivery of sensed despite the fault condition. Some algorithms also focus on detection of fault condition itself and then managing it.

Majority of these solutions propose algorithms with the ability to dynamically find alternate network paths in case a given network becomes un-usable because of a fault condition. Many algorithms present methodology where the data sensed is transferred on multiple paths simultaneously to begin with thus ensuring delivery of information at sink despite some path becoming useless because of node failure (Ali, S., Fakoorian, A. & Taheri, H. 2017).

These methods compute reliability for multiple paths depending upon status of nodes and send data on most reliable path or broadcast data on multiple routes with probabilistic approach for success. Some algorithms use Centralized or distributed approach to fault detection based on historic data or location-awareness. However, these set of methods tend to be highly memory & computational intensive thus putting significant load on the nodes and increasing the computational delays and reduction in life of the node (Alsbouí, T.A. 2019). The tolerance to node fault is high and performance of the network does not de-grade significantly with limited node failure. Focus is on many-to-one aggregation of information while distributed approach towards data aggregation is not addressed. The focus of design of algorithms is on ensuring that the packet origination from a deployed node is not dropped at any stage and if it does drop then some alternate methodology is available for recovery of the same (Akyildiz, I. F. 2020). Multiple approaches including exploiting spatial and temporal redundancies, hop-by-hop recovery, end-to end reliability, tree based optimization of routing path, multipath [joint or disjoint] approach, quick fetching of lost packets from neighbours are used. Concept of handshaking is also used in some cases using RTS/CTS or NACK as handshaking confirms success of transaction. These approaches focus on ensuring the delivery of information sensed, however, a time-constraint is not generally defined for these cases (Akyildiz, I. F. & Kasimoglu, I. H. 2014). Besides high control overheads are observed because of the effort involved in ensuring packet delivery. Algorithms are available for many-to-one and many-to-few approach for data aggregation. However, the reliability here is at a cost of end to end delay time. The focus in this approach is on using one or multiple QoS parameters for making decisions regarding the actions to be taken by the nodes in context of delivery of sensed information. In particular, the focus is on satisfying. Some QoS aware mechanisms are topology aware or geographically aware (Akan, Ö. B. & Akyildiz, I. F. 2015). Shortest path with least energy being used is the criterion generally used which may be applied at local (node level) or network wide level. Some algorithms use the end-to-end delay constraints as a part of the performance metric but the network overheads tend to be high and the delay may not meet the application criterion in many cases. Generally prescribed for many-to-one collection of data and are not suited for many-to-few criterion applicable in Multi-Actuator distributed scenario (Alimohammadi, H. & Jassbi, S. J. 2019). This approach utilizes the knowledge about the location of the nodes to deliver the information sensed through the shortest, least congested path. Geographic knowledge also enables the nodes to be scheduled for sleep patterns since the path for forwarding the information packets can be ascertained in terms of geographic location of nodes (Ali, A. 2018). However, location knowledge of nodes is generally not easily available or prevalent and actually ascertaining the same is highly energy intensive. This approach uses the Swarm Intelligence techniques like ant colony Optimization, Foraging Bee hive, for arriving at optimized decisions for reliable communication (Al-Awami, L. & Hassanein, H., 2018). Suitable for distributed problem solving approach, however, tend to have communication and energy overheads. Characteristics and constraints that apply to Wireless Sensor Networks have also been shown. The evolution of Wireless Sensor & Actuators has then been discussed along with information about the standard configurations and additional constraints that apply to it. The concept of reliability has also been traced to its traditional definition and then the application of the concept of reliability, in context with Wireless Sensor Networks, has also been presented. The field has also been presented briefly followed with the information about the contribution of this chapter.

LITERATURE SURVEY

The issue of reliability of Data Acquisition is a central part of the core activity for the reason for deployment of a WSN or a WSAN. However, there are different directions and contexts in which research work has been pursued in this regards. The issue of definition of reliability itself has been stated in different contexts (Alam, M. M., & Hong, C. S. 2019). The approaches tend to vary between the methodologies used for delivering the information collected at the deployed nodes with some techniques focussing on providing multiple paths, some on identifying multiple paths but using a single path, some on using QoS metric for identifying the best path while others focus on reduction of energy use as the main criterion. The different approaches, proposed, can be broadly categorised as below: Event Detection Reliability and Packet Delivery Reliability Achieving reliability of data acquisition can be done by using various techniques Which are essentially categorised. This Categorization is as follows: End to End Reliability and Hop by Hop Reliability. Figure.1. illustrates the categorization of the concept of Reliability in WSN based on the various techniques proposed in literature. One of the most straightforward ways to guarantee the accuracy of the data collected by a deployed Wireless Sensor Network is to make sure that the packets sent by the sensor nodes include the detected information and make it to the sink, where it can be used to reconstruct the events at a distance. One of the easiest ways to guarantee that no data is lost at any point along the multi-hop path from the source node to the sink is to have the source node retransmit the lost packet.

The definition of some common terms given in the context of reliability is given below:

Event Detection Reliability

The objective here is to ensure that sufficient packets reach the sink to enable it to detect (Atanasov, A., Kulakov, A. 2010). It is implicit that it may not be necessary for all the packets sourced by each node to reach the sink, thus a significant amount of retransmission of packets may not be needed.

Figure 1. Categorization of reliability techniques for WSN [Akyildiz, I. F. 2020]

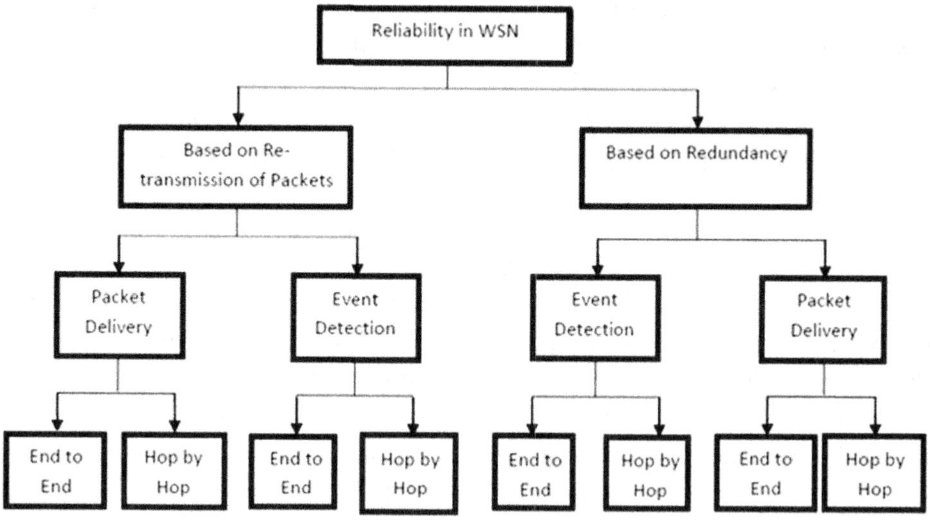

Packet Delivery Reliability

By virtue of the definition itself it is implicit that significant amount of re-transmission of packets, partial or complete, may be required.

Hop by Hop Reliability

Each node, which re-transmits the packet in between the source and destination node, is able to detect the loss of a packet that it re-transmitted (Ayadi, A. 2018). Further the re-transmission of lost packet can be done by the most recent node rather than by the originating node. Thus the detection as well as re-transmission of the lost packet can be done in the shortest period of time. The obvious disadvantage is that all nodes need to store information about each packet re-transmitted by them irrespective of whether they are the originating node or not.

Explicit And Implicit Acknowledgement

The loss of a transmitted packet is detected easily by the usage of explicit or implicit acknowledgement mechanism. In case of explicit acknowledgement the receiving node transmits an acknowledgement packet to the transmitting node on successful receipt of the packet (Baghyalakshmi, D., Ebenezer. 2010). This re-transmission is also heard, since it falls within the radio range and is accepted as an implicit acknowledgement of the receipt of packet by the previous node. The implicit acknowledgement mechanism is less energy consuming since a special acknowledgement packet need not be transmitted each time a packet is received. In case the acknowledgement packet itself gets lost, the transmitting node will read the situation as a loss of the transmitted packet and will re-transmit even though the next hop node has actually received the packet. This unnecessary retransmission will result in a duplicate packet being introduced in the network thus entailing consumption of energy to ensure that the redundant packet reaches the destination (Baronti, P., Pillai, P. 2017). A mechanism at the sink will be required to reject the duplicate packet, however, the loss of energy in the network for conveying the duplicate packet to the sink can only be prevented by having a mechanism within the network to detect and reject an apparent duplicate packet. Each node keeps track of how to get to its nearest neighbours, any of which could go to the sink. Also, the current buffer status at each node is taken into account while choosing the path. The authors provide theoretical proof that, in the absence of noise and malfunctioning environment, the mechanism will guarantee delivery of each packet created by the nodes to the sink, provided the network is a linked network. In addition to supporting route updates, the method also uses the RTS, CTS mechanism included in the IEEE 802.11 MAC to do so.

Brief discussion regarding some of the algorithms falling under the above mentioned categories is given below along with their pros and cons. These factors are very lightly affected by failure rate of nodes in terms of communication or physical node failures (Bein, D., Jolly, V. 2015). The algorithm has low control overheads but does not support flow of messages from the sink to the nodes; further the issue of loss of energy in nodes closest to the sink is also not considered. Further the task of detection of loss of a packet is done by the MAC layer mechanism rather than a formal explicit mechanism. Simulation of the mechanism indicated that the delivery ratio showed slight improvement while the delivery time reduced significantly viz a viz AODV. The route discovery overhead also shows significant reduction however, the maintenance overheads are quite significant (Bernard, T., Fouchal, H and Linck. 2018). The mechanism is designed for multiple-sink environment but has significant control and maintenance overheads for ensuring higher reliability of delivery and reduction in network latency.

Reliability of Data Acquisition is a key concern in a WSN since the whole purpose of deploying a WSN is to be able to acquire information, at a remote location, about the environment in which the WSN is deployed (Bolot, J.C. 2015). The substantial literature survey indicates that the issue is important and various algorithms, protocols and mechanisms. The issue of reliability assumes even more stringent proportions Wireless Sensor & Actuators the acquired information is expected to be the foundation for the decision making process and for actuation of the control

Figure 2. Explicit and implicit acknowledgement

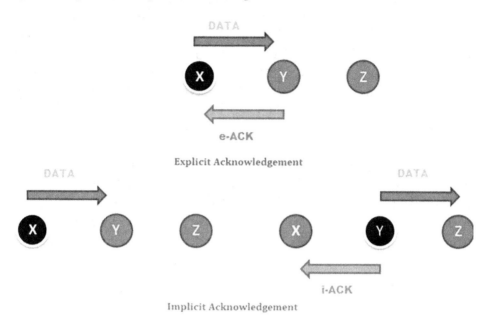

elements. However, this is where the current research work appears a little thin as the literature survey indicates that most of the work in the field of reliability appears in context of Wireless Sensor Networks and not as much in WSAN. However, it could be argued that the aspect of reliability of data acquisition is studied in the sensing part of the WSAN rather than in the actuation part (Bourdenas, T. & Sloman, M. 2019). As observed in the literature survey various types of mechanisms are designed to cater to various aspects of reliability in WSN including Reliability of Delivery, Reliability of Secure Data, Latency time etc. These include protocols to improve reliability of providing security to the network against hostile attacks by malicious nodes, mechanisms to reduce the Packet Drop rates by the transport protocols, improve the reliability of coverage of the sensed area by novel deployment schemes, methodologies to make the network more tolerant (Boukerche, A. and Pazzi, R. 2018). A significant number of algorithms focus on improvement of reliability of delivery mechanism by focusing on providing multiple paths for delivery of information. Most of the work in the field of WSAN is found to be focused on the Actuator- Actuator Communication and coordination, Actuator –Sensor communication, stability of the control action etc. However, the network latency constraints imposed upon the sensing part of the network, in view of the impending control action by the Actuator part of the network, is not particularly addressed as a focus point in an algorithm towards achieving reliability of data acquisition (Cardei, M., Yang, S. & Wu, J.

2018). Some algorithms have taken on the issue of network latency and have taken steps to achieve the same but the steps taken are not particularly aggressive and the simulated results are not likely to prove beneficial in real context. This is because most of the simulation results shown are using IEEE 802.11 as the PHY & MAC layer with channel transfer rate of 2 Mbps and it has been observed that network latency is in the range of a few milliseconds in this context. However, IEEE 802.11 is not suitable for use in WSN in any real-life implementation, this is primarily because WSN applications generally do not require such high data rates and can't afford the associated high energy budget required by IEEE 802.11. Therefore any real-life implementation is unlikely to use IEEE 802.11 protocol; rather the usage of IEEE 802.15.4 or some other variant similar to this is more likely e.g. Zigbee™ uses IEEE 802.15.4 PHY & MAC as does Wireless HART™. IEEE 802.15.4 has a theoretical bandwidth of 250 kbps which is almost 8.3 times lower than the 2 Mbps used during simulation(Camilo, T., Carreto, C. 2016). Therefore the simulated latency periods of a few milliseconds are not likely to be met or observed in real-life situations. Therefore there is a need to design and test mechanisms for Reliable Data Acquisition in WSN which use IEEE 802.15.4 PHY & MAC layers and are optimized for the constraints and functionalities present in this protocol. Any design of mechanism for improving reliability of Data Acquisition in WSAN must address the following issues: Data Acquisition mechanism must be reliable in terms of ensuring that sufficient information regarding the event sensed in the deployment environment must be made available to the sink (Chen, J., Cao, X. 2018). Data Acquisition mechanism must meet the time-constraints as may be required by the actuation system i.e. the network latency must be low. Data Acquisition mechanism must be fault tolerant i.e. it must be able to maintain reliability of the acquisition process even if some nodes fail during the data forwarding process. However, the protocol's dependability drops as low as 20% with fewer than 100 participating nodes and rises to values more than 80% only with more than 200 distributed nodes. In addition, the fault tolerance mechanism is weak, and the computing cost of the method is considerable. There is currently no way to determine when the network is congested. Each unlocalized node may learn its location if it is in range of a moving anchor that broadcasts its position through a beacon, and the authors provide a straightforward technique for doing so. After receiving three of these broadcasts, a typical node can determine its exact location. Due to the fact that every cluster is aware of a node's joining and departing as well as its location inside the cluster, localization data is very valuable. The technique is also fault-tolerant since it can identify when a node has failed. It has large control overheads, does not account for network delay, does not have a means to detect network congestion, and does not allow numerous sinks. The protocol supports three distinct forms of real-time communication: one, known as real-time unicast, is used to relay information from one region of the network to another. Second, real-time area-multicast is designed for simultaneous communications from several nodes to a central hub, and third, real-time area-anycast is for reporting events from any node in the event that the data is highly redundant. The protocol exploits nearby node locations to minimise data storage needs. Instead of sending probes specifically for the purpose of delay estimate, time stamps on the data are used instead. SPEED employs a congestion management strategy that operates on both the network and MAC layers simultaneously. When compared to AODV and DSR, the reported end-to-end latency of around 150 ms is much lower. This is true even for networks with congestion levels of up to 90%. The SPEED protocol uses less energy than AODV and DSR because it does not rely on flooding to establish connections between nodes for secure data transmission. The authors present a thorough theoretical study, simulated tests, and a working implementation using Berkeley motes to back up their claim. All of the nodes' clocks are assumed to be in phase with the central station's for the sake of this technique. Once the node has received an acknowledgment from the sink in response to the Session Initiation Packet, only then will it deliver the real data packet.. The protocol allows the programme to choose its own level of trustworthiness. In addition to the data-forwarding mechanism, a congestion-detection method is built in, with the result that each sent data packet includes information on the congestion state of the node that passed it. When this happens, the sink will let the source know so that it may change its route or slow down its packet transmission.

DATA ACQUISITION MECHANISM

In a wireless sensor network, the occurrence of an event is sensed by nearby nodes which report their observations to the sink/actuator in a multi-hop fashion over the wireless channel. As the sensors are located at a distance from the event, their observations are noisy and error prone. To obviate this limitation, sensor nodes in a region are densely deployed(Chen, X., Dai, Z. 2018). Sensor nodes sleep and wake up to sense their surroundings as per a prefixed schedule. Usually, at any point of time, the sensed region is covered by a number of sensor nodes depending on the 'Sleep-Awake cycle' schedule. An event persists for a finite period of time and triggers the generation of a number of packets by the nodes which are reporting the event. These packets traverse towards the sink, they are lost due to wireless medium characteristics like high attenuation and multi-path fading induced noise. Moreover, there is an additional loss and delay due to channel contention during broadcast over the shared wireless channel (Iyer, Y. G., Gandham. 2015). The attenuation varies non-linearly with distance necessitating multi hop transmission in lieu of direct broadcast to the sink for energy conservation at resource scarce nodes. However, multi-hop transmission requires processing at every en-route node. This increases the network latency significantly. The *energy-latency tradeoff* necessitates optimization of the number of hops packet transversal from a source node to the sink. data acquisition mechanism requires the sensors to sense data locally, communicate the measured data to the sink quickly, the sink makes quick control decisions, and then directly communicate with the actuators for them to take control actions. It must ensure that sufficient data pertaining to an event reaches the sink for it to be able to infer the state of the environment for any control action (Johnson, D. B. 1996). The requirements of particular application can be stated as follows: Reliability: Defined in terms of effective throughput and latency in terms of delay experienced by packets in the network.

False Positive: This is required to minimum spurious control action by the system. This is determined by the accuracy of event estimation which is a function of the number of sensors sensing the event, number of observations and noise in the sensing process. Availability of Data Acquisition Service by the network: LTCRDM should make best effort to provide a long Network-lifetime with reasonable reliability in-spite of not having least distance neighbour communication.

Fault tolerance against communication disruption or node failure. The remaining requirements pertain to maximization of operational lifetime of the sensor network Low Computational delays & Memory resources utilization at Sensor Nodes: Graceful de-gradation in Network performance with loss of nodes.

LTCRDM DESIGN

LTCRDM endeavors to achieve the application dependent requirements like latency and throughput based reliability, availability etc. together with network oriented requirements through the following design considerations. For the application chosen in the present study, LTCRDM implementation uses IEEE 802.15.4. Environment induced communication disruptions in a WSN, usually temporary, is local and affects only a small portion of the network. However, these disruptions cause severe loss of packets and erode reliability of the data delivery mechanism (Johnsort, D. B. 1994). LTCRDM uses a multi-path approach to packet delivery mechanism thus ensuring an increased packet delivery rate of packets to the Sink/Multiple Actuators along with high availability. Alternate path mechanism implemented to counter loss of communication between a set of neighbor nodes of a specific node. This also ensures reliable delivery of information in case of failure of substantial number of neighbors of a specific node. It also maintains time coherence among the packets containing the information sensed and transmitted by each deployed node as events are reported in the order of being sensed by the nodes. The nodes in the WSAN are static. This allows prefixing of data paths from sensor nodes to the sink/actuators. The table based approach for storing neighbor information makes the nodes aware of the local and contextual topology of the network. Fixed data forwarding path reduces the control overhead involved in multi path discovery and maintenance. This, in turn, decreases the overall network latency. This is accompanied by a corresponding reduction in the delay experienced by a packet from source to the sink. Moreover, increase in the number of hops may lead to increased contention increasing the time of flight of a packet (Juwad, M. F., & Al-Raweshidy. 2018). To achieve the critical objective of low latency with energy efficiency, a typical packet must reach the sink in optimal number of hops. For this, the sensed region is divided into disjoint region called layers. A packet is transmitted from a node in one layer to the nearest node in the next layer closer to the sink. There may be multiple nodes of the next layer in the communication range of a node in the previous layer. Ant Colony methodology is employed for increasing of probability of packet delivery by identifying high success rate neighbors. In this methodology, the node which maximizes successful transmission is chosen as the preferred forwarder to maximize the packet delivery ratio and minimize latency by reducing retransmissions. This ensures that the action to be taken by the actuators does not suffer from network lag and the decision taken for the control action is in coherence with the sensed information/event. Moreover, the mechanism automatically performs decongestion of congested nodes relieves nodes which start showing more congestion and start acting as bottlenecks. As the forwarding nodes are prefixed, there are no control computational overheads, the design works with limited computational and memory capabilities of the sensor nodes while significantly reducing computational delays. Finally, the improved packet delivery ratio obviates the need for high degree of coverage (Liang, C.M., & Terzis. 2018). With this, the requirement of having larger number of nodes to be ON simultaneously for detection of an event is reduced. The sleep-awake schedule cycle can be designed such that more nodes can be put to sleep for longer periods of time. This increases the overall network operational life-time without compromising the reliability and availability of the data acquisition service of the deployed WSAN. Packet loss over wireless channel is significant in WSN due to high attenuation, multipath fading and the lack of powerful error correction due to energy intensive nature of communication. Hence, more number of packets are required in constrained time interval for higher reliability.

PHASES IN LTCRDM

The Network Set-up Phase and the Data Acquisition phase are mandatory while the Event Detection Phase or Query Dissemination phase will depend upon the mode of operation desired for a particular application.

Network Set-Up Phase

Occurs only once in a network at the startup. Assumption is made that the nodes are static in nature and will not experience mobility subsequently as well. This phase can be divided into the following two sub-phase.

Layer Assignment Sub-Phase

During this phase the sensor nodes are divided into Layers and assigned layer numbers based on an estimate of the gap between sink & nodes and subsequently between nodes themselves. In case of Distributed Acquisition the sensors are divided into layers and assigned Layer numbers based on the gap between each Actuator and the nodes.

Neighbor Discovery Sub-Phase

During this phase each sensor node initiates the process of finding its neighbors, however, the neighbors are not the immediate neighbors as in the traditional sense. The objective is to find out four types of neighbors for each node. The neighbor information of the four types is stored in four tables along with relevant information regarding the neighbors including the priority of the neighbors.

Data Acquisition Phase

This phase is mandatory and is at the heart of LTCRDM. The objective of this phase is to ensure delivered at the Sink. If it has sensed some information which it needs to send to sink / Actuator, it initiates the process of transmitting the information gathered by it to the sink / Actuator node by passing it to the neighbour nodes (Loh, P. K. K., Hsu. 2017). A hand-shaking mechanism is utilized to improve reliability of data transfer between the nodes. If it receives some information from some other node and it is required to simply forward the same then it initiates the process of forwarding the same to its neighbor nodes towards the Sink / Actuator while simultaneously completing the hand-shaking process.

Query Dissemination Phase

This phase is initiated only in applications where data from the deployed node is to be acquired only if a specific query regarding the parameter required is initiated by the Sink / Actuator node. This phase is initiated by the Sink / Actuator node seeking information. The network nodes on receiving the query perform the following two tasks: If the nodes have the sensed data relevant to the parameter asked for in the query, then they pass on the information to the Sink / Actuator node seeking the information by invoking the Data Acquisition phase. Further the nodes re-transmit the query so that other nodes in the network are also aware of the query. However, this is done in an efficient manner such that it does not lead to network congestion or large scale duplication of the query. Network Latency Time TNLT is the time elapsed between the point when the information regarding the sensed event was transmitted by the deployed node and the point when this information was received at the sink. In case of a Wireless Sensor & Actuator Network, this factor becomes critical since a significant delay between sensing of event and its eventual receipt at the centralized sink for the purpose of decision making could have an adverse impact on the decision making process. There are multiple reasons that could result in a large value of the Network Latency Time including the number of hops, rate of packet collisions, failure of neighboring nodes leading to lack of routing path etc. Typically the energy required for these activities is almost 3-10 times more than the energy required for sampling and processing of information at the node. In any network, raw data is framed and packetized for communication. These additional bits constitute costly overhead in terms of both energy and delay. The MPDU derived from the MAC format is the PSDU i.e the payload for the Physical layer. The frame Payload for the MAC frame is the information from the Network Layer which will also contain the actual data sensed by the node which it desires to transmit. As can be observed the packet formation of PPDU requires addition of approximately 5 additional bytes at Physical Layer (SHR=4 bytes & PHR=1 bytes) and addition of about 39 bytes at MAC layer. In case the actual data at network layer is only x bytes then at-least 44 additional bytes are added to the data as packet formation overhead. The actual packet being transmitted is much larger than the data sensed and therefore will take longer to transmit. As these packets traverse over the network from a source to the sink, delay and energy penalty incur.

DATA TRANSFER IN WSN

It can be conjectured that in the absence of nodes at ideal distance, we can look for nodes in a small circular region with center at a distance from a transmitting node .

This leads to the idea of layers for minimizing both delay and energy expenditure. As shown in the Figure.3., Node A is at a distance d from the Sink S while Node B is at a distance d/2 from the Sink. Node A has the transmission capability to transmit to the Sink directly. The impact on time and energy conservation in the single hop approach and multi-hop approach is presented here. Assume that Node A has 10 bytes of sensed data (at time T1) that is to be reported to the sink. The size of the PPDU is 54 bytes. Node A transmits at Power *Pta* and this power is sufficient for the signal to be received at Sink. Further assume that channel 2.4 GHz is being used for communication where the data rate is 250 kbps. CSMA/CA protocol, which is a part of IEEE 802.15.4 MAC is being used for packet collision and collision avoidance with default value of 3 for *macMinBE*.

For Single Hop Approach

Node A transmits the PPDU of size 54 bytes at Time T1 at power *Pta*. If the channel is clear then the transmitted data will reach Sink S on the first transmission attempt at Time T2.

For Multi-Hop Approach

Node A transmits the PPDU of size 54 bytes at Time T1 at power *Pta/4* which will be received by Node B. Node B will then re-transmit the same to Sink. If the channel is clear then the transmitted data will reach Sink

S on the first transmission attempt at Time T2. As can be observed the time taken for transferring the packet from Node A to sink is slightly more than double if one additional hop is added. Based on the calculations it can clearly be seen that the communication cost is the highest in terms of the overall energy budget. The size of the first layer is kept larger than the other layers since the nodes in this layer are closest to the sink therefore maximum routing of the packets will be done by the nodes in this layer. This will put a severe strain on the energy capabilities of nodes in this layer and the problem can only be solved by having a higher node density in this layer so that the distribution of communication load is more even in this layer. DBMLAM uses Link Quality Indicator (LQI) based readings to perform the task of categorization of nodes into layers. IEEE 802.15.4 PHY layer provides a LQI value for every packet received. The structure of the Packet used for Layer categorization is given in Figure.4. As can be seen the packet consists of three main fields, viz. Packet type, Layer No. & Destination address. Since this packet is broadcast therefore the Destination Address field becomes irrelevant.

Figure 3. Single hop & multi-hop approaches

Sink
Node A
Node B
Distance d/2
Distance d
Sink
Node A
Node B
Distance d/2
Distance d

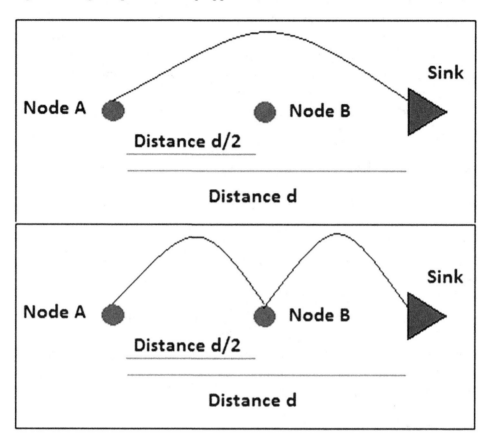

Figure 4. Structure of payload packet [Bernard, T., Fouchal, H and Linck. 2018]

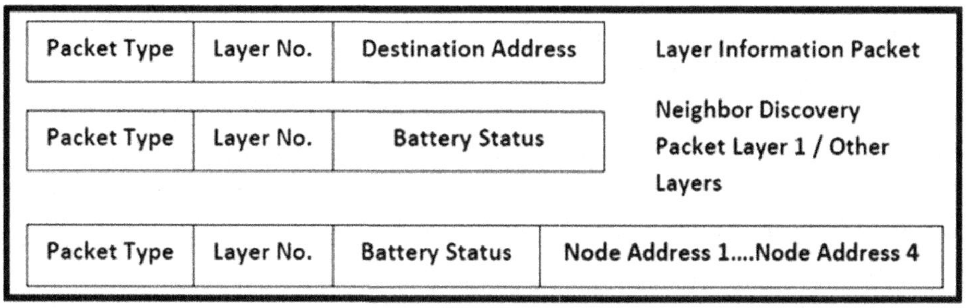

Packet Type
Layer No.
Destination Address
Layer Information Packet
Packet Type
Layer No.
Battery Status
Neighbor Discovery Packet Layer 1 / Other Layers
Packet Type
Layer No.
Battery Status
Node Address 1....Node Address 4

On completion of this phase, each node has complete information regarding all its neighbours in all possible layers and directions i.e. the topology of the network, relevant from each node's perspective, is fully established. It is pertinent to note here that the Node is not storing the topology information about the whole network but only related to its own immediate neighbours. This ensures minimum requirement of memory resource for storing the topology information. The two factors Success Ratio (SR) and Recent Failure (RF) are quite important since they help in improving the reliability of communication and also have a positive impact on reducing the Network Latency Time. Every time a node forwards a packet to its neighbor it updates the Packet Success field for that node in the Neighbor Table. Once acknowledgement of receipt of packet is obtained from the neighbor, the Success Ratio field in the format is updated and this leads to a change in the Priority of the neighboring nodes. In case there is a failure of getting the acknowledgement then the Recent Failure field (RF) is updated and this also impacts the priority of the neighbor node in the Neighbor Table. These two factors, along with priority, ensure that the neighbor node which has a higher probability of successful receipt of packet is accorded a higher priority thus improving the reliability of data acquisition. This is best explained with an example. As can be seen in Figure.5.Node 9 is the originating node with 4, 5, 6 as Upstream Neighbors, 8 as Adjacent Neighbor and 3 as Layer Bypass Neighbor. Similarly Node 8 has 9 as its adjacent neighbor and 6 & 7 as Up-stream neighbor. Similarly 3 & 1 are Upstream neighbors of 5 while 4 is its adjacent neighbor. 2 is Up-stream neighbor of 6 while 2 and 3 are adjacent neighbors of 1.

Node is Forwarding Node

In this scenario, a packet may be received from another node from either the down-stream layer or from the same layer. This is because the node is viewed as a Upstream, Adjacent or By-pass layer neighbor of the node which has sent the packet. In line with the specifications mentioned in IEEE 802.15.4, the MPDU size, with the current payload of the format of Data Packet (10 bytes) and with no security related overheads is 35 bytes. This translates into a PPDU of 41 bytes. The data being sent is assumed to be 2 bytes long and the overall PPDU size could change if the data size is enhanced. Fourteen bits have been kept for the Sequence number of the packet thus putting a limit of 16385 on the number of packets per node. Again this could be changed if required.

Figure 5. Packet routing and delivery mechanism

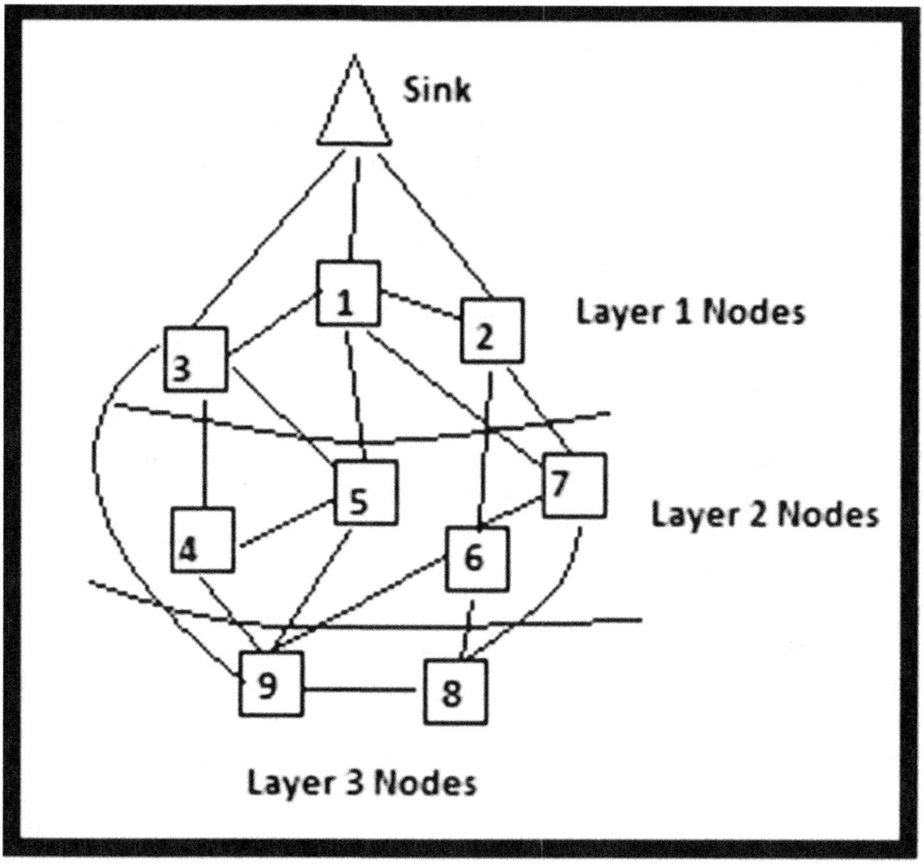

METHODOLOGY USED IN LTCRDM

Using & implementing non-disjoint Multipath Delivery Mechanism for reliable delivery of sensed packets at the sink / Actuators. LTCRDM uses a multi-path approach to packet delivery mechanism thus ensuring a sufficiently high packet delivery rate of packets to the Sink / Multiple Actuators. It also maintains time coherence among the packets containing the information sensed and transmitted by each deployed node. Adopting Table based approach for storing neighbor information i.e. making the nodes aware of the local and contextual topology of the network. This is done with the assumption that the nodes in the WSAN are static and mobility of nodes is ruled out. Deviating from the traditional approach of classifying and categorizing neighbors and suggesting & implementing a novel approach for defining neighbor with the view of achieving the objectives set forth for the mechanism. Reducing the number of hops between the sensing node and the sink / Actuator for meeting the TNLT time-constraint by implementing concept of Layers. This ensures that the action to be taken by the actuators does not suffer from network lag and the decision taken for the control action is in coherence with the sensed information/event. Alternate path mechanism implemented to counter loss of communication between a set of neighbor nodes of a specific node. It also provides a methodology for reliable delivery of information sensed by this node in case of failure of substantial number of neighbors of a specific node. Adopting limited implementation of Ant Colony methodology for strengthening the success rate of packet delivery by identifying high success rate neighbors. This mechanism automatically relieves nodes which start showing more congestion and start acting as bottlenecks. This leads to auto de-congestion of congested nodes. Usage of both explicit and implicit acknowledgement mechanism to ensure that un-required re-transmissions are avoided in face of loss of the ACK packet itself. The methodology designed for attaining a high reliability of data acquisition has been done with a consistent eye on reducing the complexity of computation to be done by the sensor nodes and the utilization of the constrained memory resource. Typically a geographical area of 40 m x 40 m has been taken as the deployment area for the sensor nodes. Beacon is switched off so that contention is resolved using CSMA/CA. The nodes are deployed randomly while the number of nodes deployed is varied from 20 to 70, depending upon the parameter being evaluated. Antenna used is isotropic with 0.0db transmitting power. The probability of packet loss is highest when all the nodes attempt to transmit simultaneously. All the simulations are carried out with this worst case scenario, where all the nodes transmit simultaneously rather than in randomized manner which could alleviate the situation. The objective of this test is to establish the impact of using the concept of layers and layer based neighbors on the Network Latency Time (TNLT) in a given deployment scenario. Comparison is made by replacing LTCRDM with AODV & DSR in the same deployment scenario to compare the Network Latency Time. Results sown in Figure.6.

The objective of this test is to evaluate the impact of the increase in the number of nodes and the resulting packet traffic, on the Effective Reliable Delivery Ratio capability of the network while using LTCRDM viz a viz the traditional methods. With increase in congestions, the number of re-transmissions and reliance on alternate paths increases to ensure delivery of packet at the sink. However, this also significantly reduces the effective reliability as the constraint on Network Latency time may not be met by large number of packets ultimately reaching the sink. The application considered is that of a human being in a space capsule where various types of sensors measuring temperature, pressure, oxygen level, CO_2 level, air toxicity level, light intensity, heart-beat, temperature & blood pressure and more are deployed in comparatively small vicinity. The information received from the

Figure 6. Impact of layer mechanism

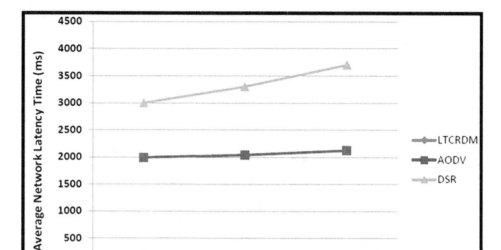

various sensors is aggregated at the sink and is transmitted to the ground station. Based on the sensed values the ground station may send control signals to the capsule thus controlling the various parameters. One of the key requirements is that the data sensed by the nodes must reach the sink not later than 500 ms i.e. the upper limit on the network latency time TNLT max = 500ms. The objective is to analyze the impact of number of Neighbors i.e. Adjacent & Upstream neighbors on the Effective Packet Delivery Ratio in LTCRDM. For Layer 1 nodes, only the number of Adjacent numbers is changed while for nodes in other layers, both adjacent and upstream neighbors are changed. The change in number of neighbors in Lateral & Upstream layer is controlled by setting values for the configurable *Amax* & *Umax* parameters in the software running at each node. Having a larger number of neighbors is advantageous since it increases the number of multi-path available between the source and destination nodes. This

has a direct favorable impact on the probability of successfully delivery of a packet to the sink. In case of LTCRDM, having a larger number of upstream and adjacent neighbors is expected to increase the Packet Delivery Ratio since multipath decisions can be taken at every node while forwarding a packet to the sink. As can be seen from Figure.7. & Figure.8., the Packet Delivery Ratio is steadily increasing as the number of neighbors increases. However, the Effective Packet Delivery Ratio depends not only upon the successful delivery of the packet but with an additional

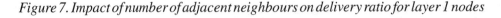

Figure 7. Impact of number of adjacent neighbours on delivery ratio for layer 1 nodes

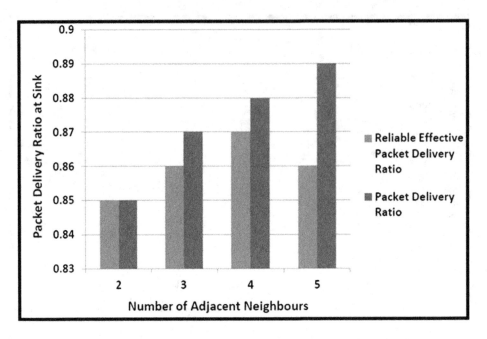

condition that the packet must be delivered within a specific time constraint. It is observed form the results that although the Packet Delivery Ratio is improving steadily, the Effective Packet Delivery Ratio peaks with a specific number of neighbors and then actually reduces with an increasing number of nodes. This is because with an increase in the number of neighbors, the network latency may suffer because of the larger number of hops involved.

LTCRDM was subjected to a series of evaluation tests to ascertain the performance of the mechanism against the objectives set forth for it. Comparative analysis was also done in specific cases against established protocols and algorithms to get a better understanding about the performance of LTCDRM. A summary of the variety of the test conducted, parameters changed and results and findings are provided below:

Impact of Layering on Network Latency Time

It has been shown that the theoretical assumption made regarding the reduction in the Network Latency Time because of reduction in the number of hops and by changing the traditional definition of neighbors has been proven to be correct by the simulation done.

Figure 8. Impact of adjacent and upstream neighbours on delivery ratio

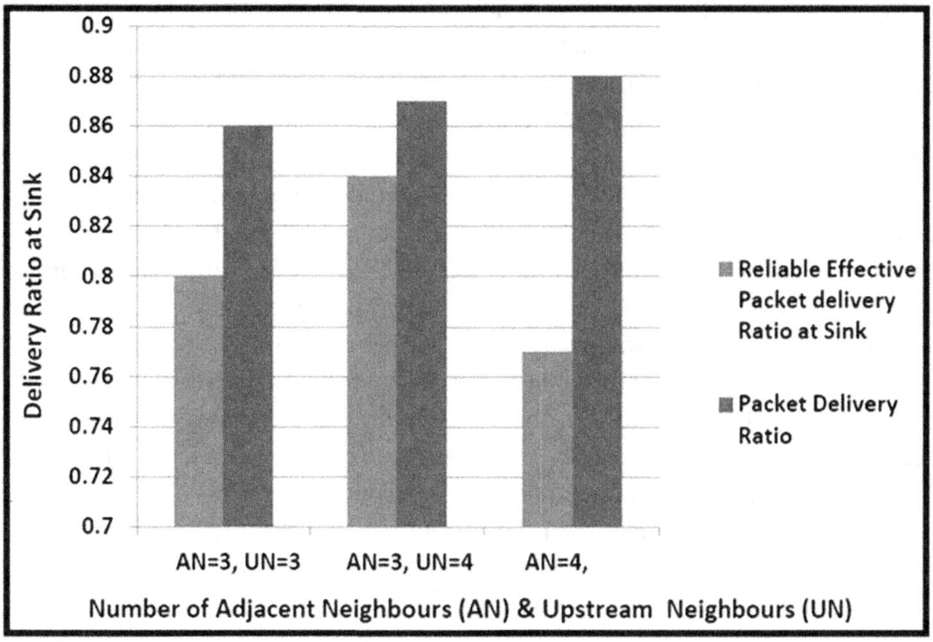

Effective Reliability of Data Acquisition

Performance evaluation of LTCRDM was conducted to check the Effective Reliability of Data Acquisition at Sink by including the Network Latency Time (TNLT) time constraint. Thus the packets received at the Sink beyond the TNLT time constraint were treated as dropped or lost packets. A series of evaluations was done by varying parameters including Number of nodes deployed, percentage of nodes deployed transmitting information at a given time, rate of transmission of information in case of periodic data acquisition to check the impact on the Effective Reliability of Data Acquisition at the Sink. Comparison analysis was also done by checking performance of the Packet Delivery Ratio of LTCRDM against known standard transport protocols like AODV & DSR. AODV with Binary tree is the choice of protocol used in Zigbee™ protocol which has been designed for commercial grade data acquisition applications. It was observed that LTCRDM outperformed in comparison to others on variation of the parameters mentioned with the achieved Effective Reliability of data acquisition as high as 0.9 in some scenarios. Performance evaluation of LTCRDM was done to ascertain the impact of the number of Adjacent neighbor nodes allowed for Layer 1 nodes and adjacent neighbors and the number of Upstream neighbors allowed for nodes lying on layers beyond 1st layer. Various combination of the two were tried and it was observed that a simple enhancement in the number of maximum allowed Adjacent neighbors and Upstream neighbors does improve the effective reliability of acquisition to a limit; however, beyond a specific point further increase in the number of such neighbors was actually counter-productive since the TNLT constraint was repeatedly violated leading to a drop in the Effective Reliability of Acquisition. A comparison was also made with Reliability of acquisition in the same case i.e. without taking the Network Latency Time into consideration, to establish the percentage of packets being lost due to the time-constraint. Performance evaluation of LTCRDM was done to ascertain the impact of the failure of nodes, present in the network, on the Effective Reliability of Acquisition. Node failure was introduced in the network @ 10% Node Failure, 20% node failure and so on till 40% node failure. The impact on the effective reliability of Acquisition was then studied. It was observed that the effective reliability of acquisition reduced with increase in the number of failed nodes; however, the rate of reduction in the effective Reliability of Acquisition was less as compared to the other scenarios. Evaluation was done to ascertain the impact of Effective Reliability of Acquisition on the number of nodes that could be put into sleep mode thus having a positive effect on network life. It was observed that the number of nodes that could be put in the sleep mode increases as the effective reliability of the data acquisition mechanism increases.

There are distinct advantages of the distributed decision making approach as in many applications, the impact area of an actuator is rather localized and does not have an effect on the overall area covered under the complete wireless sensor network. Thus, for decision making, it needs data collected by sensor nodes in its vicinity rather than the data from the complete sensor network. Further the QoS requirements are better met when the number of nodes and the geographical area in which they are deployed is small. The objective of this phase is to categorize the deployed nodes into Layers based on the tentative proximity of the nodes from the various actuators available in the WSAN. The overall network is sub-divided into smaller networks with an actuator acting as the sink in each sub-network. Therefore, within each sub-network the problem becomes similar to the situation where multiple nodes exist with only one sink in the network. Figure.9. illustrates the situation.

Each sub-network created by virtue of presence of Actuator Node behaves as if it is a single network with the Actuator as Sink. This behaviour is identical to the Centralized Decision Making Scenario with a single sink, which in this case is the Actuator. Therefore the Data Acquisition process, within the sub-network of an Actuator, in the case of Distributed Decision Making Scenario is identical to the Data Acquisition mechanism in the case of Centralized Decision Making scenario. However, each sub-network is identified by the Actuator No. of that sub-network which corresponds to the Actuator in that sub-network. The nodes, once layered

Figure 9. Layer formations in wsan with distributed decision making scenario

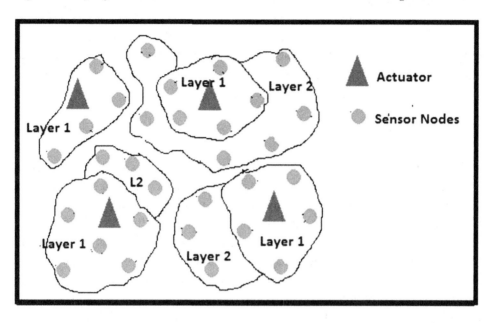

and linked to one actuator will send their packets to only that actuator and will not send their data to the actuator of the other sub-network. In case the information from one actuator based sub-network is to be shared with the actuator of some other subnetwork then the same will require communication between the actuators. The LTCDRM mechanism in context of Distributed Decision Making Scenario is presented in this chapter and it is shown that a Distributed Decision Making Scenario can actually be seen as a collection of disjoint and mutually exclusive sub-network centred on each Actuator. Each such sub-network is identified by the Actuator No. of the actuator in that network. In this context then the LTCRDM algorithm can be used with modifications in the Packet formats and Tables by incorporating the Actuator Number as a parameter for identifying the sub-network. This ensures that the nodes within the sub-network do not interact with nodes of other sub-network. The performance of the LTCRDM in the Distributed Decision Making scenario is equally efficient and the sub-network approach yields better results as compared to the centralized approach especially as the number of nodes deployed increases. A query mechanism has been designed to work in the Layer Based architecture proposed as a part of LTCRDM. The objective of the algorithm was to ensure that a query arising out of the sink is disseminated to all the nodes in the network irrespective of their layer position has been met. At the same time, the benefit of the Layer Based mechanism in terms of the time taken for the dissemination of the query has also been met. The algorithm also ensures that each node has the ability to reject duplicate queries which may reach it and it neither re-sends the query nor does it respond to the same query again. This is possible because of a unique Query Sequence number assigned to each query generated by the Sink and the availability of Query Storage Table. The simple structure of the Query Storage Table ensures that it does not impose severely on the memory storage capability of the node.

CONCLUSION

According to the findings, the Distance Based Multi-Layer Assignment Mechanism (DBMLAM) is able to classify deployed nodes into layers for both the centralised and decentralised decision making approaches (single sink and multiple actor/sink). Since nodes are deployed randomly, their layer sizes and shapes vary, suggesting that RSSI is impacted by both time-varying channel properties and random node placement. Because of this, there may be fewer reliable neighbours to choose from. The results also show that the Layer Based Neighbours (LBN) method is effective, with installed nodes being correctly identified as Downstream, Upstream, nearby, and Layer Bypass neighbours. The first observation, however, also revealed that some nodes were left as orphans because they lacked neighbours who could be recognised. Multiple packet exchanges were added to the system to guarantee that all nodes were correctly recognised as neighbours. To create a gradient between the nodes and the sink in the Centralized Decision making approach and between the actuators and the nodes in the Distributed decision making approach, LTCRDM employs the layering technique. However, the present implementation of LTCRDM cannot transmit data to a user-specified destination node other than the sink or actuators.

ACKNOWLEDGMENT

We would like to acknowledge the support given by BMS Institute of Technology and Management, Bangalore, India.

REFERENCES

Abroshan, S., & Moghaddam, M. H. Y. (2014, May). SESRT: Score based Event to Sink Reliable Transport in wireless sensor networks. In *Electrical and Computer Engineering (CCECE), 2014 IEEE 27th Canadian Conference on* (pp. 1-6). IEEE.

Adelstein, F., Gupta, S. K., Richard, G., & Schwiebert, L. (2005). *Fundamentals of mobile and pervasive computing*. McGraw-Hill.

Akan, Ö. B., & Akyildiz, I. F. (2015). Event-to-sink reliable transport in wireless sensor networks. *IEEE/ACM Transactions on Networking, 13*(5), 1003–1016. doi:10.1109/TNET.2005.857076

Akyildiz, I. F., & Kasimoglu, I. H. (2014). Wireless sensor and actor networks: Research challenges. *Ad Hoc Networks, 2*(4), 351–367. doi:10.1016/j.adhoc.2004.04.003

Akyildiz, I. F., Su, W., Sankarasubramaniam, Y., & Cayirci, E. (2020). A survey on sensor networks. *Communications Magazine, IEEE, 40*(8), 102–114. doi:10.1109/MCOM.2002.1024422

Al-Awami, L., & Hassanein, H. (2018, October). Energy efficient data survivability for WSNs via decentralized erasure codes. *IEEE 37th International Conference on Local Computer Networks*, 577–584.

Alam, M. M., & Hong, C. S. (2019). CRRT: Congestion-aware and ratecontrolled reliable transport in wireless sensor networks. *IEICE Transactions on Communications, 92*(1), 184–199. doi:10.1587/transcom.E92.B.184

Ali, A., Latiff, L. A., Sarijari, M. A., & Fisal, N. (2018, June). Real-time routing in wireless sensor networks. In *Distributed Computing Systems Workshops, 2008. ICDCS '08. 28th International Conference on* (pp. 114-119). IEEE.

Ali, S., Fakoorian, A., & Taheri, H. (2017, October). Optimum Reed-Solomon erasure coding in fault tolerant sensor networks. *4th International Symposium on Wireless Communication Systems (ISWCS)*, 6–10.

Alimohammadi, H., & Jassbi, S. J. (2019). A Fault Tolerant Multipath Routing Protocol for Mobile Ad Hoc Networks. *European Journal of Scientific Research, 85*(2), 317–326.

Alsbouí, T. A. A., Hammoudeh, M., Bandar, Z., & Nisbet, A. (2019). An Overview and Classification of Approaches to Information Extraction in Wireless Sensor Networks. *Proceedings of Fifth International Conference on Sensor Technologies and Applications (SENSORCON)*, 255-260.

Atanasov, A., Kulakov, A., Trajkovic, V., & Davcev, D. (2010). Testbed Environment for Wireless Sensor and Actuator Networks. *Fifth International Conference on Systems and Networks Communications*, 1,6, 22-27. 10.1109/ICSNC.2010.8

Ayadi, A. (2018). Energy-efficient and reliable transport protocols for wireless sensor networks: State-of-art. *Wireless Sensor Network, 3*(3), 106–113. doi:10.4236/wsn.2011.33011

Baghyalakshmi, D., & Ebenezer, J. & SatyaMurty, S.A.V. (2010, Jan). Low latency and energy efficient routing protocols for wireless sensor networks. *International Conference on Wireless Communication and Sensor Computing,* 1,6. 10.1109/ICWCSC.2010.5415892

Baronti, P., Pillai, P., Chook, V. W., Chessa, S., Gotta, A., & Hu, Y. F. (2017). Wireless sensor networks: A survey on the state of the art and the 802.15. 4 and ZigBee standards. *Computer Communications, 30*(7), 1655–1695. doi:10.1016/j.comcom.2006.12.020

Bein, D., Jolly, V., Kumar, B., & Latifi S. (2015). Reliability Modeling in Wireless Sensor Networks. *International Journal of Information Technology, 11*(2).

Bernard, T., Fouchal, H., Linck, S., & Perrin, E. (2018). Impact of routing protocols on packet retransmission over wireless networks. *2013 IEEE International Conference on Communications,* 2979,2983.

Bolot, J. C. (2015, September). End-to-end packet delay and loss behavior in the internet. *ACM SIGCOMM Conference on Communications Architectures, Protocols and Applications (SIGCOMM),* 289–298.

Boukerche, A., Martirosyan, A., & Pazzi, R. (2018). An inter-cluster communication based energy aware and fault tolerant protocol for wireless sensor networks. *Mobile Networks and Applications, 13*(6), 614–626. doi:10.100711036-008-0093-x

Bourdenas, T., & Sloman, M. (2019, June). Towards self-healing in wireless sensor networks. In *Wearable and Implantable Body Sensor Networks, 2019. BSN 2019. Sixth International Workshop on* (pp. 15-20). IEEE.

Camilo, T., Carreto, C., Silva, J. & Boavida, F. (2016). An energy-efficient antbased routing algorithm for wireless sensor networks. *Ant Colony Optimization and Swarm Intelligence,* 49-59.

Cardei, M., Yang, S., & Wu, J. (2018). Algorithms for fault-tolerant topology in heterogeneous wireless sensor networks. *Parallel and Distributed Systems, IEEE Transactions on, 19*(4), 545-558.

Chen, J., Cao, X., Cheng, P., Xiao, Y., & Sun, Y. (2018). Distributed collaborative control for industrial automation with wireless sensor and actuator networks. *Industrial Electronics, IEEE Transactions on, 57*(12), 4219-4230.

Chen, X., Dai, Z., Li, W., & Shi, H. (2018). Performance Guaranteed Routing Protocols for Asymmetric Sensor Networks. *Emerging Topics in Computing, IEEE Transactions on, 1*(1), 111-120.

Iyer, Y. G., Gandham, S., & Venkatesan, S. (2015). STCP: a generic transport layer protocol for wireless sensor networks. In *Computer Communications and Networks, 2015. ICCCN 2015. Proceedings. 14th International Conference on* (pp. 449-454). IEEE.

Johnson, D. B., & Maltz, D. A. (1996). Dynamic source routing in ad hoc wireless networks. In *Mobile computing* (pp. 153–181). Springer US. doi:10.1007/978-0-585-29603-6_5

Johnsort, D. B. (1994, December). Routing in ad hoc networks of mobile hosts. In *Mobile Computing Systems and Applications, 1994. WMCSA 1994. First Workshop on* (pp. 158-163). IEEE. 10.1109/WMCSA.1994.33

Juwad, M. F., & Al-Raweshidy, H. S. (2018, May). Experimental Performance Comparisons between SAODV & AODV. In *Modeling & Simulation, 2018. AICMS 08. Second Asia International Conference on* (pp. 247-252). IEEE.

Liang, C. M., & Terzis, A. (2018). Typhoon: a reliable data dissemination protocol for wireless sensor networks. *5th European conference on Wireless Sensor Networks*, 268-285.

Loh, P. K. K., Hsu, W. J., & Pan, Y. (2017). Reliable and efficient communications in sensor networks. *Journal of Parallel and Distributed Computing, 67*(8), 922–934. doi:10.1016/j.jpdc.2007.04.008

Chapter 11
Minimize the Energy Consumption for Communication Protocol in IoT

Manjula Gururaj Rao
N.M.A.M. Institute of Technology (Deemed), India

Sumathi Pawar
N.M.A.M. Institute of Technology (Deemed), India

Priyanka H.
(iD) https://orcid.org/0000-0002-4155-0418
People's Education Society University, India

Hemant kumar Reddy
VIT-AP University, India

ABSTRACT

Any internet of things (IOT) deployment must have connectivity; this is accomplished by WSNs (wireless sensor networks). A few factors need to be taken into account when choosing a wireless technology for an IoT device: the maximum throughput, the distance range, the availability in the deployment zone, as well as the power consumption. The aim of this research is to maximize the lifetime of the nodes of WSN and to reduce the energy consumption. The system is also focused on managing WSN nodes with huge residual energy, small routing distance, and with maximum number of neighbors. This system makes use of LEACH protocol, and this protocol is hierarchical clustering protocol and also energy efficient. The cluster is constructed in such a way that average dissipation of energy in each node is minimized, and speed of the network is increased. To connect sensor networks with gateways and transfer data from these sensor networks, other technologies such Sub-1 GHz are Zigbee can also be employed.

DOI: 10.4018/978-1-6684-4974-5.ch011

INTRODUCTION

Communications refer to exchanging information between various entities, such as humans, computers, organizations, agencies, and firms. During digital communication, data is transmitted between several computing devices. Here data transmission (datacom) takes place over a wired or wireless radio signal, fiber optic cable, or telephone line. A computer network which consists of group of digital devices enables gadgets to "speak" to one another according to their connection. Devices for communications use both wired and wireless connections. In communication over a wired connection, the devices are physically connected to one another. Devices used for wireless communication are not physically connected to one another during communication. Data and voice transmissions take place during the communication. The availability, accessibility, and efficiency are all improved through wireless transmission. The coverage area is increased by wireless networks. Despite the fact that wireless networks offer greater opportunities, there are often installation-related disadvantages with wireless communications.

WSN (Wireless Sensor Network)

A wireless sensor network (WSN) comprises many sensor nodes that gather environmental data and deliver it to a sink. WSNs also refer to a collection of specialized sensors that are distributed spatially and are used to track environmental and physical conditions, record them and organize the information at a central point. WSN is used in various industries including the military, the environment, healthcare, home and other businesses. Numerous elements such as fault tolerance, scalability, cost of manufacture, operational environment, transmission medium and power consumption impact the design of sensor networks. A wireless network composed of various wireless sensors called the Wireless Sensor Network (WSN). In WSN, sensor nodes with an embedded CPU regulate and monitor the environment in a given space. They are connected to the Base Station (BS), which acts as the central processing node for the WSN System. To share data, the BS of a WSN system connected to the internet. Figure 1 depicts the user connection, internet, and WSNs (Geeks for Geeks, n.d.; Sharma et al., 2015).

WSNs are typically set up in a environment to watch over either static or dynamic events. Measuring static events (like temperature and humidity) is very simple. The drive of an unwelcome vehicle on a battlefield or the motion of whales in the ocean are examples of dynamic occurrences, which are often non-cooperative. It is difficult to monitor and unstable as they fluctuate (Fantin Irudaya Raj & Appadurai, 2022; Neelakandan et al., 2022; Thilakarathne et al., 2022). A certain protocol is needed for sensor networks to function well. For instance, a protocol can take the form of a particular application that aggregates data in a precise order and optimizes energy use. Hierarchical routing is the name given to this type of protocol.

Additionally, it has a data-centric routing protocol that defines a network environment. A sensor node also uses a data-centric method to execute sensing applications to find the best route from several sources to a single destination. A list of attribute-value pairs can define data from every node in a network termed attribute-based addresses, which explain how a node can reveal its availability to the entire sensor network. However, as the energy allotted for sensor nodes are often quite limited, it is imperative to improve wireless sensor networks' energy efficiency. Additionally, the complexity of wireless sensor network technology will expand as society becomes more dependent on it.

Components of WSN Elements

1. **Sensors:** In a WSN, sensors are utilized to collect data and capture ambient variables. Electrical signals are created from sensor signals.
2. **Radio Nodes:** It takes in and transmits data generated by the sensors to the WLAN access point. It comprises a transceiver, external memory, microprocessor, and power source.
3. **Access Point for WLAN:** It accepts wirelessly sent data from radio nodes, typically via the internet.
4. **Analysis Software:** A program known as Evaluation Software processes the data that WLAN Access Point receives, to provide the users with a report that may be used for data processing, analysis, storage and mining (Elprocus, n.d.).

WSN and IoT

An IoT system's sensors transmit data directly to the internet. For example, a sensor could be used to monitor the temperature of a body of water. For a server to process and show the data in front - end interface, the data will be delivered promptly or repeatedly over the internet.

On the other hand, a WSN lacks a straight internet connection. Instead, a router or other central node is connected by numerous sensors. The data can then be forwarded from the router or central node. An IoT device can use a wireless sensor network to gather data, by corresponding with its router. A wireless sensor network might be considered more as a collection of sensors or "a huge sensor" and less as a "rival" or "competitor" to the IoT.

Figure 1. Wireless sensor networks and user.

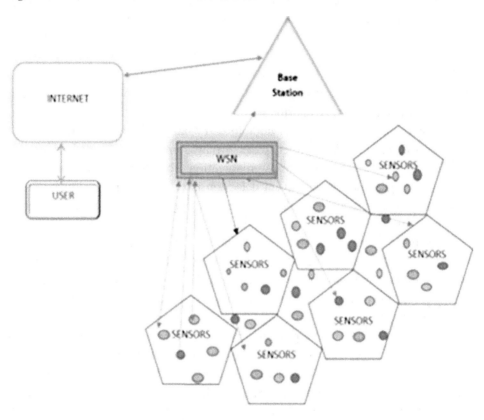

WSN as an IoT Subset

The IoT is more advanced than WSN. In other words, WSN is a technology that is frequently utilized in IoT systems. In an IoT system, a big group of sensors, like those in a mesh network can be utilized to collect data on their own and transfer it via a router to the internet. It's also crucial to remember that "the internet of things" encompasses far more than a "wireless sensor network." WSN is a network made up entirely of wireless sensors. The network could no longer be referred to as a "wireless sensor network" if it had a wired sensor. IoT is not like this. IoT devices are the gadget that connects to the internet. As a result, an "IoT system" can be thought of as a collection of numerous IoT devices (Shiverware, n.d.).

PCs, laptops and smartphones are usually left out of the process of connecting devices to the internet in the IoT. A standalone internet-connected gadget that can be monitored and operated from a distance is also referred to as an IoT device, according to Business Insider. A wireless sensor network is a collection of wireless sensors that may or may not be linked to the internet (Shiverware, n.d.).

WSN for IoT:

When a WSN is enabled, several tasks are carried out to build the network's essential infrastructure, including the distribution of sensor nodes and the data transmission routing that will permit the sensor node to carry out its functions normally. Each node must determine which other nodes it may communicate directly and have sufficient radio power to maintain communication. Therefore, researching energy-saving techniques for dynamic event detection is very challenging. For instance, a target tracking program uses a dynamic monitoring technique, whereas a forest monitoring application uses a static monitoring strategy.

Since the nature of the WSN, sensor nodes are often powered by batteries, which results in a highly limited energy budget (Sharma et al., 2015). The WSN can work either single-hop or multi-hop. In the single-hop WSN, the base station (BS) is connected directly to all the sensor nodes. This leads the sensor node to be active all the time. Because this energy level in the sensor node will be decreased very fast, to avoid this type of problem additionally, peer nodes and cluster heads are employed in multi-hop WSNs to convey the information to conserve energy.

Due to its architecture, hierarchical routing offers superior scalability and energy efficiency. This kind of protocol divides the entire network into clusters and distinct nodes are selected based on a list of standards. These unique nodes, called cluster heads (CHs), gather, assemble and compress the data obtained from neighboring nodes that formerly transmit the BS with compressed data. Since it provides additional services to other nodes in the cluster, the CH consumes more energy than other nodes in the cluster. A common tactic for balancing the cluster's energy loss is cluster rotation (Singh et al., 2017).

Routing Protocol Analysis for IoT

For routers to be able to choose routes between any two nodes on a computer network, they must be able to communicate with one another by a routing protocol. A routing protocol needs to satisfy certain criteria and use various tactics to address the limits of IoT systems. Such a protocol must be adaptable regarding power consumption and the traffic flow in the deployment region. As well as scaling in terms of memory and performance, it needs to handle sparse location changes. An IoT routing system is also necessary to identify and avoid one-way networks and use transmitter energy sparingly (Iqbal & Kim, 2022). Support for IPv6 and mobility are key traits. However, they are not the last. The techniques employed include reactive routing, which involves searching the routes as needed and proactive routing, which aims to maintain an overall picture of the entire network topology at all times. Since nodes in an IoT network serve as hosts and routers that provide data to the gateways, routing is crucial. The power forwarding nodes use how data is sent from a source to a final destination.

LEACH (Low-Energy Adaptive Clustering Hierarchy) Protocol Overview

The LEACH protocol is a hierarchical cluster-based routing protocol. LEACH centralized, single-hop, multi-hop and mobile Leach protocols are different types of this protocol. With this approach, the clusters' heads will randomly alternate between different nodes. The dynamic network enables robustness, fewer sinks and less bandwidth usage. Leach accomplishes data aggregation at the level of the CH using localized coordinates. The CH is chosen using the network's setup method, in which all nodes are involved. With a non-persistent protocol, the elected cluster leaders disseminate an advertisement to other nodes. When a cluster member learns about the cluster head, it gathers information and sends it to the CH. After that, the cluster head sends this data to the BS. The data is transferred from the cluster head to the base station using single-hop communication. The CH maintains the network in a steady state by fusing data and forwarding it using an inter-cluster approach. This protocol's benefits include a longer lifetime, less base station-based head control and less communication between CH and members. This has a substantial impact on lowering the system's energy consumption.

BACKGROUND

There are several other LEACH method adaptations that researchers have created. Using LEACH protocol replacements in single-hop and multi-hop configurations is now possible. Due to the WSN's application in the IoT nowadays, extensive research has been conducted (Albert et al., 2022; Majid, 2022; Pushparaj et al., 2022).

S. K. Sigh et al. (Singh et al., 2017) proposed the reviews on LEACH routing protocol variations that have been put forth to date and discussed how they could be improved. Researchers have produced many other LEACH method variations. In single-hop and multi-hop scenarios, LEACH protocol successors are now available. Much research is being done because the WSN is now used in the IoT. A temporal comparison analysis employing nine distinct characteristics, including scalability, overhead, energy efficiency, complexity, etc., has been provided. The article also addresses the advantages and disadvantages of every LEACH version.

The utilization of WSN energy by using the round-robin technique to select the CH is proposed by R. Sharma et al. (Sharma et al., 2015). Energy can be saved in this way to allow for longer iterations. Clustering technique remains flexible and is used with stationary nodes. Here, depending on the round-robin approach, every node serves as a cluster head at least once.

Improving the energy level in the sensor node increases the network lifetime. This concept is used by improving the LEACH protocol by Z. Peng et al. (Peng & Li, 2010). The suggested model extends the LEACH stochastic cluster-head selection algorithm by altering the round time to account for sensor network conditions. Variable-round LEACH is its name. The outcome is simulated by NS2.

Khediri et al. (Khediri et al., 2014) propose energy-saving techniques that have frequently used clustering algorithms. In this situation, choosing a cluster is the most important aspect of this topology. Alternatively, choosing a cluster is one solution. However, this decision needs to consider energy as a crucial factor in publications. This work proposes a new method to lower energy consumption to optimize the Low Energy Adaptive Clustering Hierarchy (OLEACH). It enhances it by dynamically selecting clusters depending on the node residual energy in LEACH and LEACH- C.

Fu C et al. (Fu et al., 2013) propose a new, enhanced LEACH protocol method (LEACH-TLC). This aims to balance the network's overall energy consumption and increase network longevity. The implementation is conducted in MatLab 7.

To improve the hierarchical routing protocol, LEACH is proposed by J. Xu et al. (Xu et al., 2012). They provided the E-LEACH cluster routing algorithm. The initial selection method for the cluster heads in the E-LEACH algorithm is random and the round time for the selection is fixed. The author designed the E-LEACH algorithm, which changes the round time depending on the appropriate cluster size, taking into account the remaining power of the sensor nodes to balance network demands.

Ran, G et al. (Ran et al., 2010) propose that the LEACH-FL protocol, which considers battery level, distance and node density is improved in this research utilizing fuzzy logic. By employing Matlab comparative simulations, it has been demonstrated that the suggested strategy makes a better choice.

According to Mosorov, V. et al. (Mosorov et al., 2015), objectives are to raise awareness of wireless sensor networks and provide a new approach to self-organizing nodes in sensor networks. By assigning the role of the root node to each node within the cluster, the suggested idea under development decreases the occurrence of excessive traffic during the first phase of selecting the root nodes and lengthens the life of each node.

Ahlawat et al. (Ahlawat & Malik, 2013) introduced an improved version of the leach protocol, dubbed VLEACH, to lengthen network lifetime. The usual clustered Routing Protocol-LEACH and its flaws were thoroughly examined in this paper before an enhanced v-leach was suggested. Improvements must be made to the v-leach protocol for choosing the vice cluster head. The vice cluster head is the backup head that only functions if the cluster head passes away. The method of choosing a cluster head is based on three criteria: minimum distance, maximum residual energy and minimum energy. The suggested method will lengthen the network's lifespan because no cluster head will fail. When a cluster head passes away, its vice-cluster head will take its position. Numerous simulations revealed that the newly upgraded v- LEACH version outperforms the old leach protocol by extending network lifetime by 49.37%.

In this study (Mahapatra & Yadav, 2015; Tyagi & Kumar, 2013), the authors give a timeline and a recent survey of LEACH's legacy hierarchical routing technology. Additionally, these techniques are being compared based on several suppositions.

According to energy efficiency and power consumption, Mahmoud et al. (Mahmoud & Mohamad, 2016) gives a complete review of IoT technologies, including Low-Power Short-Area Networks (LPSANs) and Low- Power Wide-Area Networks (LPWANs). Existing consumption models and energy efficiency techniques are categorized, investigated and disputed to highlight the important trends presented in literature and standards toward developing energy-efficient IoT networks.

Since every node tends to relay its data to the base station via a randomly chosen cluster head every round, noticed by L Yadav et al. (Yadav & Sunitha, 2014). The observation is that the first node in the non-hierarchical arrangement dies more quickly. During the simulation round, the elected cluster heads' limited load significantly and quickly depleted their energy.

In the smart city, energy consumption and minimization are very much important criteria and it is done by using the predictive optimization and time management models proposed by Iqbal N et al. (Iqbal & Kim, 2022).

The round-robin technique employed to choose the cluster head to conserve node energy is proposed by Sharma et al. (Sharma et al., 2015). Energy can be saved in this way to allow for longer rounds. The clustering technique is flexible and used with static nodes. Here, depending on the round-robin approach, every node serves as a cluster head at least once.

The literature survey reviewed that many attempts are made to increase the energy of the WSN. Most of the time protocol used is LEACH. This chapter suggests a novel strategy, based on the LEACH hierarchical protocol in the following section.

PROPOSED MODEL

The extensive use of WSNs and the requirement for an energy-efficient strategy call for effective network topology organization to balance the load and increase network lifetime. It has been demonstrated that clustering can extend the network

Figure 2. The connection between the cluster and IoT.

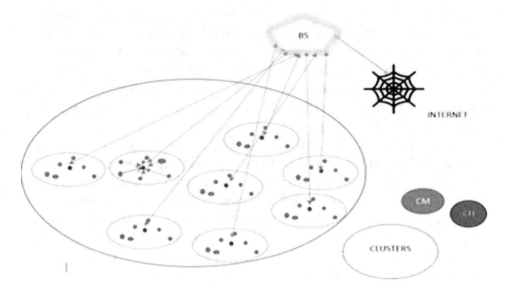

lifetime and offer the necessary scalability. A sensor network loses contact with the base station due to the bottleneck phenomenon in WSNs, wasting the residual energy resources of the functional nodes.

The proposed model has the input/output unit, energy consumption unit-LEACH and base station unit. The main intention of the model's design is to minimize the energy consumption in sensor nodes present in the wireless network. Minimizing the energy level in the sensor node can be done by using the multi-hop transmission and reducing direct transmission from the sensor to the BS. This will be achieved using the LEACH protocol and explained in the following sections.

- Base Station (BS): In wireless communications, it is a transceiver linking several other devices to one another and a larger area, whereas it is a GPS receiver at a specified place. It serves as the link between mobile phones and the larger telephone network in cellular telephones. A transceiver serves as a switch for computers in a computer network, potentially linking them to the internet and other local area networks. In our model, the cluster head present in the clusters of the WSN will directly communicate with it. From there, the BS is connected to the internet. The BS is also called as sink.
- Input/ Output Unit: This is present in each sensor unit, which contains the information regarding itself and all other nodes or the cluster head.
- Energy Consumption Unit: The nearby nodes will be combined and become a cluster. In the cluster, one node will become the head called cluster head

(CH). The sensor node contains information regarding all other sensor nodes present in the cluster. It contains the attributes in the specialized table, which might be the battery level and sensor network id. Choosing and forming of the cluster are done using the LEACH protocol.

LEACH PROTOCOL

For the design of WSNs, this hierarchical clustering-based routing protocol is employed. This protocol randomly switches the cluster heads and one by one all the nodes take over as CHs. LEACH aggregates data at the cluster head level for robustness and scalability. This method employs localized synchronization for the dynamic network. As a result, bandwidth is conserved while less data is delivered to the sink.

In the setup step of the clustering system process, all nodes choose CHs using an indicator function. Using the non-persistent protocol, the selected cluster heads then broadcast an Advertisement (ADV) message that contains the CH's ID (also called CHID).

The LEACH Protocol's Operation

- The hierarchical nature of the LEACH protocol makes the majority of nodes broadcast to CH, who then compiles and passes the data to the BS. The CH then develops a transmission schedule for each node in the cluster.
- The LEACH protocol employs the distributed CH election technique, in which some network nodes are randomly selected as CHs while others are designated cluster member nodes.
- When a node broadcasts that it has become a CH, the nodes select the CH with the strongest signal received to join to create a cluster.
- Through multi-hop communication, the cluster member node gathers data and sends it to the CH, receiving it and sending it to the BS. The CHS are responsible for doing the work tasks, such as managing the cluster member nodes, compiling the data received by the member nodes, data fusion and inter-cluster forwarding. Therefore, the CHs cycle and the cluster topology are often modified to balance the nodes' energy usage.

Two stages comprise the LEACH Protocol: The initialization phase and the steady-and-round-state phase.

During the setup step, the cluster head is chosen. During the steady-state phase, data is transmitted between nodes when the cluster head is established.

Algorithm for Initialization Phase

Choosing the Cluster and Cluster Head (CH)

Step 1: Initially, any sensor node (P) with an energy-level $>=$ to the threshold will be randomly chosen as the CH. The CH can be chosen using the RCC (Random Competition-Based Clustering).

Step 2: Now node P (become the CH) will transmit its status and energy information, the proximity L from BS and proximity M between a new node and CH, to all the member sensor nodes in the cluster. This is repeated for all the nodes present in the adjacent nodes, up to some minimum distance.

If the node's energy level is better than the CH and the distance is $<$ L, then the new node will become CH.

Step 3: Each node's data transfer schedule is made by the Cluster Head using TDMA. Each node in the cluster is informed of the TDMA schedule by the Cluster-Head.

Step 4: The clusters are formed using steps 1 to step 3 for the sensor nodes belonging to the WSN.

Algorithm for the Steady-and-round-State Phase

Step 1: Any node (P) present in the cluster senses data and transfers to the CH for every time- interval.

Step 2: Now, the CH gathers data from all of the cluster's nodes and transmits it to the BS at regular intervals.

Step 3: The CH checks the cluster members' energy levels after some time has elapsed. Depending on factors, including their proximity to the BS and energy level, CH selects one of the member nodes to serve as the next CH. The initialization procedures from steps 2-4 are then repeated.

The system proposed the Round-Robin with the priority in the two algorithms while choosing the CH. The priority attributes are the energy level and proximity of the sensor node to the BS. If the energy level is higher and the proximity to the BS is less, the member node will become CH.

IMPLEMENTATIONS

Clustering in WSN

Using a technique called clustering, sensor nodes are hierarchically arranged according to how close they are to one another. An efficient and consistent method for sending the data acquired from the physical environment to the BS via the sensor nodes is created by the hierarchical energy consumption (sensor nodes clustering). By grouping sensor nodes, the routing table can be compressed. Moving through the discovery phase between sensor nodes more quickly is possible. Clustering can also utilize less communication bandwidth since it limits the scope of interactions between clusters and minimizes the repeated exchange of messages across sensor nodes. Before sending the data to the cluster head, each sensor node will check the routing table for the CH in its region. The CH performs a route-finding estimation based on the shortest distance to a recipient CH closer to the BS or straight to the BS. Link information is periodically transferred between sensor nodes to keep track of the routing table to adjust changes in energy for each node that needs to transmit data.

Cluster Formation and Rotation

Clustering offers an effective way to manage sensor nodes to increase their longevity in light of the changing trend in the application and management of WSNs. Several clustering formation methods have been created, such as random competition-based clustering (RCC). The RCC technique uses a random timer and the First Declaration Wins Rule is used to choose nodes for cluster formation. According to this rule, any node that initially identifies a CH to other nodes in its radio network is given governorship status.

An initiator is a name for this node that was chosen at random. All initiators broadcast a cluster advertisement message to all networked sensor nodes. Any node in the network that is not an initiator, sends a message to the initiator from which it received an advertisement message, if it receives one within the cluster. In addition to responding, it will forego accepting further cluster advertisement messages for that simulation cycle. However, such a sensor node will link this initiator's cluster.

Multi-hop broadcasting, on the other hand, uses a specific transmission range to deliver a cluster advertisement message to the sensor nodes. All sensor nodes within the receiving node's transmission range must continue to receive the cluster-advertised message. Since these strategies leverage the principle of lowest communication energy, the sensor node that is easiest to reach will be a part of the initiator cluster. Additionally, dynamic cluster creation involves regular restructuring.

Figure 3. Formation of the different nodes

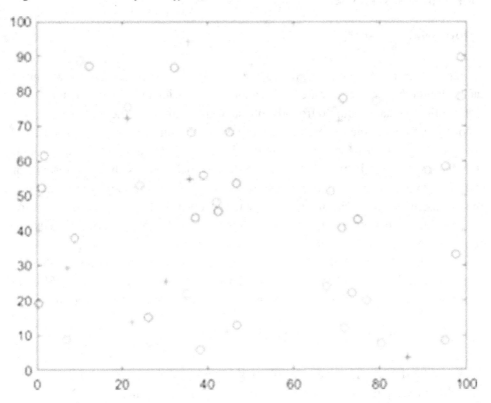

The problem of energy utilization is significantly reduced by multi-hop transmission. This is because there is a limit to how much energy may be required to transmit and the minimal transmission energy of nearby sensor nodes is spent. As a result, there will not be a requirement for distant sensor nodes to communicate directly with one another. Its drawback is that, compared to the earlier method of broadcasting, it has a longer latency. This is because multi-hop broadcasting necessitates the processing of data by individual sensor nodes along the multi-hop path, this delays cluster formation. However, if the latency issue is present, the multi-hop is significantly superior to the direct broadcast. The placement of nodes in the WSN is depicted in Figure 3.

Cluster Head Selection

Following cluster establishment, CHs are chosen to serve as cluster leaders. The cluster leaders are responsible for data gathering and information routing for the data of the cluster members to the BS. Clusters among many nodes have a greater

Figure 4. Cluster formation and cluster head selection

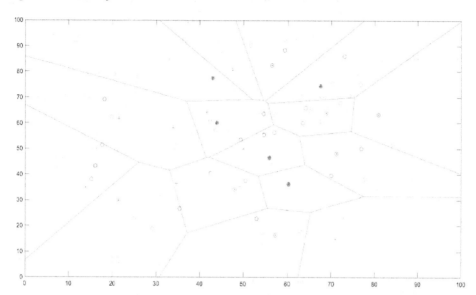

burden than clusters with fewer nodes because CHs for large-sized clusters must receive, aggregate and distribute more data than CHs for clusters with fewer nodes.

In the beginning, a CH may be chosen at random. It is also possible to choose a CH by considering the cluster's nodes' remaining energy. The job of the CH is rotated to distribute the burden because it is known that CHs have larger burdens than member nodes, extending the usable lifetime of those clusters.

The rotation happens regularly. In contrast, the sensor node with the greatest energy level in the cluster is chosen as CH in the residual energy selection method. Up until the interval time slot, it will remain the CH. The total energy of the sensor network will be uniformly distributed due to the CHs' rotation. The network lifetime progressively gets better with this method. Only the member clusters near the BS and have a level greater than the threshold energy are considered when choosing the cluster's next CH. Only the energy level and the separation between the sensor node and the BS are considered. Figure 4 depicts the creation of the clusters and the CH selection.

Data Transmission

Data transmission can begin as soon as the clusters are created and the TDMA schedule is configured. Assuming nodes always have information to provide, they do it during their assigned transmission window to the CH. This communication uses a very small amount of energy. To reduce energy loss at non-cluster-head nodes, the radios of each one can be disabled until the node's designated communication time. The CH node must keep its receiver to receive all data from the cluster nodes. After receiving the information, the CH executes the procedures for involving signals to combine all of the data into a single signal. For instance, the CH can beamform various signals to create a complex signal. The base station receives this mixed signal. The transmission requires high energy due to the distance from the BS. The LEACH networks perform this during the steady state. Next, depending on the priority, each node decides if it should serve as a CH for the consecutive round. The consecutive rounds selection of the CH node will be made after a particular time-stamp.

RESULTS AND DISCUSSION

This system is implemented using LEACH protocol with MATLAB. Some of the features of the nodes were presumptions about during node deployment. The base station is located in the zone's center. Nodes and clusters are stationary. Normal nodes send data directly to their corresponding CH within a particular cluster. CH uses multi-hop routing to transmit data to the BS. Each node has a uniform nature. With the same beginning energy, all nodes start (Mahmoud & Mohamad, 2016; Yadav & Sunitha, 2014). The list of variables used in the simulation is shown in Figure 5.

The following formula is used to determine the total data transferred by a node in the WSN:

$$Data\ Transmitte(D) = Data\ CH\ to\ BS + Data\ P\ to\ BS \tag{1}$$

where CH is Cluster Head, BS is Base Station and P is the sensor node.
The energy dissipated in the data transmission is denoted by equation 2.

$$ET(L,d) = \{\{LM + LA_{FS}d^2\} \quad d \le d0 \quad ;$$

$$\{LM + [\![LA]\!]\_MP\,d\}\,d \ge d\_0 \tag{2}$$

where L is the bit amount to transfer the data, M is the energy consumed to transmit or recieve 1 bit message.

AFS and AMP amplification coefficient of free − space and multi − path fading signal; d is distance.

The energy spent on receiving the data is written as following equation 3.

Figure 5. List of parameters for simulation

Parameters	Value
Simulation Area	200*200
Initial energy	0.5J
Base station	50m*50m and 100m*100m
Transmitter/Receiver Electronics	50 nJ /bit
Number of nodes	100 and 300
A_{FS}	10pJ/ bit/m
A_{MP}	0.0013 pJ/bit/m
Packet size	4000bits

$$ER = LM \tag{3}$$

Equation 4 represents the total energy consumed by the node as a result.

$$Total\ Energy\ (E) = ET + ER + D \tag{4}$$

The energy usage of each node with the number of iterations is depicted in the Figure 7.

Wi-Fi sensors utilize less battery due to the low-power Wi-Fi module or chip. The base for this system was the IEEE 802.15.[4] protocol used in 6LoWPAN and ZigBee protocols. These are designed for low-power applications, including low-power

Figure 6. LEACH protocol's number of dead and active nodes

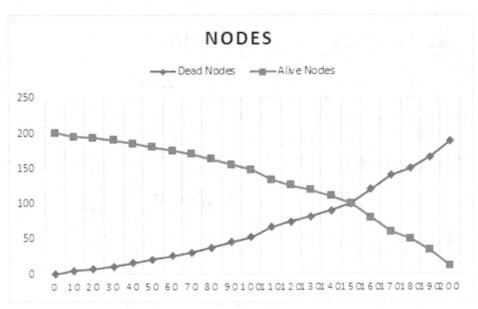

Figure 7. Energy consumption of nodes in the LEACH protocol.

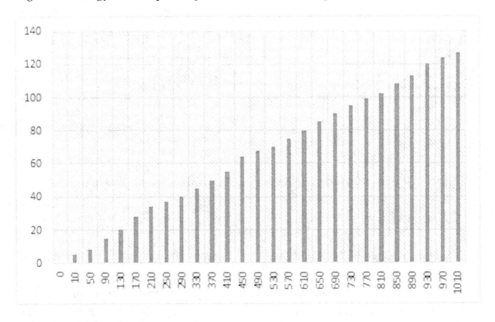

WSN. Consequently, their low power consumption makes attractive candidates for IoT applications. The range decreases with a constant transmission and receives power as the frequency rises.

Additionally, more transmission power is required to extend the distance between sensor nodes. "sub-1 GHz" refers to a low wireless communication frequency below 1 GHz. IoT communication typically uses frequency ranges like 915 MHz, 868 MHz, 433 MHz and 315 MHz. Sub-1 GHz-based sensor networks have various benefits. We can now not recommend a specific module for IoT applications in terms of distance and power utilization because proximity depends on the type of application (Psiborg, n.d.).

FUTURE RESEARCH DIRECTIONS

The research can be carried out to minimize the energy during the transfer of the data from the cluster to the BS in the different types of technologies.

CONCLUSION

This research aims to maximize WSN node longevity while lowering energy usage. The management of WSN nodes with high residual energy, short routing distances and a large number of neighbors is another area of emphasis for the system. This system employs the LEACH protocol, an energy-efficient hierarchical clustering mechanism. The cluster is designed to maximize network speed and decrease overall energy consumption in each node. The CH is done using the prioritized Round Robin method, where the new CH is selected in the cluster depending on the proximity from BS and the Battey node-level. A wireless sensor network based on Lora is frequently used for communication. To connect sensor networks with gateways and transfer data from these sensor networks, other technologies such as Sub-1 GHz, Zigbee are also can be employed. Sub 1Gz, ZigBee IOT communication Protocol stack is one simple network protocol stack used for IoT technologies.

REFERENCES

Ahlawat, A., & Malik, V. (2013). An Extended Vice-Cluster Selection Approach to Improve V Leach Protocol in WSN. *2013 Third International Conference on Advanced Computing and Communication Technologies (ACCT)*, 236-240. 10.1109/ACCT.2013.60

Albert, J. R., Kaliannan, T., Singaram, G., Sehar, F. I. R. E., Periasamy, M., & Kuppusamy, S. (2022). *A remote diagnosis using variable fractional order with reinforcement controller for solar-MP*PT intelligent system. In Photovoltaic Systems (pp. 45–64). CRC Press. doi:10.1201/9781003202288-3

Elprocus. (n.d.). https://www.elprocus.com/architecture-of-wireless-sensor-network-and-applications/

Fantin Irudaya Raj, E., & Appadurai, M. (2022). *Internet of things-b*ased smart transportation system for smart cities. In Intelligent Systems for Social Good (pp. 39–50). Springer. doi:10.1007/978-981-19-0770-8_4

Fu, C., Jiang, Z., Wei, W. E. I., & Wei, A. (2013). An energy balanced algorithm of LEACH protocol in WSN. International Journal *of Computer Science Issues, 10(1),* 354.

Geeks for Geeks. (n.d.). https://www.geeksforgeeks.org/wireless-sensor-network-wsn/

Iqbal, N., & Kim, D. H. (2022). IoT task management mechanism based *on predictive optimization for efficient energy con*sumption in smart residential buildings. Energy and Building, 257, 111762. doi:10.1016/j.enbuild.2021.111762

Khediri, S. E., Nasri, N., Wei, A., & Kachouri, A. (2014). A new approach for clustering in wireless sensor networks based on LEACH. Procedia *Computer Science, 32, 1*180–1185. doi:10.1016/j.procs.2014.05.551

Mahapatra, R. P., & Yadav, R. K. (2015). Descendant of LEACH based routing protocols in wireless sensor networks. Procedia Computer S*cience, 57, 1005–1014. do*i*:10*.1016/j.procs.2015.07.505

Mahmoud, M. S., & Mohamad, A. A. (2016). A study of efficient power consumption wireless communication techniques/modules for inter*net of things (IoT) appl*ications. Academic Press.

Majid, M. A. (2022, March). Energy-Efficient Adaptive Cluster*ing and Routing Protocol for Expanding the Life Cycle of the IoT-based Wireless Sensor Network. In 2022 6th International* Conference on Computing Methodologies and Communication (ICCMC) (pp. 328-336). IEEE.

Mosorov, V., Biedron, S., & Panskyi, T. (2015). Analysis of a new model of low energy adap*tive clustering hierarchy protocol in the wireless sensor network.* Восточно-Европейский журнал передовых технологий, 5(9), 4-8.

Neelakandan, S., Rene Beulah, J., Prathiba, L., Murthy, G. L. N., Irudaya Raj, E. F., & Arulkumar, N. (2022). Blockchain with deep lea*rning-enabled secure healthcare data transmission a*nd diagnostic model. International Journal of Modeling, Simulation, and Scientific Computing, 2241006.

Peng, Z., & Li, X. (2010). The improvement and simulation of LEACH protocol for WSNs. IEEE International Conf*erence on Software Engineering and Service Sciences, 500-503. 10.1109/*ICSESS.2010.5552317

Psiborg. (n.d.). https://psiborg.in/wireless-sensor-network-based-on-sub-1-g*hz/*

Pushparaj, T. L., Raj, E., Rani, E., Darwin, S., & Appadurai, M. (2022). Employing Novel Si-Over-Si Technology to Optimize PV Effect in Solar Array. Silicon, 1-13.

Raj, E. F. I., Appadurai, M., Darwin, S., & Rani, E. F. I. (2022). Internet of Things (IoT) for Sustainable Smart Cities. In Internet of Things (pp. 163–188). CRC Pres*s. doi:*10.1201/9781003219620-9

Ran, G., Zhang, H., & Gong, S. (2010). Improving on LEACH protocol of wireless sensor networks using fuzzy *logic. Journal of* Information and Computational Science, 7(3), 767–775.

Sharma, R., Jain, G., & Gupta, S. (2015, November). Enhanced Cluster-head selection using round-robin technique *in WSN. In 2015 International Conference on Comm*unication Networks (ICCN) (pp. 37-42). IEEE. 10.1109/ICCN.2015.8

Shiverware. (n.d.). https://shiverware.com/iot/iot-vs-wsn.html

Singh, *S. K., Kumar, P., & Singh, J. P. (2017). A Survey on Successo*rs of LEACH Protocol. IEEE Access: Practical Innovations, Open Solutions, 5, 4298–4328. doi:10.1109/ACCESS.2017.2666082

Thilakarathne, N. N., Kagita, M. K., & Priyashan, W. D. (2022). Green *internet of things: The next generation energy effi*cient internet of things. In Applied Information Processing Systems (pp. 391–402). Springer. doi:10.1007/978-981-16-2008-9_38

Tyagi, S., & Kumar, N. (2013). A systematic review on clustering and *routing techniques based upon LEACH p*rotocol for wireless sensor networks. Journal of Network and Computer Applications, 36(2), 623–645. doi:10.1016/j.jnca.2012.12.001

Xu, J., Jin, N., Lou, X., Peng, T., Zhou, Q., & Chen, Y. (2012). Impr*ovement of LEACH protocol for WSN. 2012 9th* International Conference on Fuzzy Systems and Knowledge Discovery, 2174-2177. 10.1109/FSKD.2012.6233907

Yadav, L., & Sunitha, C. (2014). Low energy ad*aptive clustering hierarchy in wireless sensor network (LEACH). Internation*al Journal of Computer Science and Information Technologies, 5(3), 4661–4664.

Chapter 12
An Energy–Efficient Keyless Approach to Home Security Using Internet of Things

Sandeep Kumar Hegde
N.M.A.M. Institute of Technology (Deemed), India

Rajalaxmi Hegde
ⓘ https://orcid.org/0000-0001-5610-8748
N.M.A.M. Institute of Technology (Deemed), India

ABSTRACT

The primary goal of the chapter is to modernize the security requirements for smart homes while building a complete home security system based on the internet of things. This system provides better security and dependability at a lower cost when compared to other security systems. Using the registered data to unlock the door increases security by preventing unauthorized unlocking. Security is ensured using two different techniques: first, the user can utilize facial recognition technology; second, they can provide access using a control app. The system provides the user with safer and more secure locking and unlocking technologies than the traditional technique.

INTRODUCTION

The Internet of Things, or IoT, is a network of connected physical things that can exchange information and communicate with one another without a person's help. Since it enables us to gather data from various sources, such as people, animals, cars,

DOI: 10.4018/978-1-6684-4974-5.ch012

and kitchen appliances, IoT has been formally referred to as an "Infrastructure of Information Society." Therefore, any physical object that can be given an IP address to facilitate data transfer across a network may be regarded as a member of the IoT system by adding electronic hardware like sensors, software, and networking tools.

The IoT differs from the Internet in that it makes it possible for everyday items with embedded circuits to interact and communicate with one another using the Internet's pre-existing infrastructure. Peter T Lewis first mentioned the concept of the "Internet of Things" in his speech to the Federal Communications Commission (FCC) in 1985, and it was then given that name. The IoT has grown significantly since that time. Currently, there are more than 12 billion linked devices, and by the end of 2020, experts expect that number to rise to 50 billion. The IoT infrastructure's real-time information collection and processing capabilities, which rely on accurate sensors and seamless connection, let people make smart decisions. Both businesses and consumers have profited from the development of IoT. By offering value-added services that enhance and lengthen the lifespan of their products or services, manufacturers may gain a greater understanding of how their products are used and how they function in the real world as well as boost their revenue. For a more personalized and effective user experience, consumers, on the other hand, have the choice to combine and control several devices.

Overview

As we all know, automation is taking over every aspect of life. The notion of automation is only one of the amazing discoveries that man, with his intellect, is always behind. As a result, the proposed paper is working to develop a marvellous solution that encourages keyless entry into your lovely homes. Deep learning and the internet of things help make keyless technology viable. Using a Raspberry Pi, this unlocking may be accomplished as a face recognition and smartphone app. The project places a strong emphasis on the concept of smart keyless home security. The most popular locking and unlocking method is the actual key, and it is all mechanical in nature. It is challenging for the resident when the key is lost or stolen. When employees at a company are required to keep numerous keys for various doors, the issue with handling keys gets worse. Therefore, keyless technology is being implemented as a solution to this issue. It is more flexible and practical because it allows users to lock and unlock the doors. The majority of keyless entry locks are just as simple to install—if not simpler—than their conventional equivalents. It gives users the option to remotely lock and unlock their houses, increasing flexibility and convenience.

Importance

It is urgently necessary to modernize our daily items and make them intelligent; also, the moment has passed when we can no longer blindly rely on antiquated and traditional security methods, particularly when referring to our door locks. This will boost convenience and flexibility. Any object needs to be modified and brought up to date by getting rid of its flaws and adding more capability. The system is made more secure by features like Face Detection, smartphone remote access, and often changing passcodes. A smart door lock has become the standard for home security today. Any suspect behaviour is reported to the owner through alerts. A photo of the front door is sent to the owner if someone attempts to open it inappropriately. It increases trust in the owner to be able to remotely unlock a door using a smartphone. The owner can be trusted because the door may be opened remotely with a smartphone.

Problem Statement

The following problem statements apply to the present circumstance and system:

i)   If the key is lost, the lock must be disassembled.
ii)   If two or more larger groups of people are using the lock, it is always necessary to physically carry the key or several copies of the same key.
iii) If someone breaks into the residence, there is no information or lead.
iv)   Current home security systems are relatively expensive and geared toward high-end consumers.

Objective

The following are the research objectives of the proposed work.

i)   Identifying the current home security issues that exist.
ii)   Adding features like Face Detection, smartphone remote access, and dynamic passcodes greatly improve security.
iii) Create a product prototype that is reasonably priced, dependable, and easy to use.
iv)   Current home security systems are relatively expensive and geared toward high-end consumers.

Motivation

Over the last few years, the Internet has gained a reputation as a vast data source. New forms of bidirectional communication between people and things as well as between people and other people were made possible by the integration of mobile transceivers into commonplace objects and gadgets. The Internet of Things concept, which Kevin Ashton first proposed in 1998 and has since gained increasing attention in academia and business, would provide the field of information and communication technology with a new dimension. Even while this paradigm is expanding and improving many facets of our lives, there are still pressing problems that need to be taken into account and resolved. The main concern is to make sure that users' data is secure and private. Another problem is making interactivity between linked devices possible to fully realize their intelligence. To achieve the latter, data exchange and autonomous behaviour are essential. IoT definitions vary depending on the viewpoint, but they all centre on "things" in general gathering, exchanging, and transmitting data with people and other "things" over the "internet". IoT facilitates decision-making and secures nearly everything in our environment. Through a variety of applications, the IoT vision for a smarter life promises to soon enable smart home security, allowing for everything from basic home monitoring to actual control. Monitoring and securing houses from different parts of the world is now possible through the integration of IoT and home security.

LITERATURE SURVEY

Related Work

Numerous remote control and monitoring systems can be used as platforms for study or as commercial goods. Research on these factors has been done by experts in wireless technology, and the findings have been successful. The research listed here is praiseworthy. The most popular wireless networking technologies are Zigbee, Bluetooth, and Wi-Fi. Zigbee is suggested as the primary system module in the proposed system. The system also has several other modules, including a camera, relay module, video phone, and a human detection module (HDM) to detect users at doors. The Zigbee module then retrieves a Zigbee tag to verify the user. The person at the door can communicate with the owner on the speakerphone if the user verification process is unsuccessful. Radio-Frequency IDentification(RFID) was suggested as the system's primary technology by certain systems. A Liquid crystal display (LCD) touchscreen is utilized by the system. In the "Web-based door locking system ", the user may easily access the door with a mobile device and can also check the door's status, including whether it is locked or not, with the use of a web application that is made to keep track of the door.

This method is flawed in such a way that anyone with the security code who hacks into the system may enter that room with ease. The Raspberry Pi was used to investigate the various algorithms of the facial recognition system. This study applied the LBPH (Local Binary Pattern Histogram), Fisher Faces, and Eigen Faces algorithms to examine the temporal complexity and precision of the different approaches.

In order to outperform traditional locks in terms of security and modern security systems in terms of cost-effectiveness, in (Touqeer et al., 2021) password-protected electronic lock system for smart home security is proposed. In this paper, the user is prompted to enter the security code on the LCD screen. If the user enters the code accurately, it is validated, and then the servo motor turns to unlock the door. This device's range is larger, and it has multiple levels of protection, including biometrics and ID card scanning. To learn more about the new security systems being proposed and implemented that use similar concepts, the author suggests an automatic door-locking system by doing literature reviews of numerous research articles(Pawar et al., 2018). Automatic door-locking systems for Android mobile phones can be created by using a customized app. The author suggested face recognition can be used as one of the approaches for an effective door-locking system, both in terms of design and implementation(Gopi Krishna et al., 2019). The distinguishing feature of the specific model is that it combines a variety of features while being significantly more straightforward in use. Additionally, clever usage of solenoids will soon render stepper motors obsolete. By including motor detection, tamper detection, and connecting the model to a mobile application, consumers can remotely monitor their homes as part of this process. By offering a low-cost authentication solution, the author suggests leveraging Bluetooth technology to enhance home security(R. et al., 2020). The core of this technology is the effective usage of keyless door lock systems with smartphones. The system contains several features that set it apart from competing products. It has motion detectors that make it easier to identify users. If the door is attempted to be opened, the camera will record images and email them to the owner. Additionally, the house address may be included in an automated trigger report that is submitted to the neighbourhood police station. The smart door-locking system built on Android is handy in several ways(Hoque & Davidson, 2019). The system is built on the idea of pre-determined passwords. It offers an additional layer of security to stop any unauthorized unlocking. The owner need not be concerned if they forget their password because they have the opportunity to reset it. In the future, the user will be able to control the lock system from anywhere in the world by using cloud computing(Bamnote et al., 2021). Additionally, other functions like face recognition and finger scanners can be implemented. The author suggested a

Face recognition-based door-unlocking system utilizing Raspberry Pi(D. Deshmukh et al., 2019). It limits access for intruders and unauthorized people. It has a passcode entry keypad or a facial recognition security system that allows you to unlock the door by entering the passcode. Any person with the passcode can access the entry to the house since it uses face recognition or a passcode. Since it primarily relies on face recognition modules, there is a possibility that faces won't be recognized and doors won't be able to be unlocked. Future updates may substitute fingerprints for the passcode because they are more distinctive.

The verification process should start with facial recognition and move on to fingerprints. In place of the outdated key system, home automation using digital techniques namely as a secret code, semiconductors, smart cards, and fingerprints have proposed(Nur-A-Alam et al., 2021; Syed et al., 2021). Because each fingerprint is distinct and can only be opened by the owner, it offers exceptional security. Additionally, more adaptability and extensibility are offered. It is a reliable, affordable, and simple system. Since a secret code, semiconductor, or smart card can be used to access the door, the door can be opened by anyone with the secret code to gain entry to the home. Future upgrades could include fingerprint and facial recognition to increase security and efficiency. The author proposed Automated Control of Door using Face Recognition by facial detection, which is more reliable, and efficient and uses fewer data and power than other products because it is powered by batteries and connects to the internet wirelessly using a USB modem(Abbas et al., 2021; Alamsyah et al., 2021; Shamim Hossain et al., 2016). Professional assistance is needed to install smart doors using OpenCV and Python. It will be challenging for regular people to handle. Without expert assistance, it would be challenging to open the door if the software is compromised. The communications might be readily removed without the owner's knowledge if someone managed to hijack the owner's email account. It can be made easier and more user-pleasant in the future. By lowering the danger of burglary, the author suggested a Facial Recognition Enabled Smart Door(Park et al., 2009; Roy et al., 2018)(Sheikh, 2018). If our friends or families arrive to greet us, we can also open the door without being there physically. Installing smart doors prevents scenarios where we get locked out of the house or inside the house from happening. There are limitations because the smart door makes use of facial recognition technology. Hackers can modify the database and the facial recognition technology as it is software-operated. In the modern world, criminals won't have any trouble employing cutting-edge technologies to make a replica model of the owner's face. In the future, the software can be changed so that it alerts the owner in case of an emergency if a person on the blacklist tries to enter the door.

To improve security and usability, the author proposed a digital door lock system for the Internet of Things that sends the collected images of the intruder to the owner's phone whenever an unauthorized person tries to enter the home or place of business(Park et al., 2009) (Khan et al., 2021; Meenakshi et al., 2019). If someone tries to physically harm the door, it also sends frightening signals to the owner's mobile device. The owner will be able to take the necessary action in the event of unauthorized entries because the messages are immediate and quick. In (Ngurah Desnanjaya & Arsana, 2021; Radu, 2020; Wiyanto & Oktavianti, 2021) authors have proposed deep learning and machine learning algorithms in combination with IoT to enhance the protection of intrusion detection systems. In (Abbas et al., 2021; Alfa et al., 2021; Neelakandan et al., 2022) the author has introduced blockchain-enabled deep learning techniques in the IoT platform to strengthen home security systems.

As discussed in the literature review section the current approaches proposed for home security systems have common flaws such as they are relatively expensive and aimed at high-end customers. These systems consume more power to run the setup and they are less reliable. In the proposed approach Raspberry Pi and the Internet of Things-based home security system have been implemented using openCV based haar algorithm which provides additional features such as face detection, smartphone remote access, and dynamic passcodes which greatly improves security at a reasonable price by consuming less power. To develop the proposed technique wide variety of inexpensive, power-efficient sensors have been made use. The section below explains the details methodology followed while implementing a home security system using the proposed technique.

PROPOSED METHODOLOGY

In this proposed work, a home security system is suggested that solicits input from actual people. The same thing is done with Raspberry Pi and the Internet of Things. As part of this work, a database is created for the family, and taking 30 pictures of each member are. It will be simple for the algorithm to determine whether the person coming is authorized or not because men change their appearance every few days. The openCV-based haar algorithm is applied to make this happen. The image is downloaded into the database from the real-time video streaming. The grey scale is used for conversion to turn each image into greyscale to conserve memory. The dimensions of each image for taking the input image from the video streaming are 130 pixels in width and 100 pixels in height. The Haar method uses a cascaded version of numerous inputs to identify which aspect of a face is and we can create the ideal face detection algorithm with the precise calculations of each character in the face. The camera will take a picture of any illegal individuals who approach the house, transform it to grayscale, and match it to the database. When the output is displayed, the pixels will be more accurate if the image has already been recorded in the database, however, when the input image has not been stored in the database, the pixels will match the least closely. The proposed work demonstrates the connection between the computer and the Raspberry Pi (RPi). Wi-Fi is required for the system-wide connectivity of RPI, and the Internet of Things serves as the primary gateway for this connectivity. The RPi is known as the machine-to-machine and machine-to-machine transaction of the data, where IoT aids in the transaction of the data from one device to another. We cannot take the desktop with us wherever we go because it is not portable; however, we can share the screen by using the same IP address.

Figure 1. System boundary

2.1 System Boundary

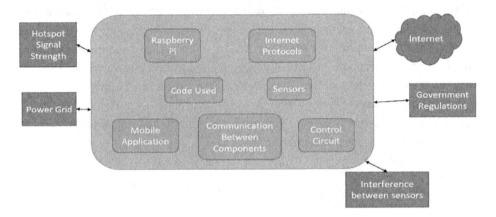

The laptop must be connected to the same Wi-Fi network after the desktop has been connected to the RPi and the Wi-Fi connection has been verified. As soon as we determine the IP address the RPi is using, we must enter the same information into the laptop's remote desktop connection and wait for the login prompt. Once the login screen displays, we must provide the proper credentials to access the remote desktop. Once logged in, the screen we are viewing is identical to the one on the desktop. We are now able to manage it from a distance. The user can also make use of the Blynk app's features, which let him remotely lock and unlock doors for himself or a person who is standing at the door.

As shown in figure 3.1 the system border outlines what is programmable within the system and what needs to be interfaced externally to the system. Everything inside the system boundaries can be readily and precisely defined and configured. The user can operate and get updates on his mobile phone using the Blynk mobile application. The Raspberry Pi's power supply is managed via a mechanism known as a control circuit, which enables the use of either a wall connector or a battery. Communication between components identifies the transmission channels being utilized inside the system, such as the Internet or physical connections, to transfer data between units. Data formatting for transmission over the Internet is described in Internet Protocols. The code utilized identifies the programming languages, including Java and Python, in which the various project components are being created. The components that send data to offer real-time information on the lock's state are represented by the sensors. The Blynk can be programmed to communicate with the selected server and operate any selected device. An effective object recognition method that makes use of Haar feature-based cascade classifiers was proposed by Paul Viola and Michael Jones in their 2001 paper "Rapid Object Identification using a Boosted Cascade of Simple Features."

A cascade function is taught by machine learning utilizing a sizable number of both positive and negative pictures. Following that, it is used to locate things in other pictures. The method initially needs a significant quantity of both positive (photos

Figure 2. Haar cascade classifier object detection

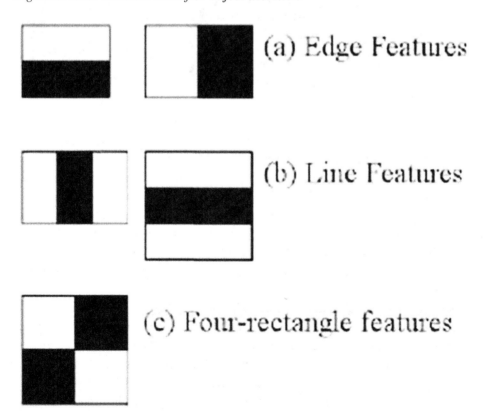

of faces) and negative (images of objects) inputs (images without faces). We must then extract characteristics from it. For this, the Haar characteristics in the picture below are used. They are a precise match for our convolutional kernel. Each feature is a single value, as illustrated in figure 3.2.

Figure 3. Haar Cascade Classifier Detection

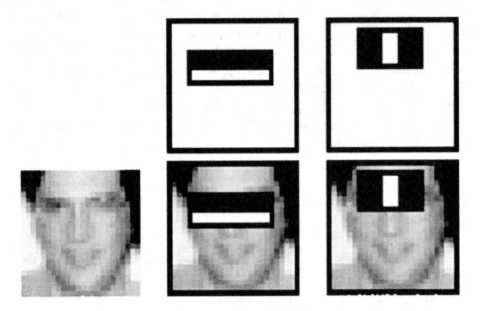

All practical sizes and placements for each kernel are now used to calculate a huge number of features. (Just think about how much computation is needed). There are about 160000 traits in only a 24x24 time frame. For each feature computation, the total number of pixels under each white and black rectangle must be determined. The major emphasis of the first attribute picked seems to be the fact that the region around the eyes is typically darker than the region around the nose and cheekbones. The second characteristic was picked because of the eyes' deeper shade than the nose's bridge. But it doesn't matter if the cheeks or any other place gets the same windows.

Every attribute for this is applied to all of the training photographs. It chooses the best cut-off for each quality to classify the faces as positive or negative. Since they can discriminate between images with and without faces the best, the characteristics with the lowest error rate are selected. Each pixel in an image is assigned a label by the Local Binary Pattern (LBP) texturing operator by thresholding its immediate surroundings and considering the result as a binary integer, as illustrated in figure

Figure 4. LBPH face recognizer

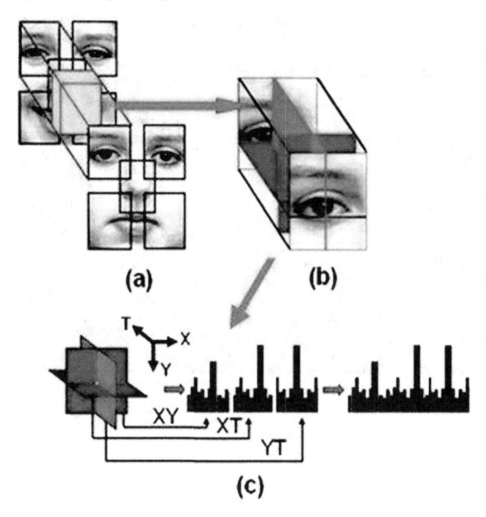

3.4. The LBP texture operator has gained prominence in numerous applications due to its potent discriminative powers and computational simplicity. It may be viewed as a unifying method for the statistical and structural models of texture analysis, which are frequently incongruent. The LBP operator's resistance to monotonic grey-scale shifts brought on by factors like lighting may be its most crucial feature in real-world applications.

Blynk was developed for the Internet of Things and is depicted in figure 3.5. It is capable of storing data, visualizing it, displaying sensor data, remotely controlling hardware, and many more exciting tasks. The platform is divided into three primary sections:

Figure 5. BLYNK

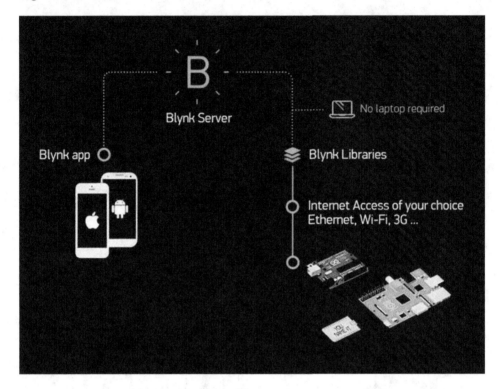

- Using a selection of the Widgets we provide, the Blynk App enables you to create beautiful user interfaces for your projects.
- The Blynk Server, which controls all communication between the smartphone and hardware.

Alternatively, by using our Blynk Cloud, you may operate your private Blynk server locally. It is open-source, can even be launched on a Raspberry Pi, and can easily manage thousands of devices.

High-Level Design Architecture

The high-level architecture of the proposed approach is shown in Figures 3.6 and 3.7 below. The owner's and other users' facial data is initially stored in the database. When the user approaches the door, the saved facial data is used to match the user's face. The door unlocks for the user when the face is recognized. A mobile application attached to the module allows the owner to permit or restrict access to the door if the

Figure 6. High-level design architecture

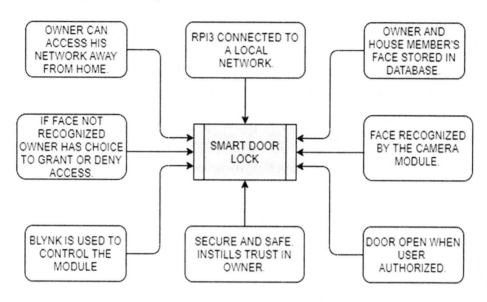

Figure 7. Sequence diagram of working module

5.2.1 Sequence Diagram /DFD

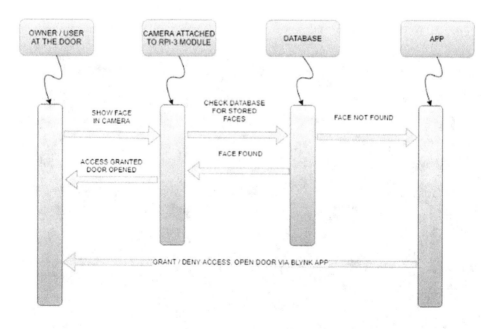

module is unable to identify the person. The owner may choose a local or remote connection for this connection.

The essential components used while conducting the experiments are listed as below.

Camera: Records the user's facial information for comparison.

Touch LCD: It shows the camera feed while also demonstrating the facial recognition technique.

Database: To locate a match, the facial data is searched across the current database.

Blynk Server: A Java server built on the Open-Source Netty platform. Blynk Server forwards messages between the Blynk mobile application and numerous SBCs and microcontroller boards (i.e., Arduino, Raspberry Pi).

Blynk Application: This is a mobile app that may be used to control the module if it is linked to the raspberry pi's Blynk server and both devices have an active internet connection or are part of the same local network. Various microcontroller boards and SBCs are used for message passing between the Blynk mobile application (i.e. Arduino, Raspberry Pi) listed below.

Applying Blynk: It is a mobile application that can be linked to raspberry pi's Blynk server and used to control the module as long as they both have an active internet connection or are connected to the same local network. The activity diagram of the Blynk server is shown in figure 3.8 below.

Figure 8. Activity diagram

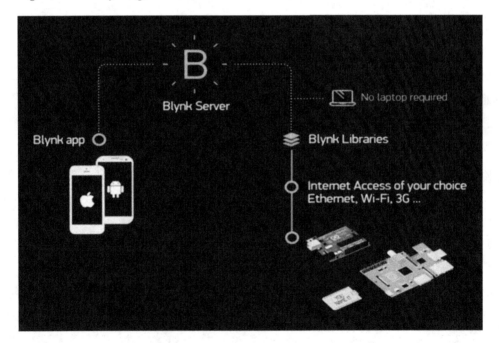

With Blynk, the owner may easily maintain control over his project. One of its important characteristics is its capacity to remotely control the proposed model securely while also giving the owner simple manageability. The use case diagram of the proposed technique is shown in figure 3.9 below.

Figure 9. Use case diagram

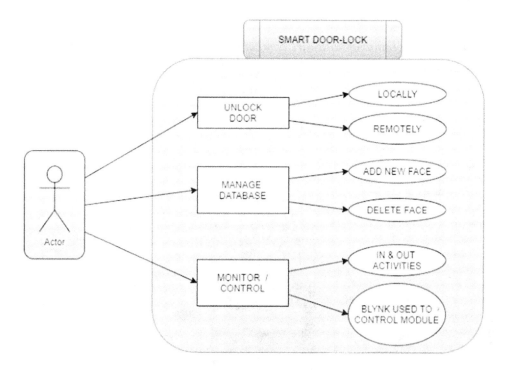

The Use case Diagram consists of the following components.

1. **Facial inputs:** The shortcomings of conventional door locks can be overcome by a smart door equipped with facial recognition technology. Since they don't have to fight with keys and unlock the door, it is a huge aid for the elderly and the visually impaired. Additionally, it lowers the chance of theft.

2. **Remote Access:** The ability to remotely open the door using a smartphone fosters owner confidence. For instance, the user can use the app to remotely unlock the door for a visitor he wants to let into the property. Due to the lack of a physical key, this method is both incredibly convenient and secure.

3. **App Notifications:** If there is an action on the system, the user will receive a notification on his app.

System Testing

The test plan calls for the team to come up with ways to trick the system into unlocking the door, whether by providing false facial information or by any other security-bypassing technique. Additionally, it would involve testing certain parts, the server, and the Android application. Figure 3.10 below shows the Integration of hardware components.

Figure 10. Integration of hardware components

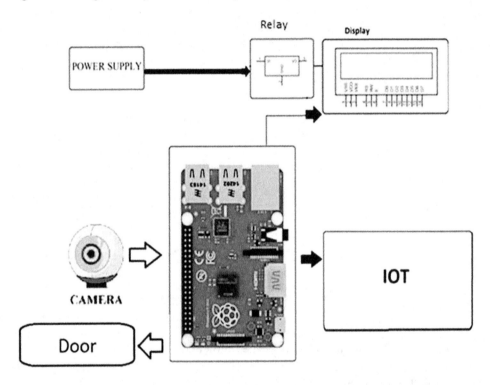

Individual software units or components are tested during the level of software testing known as unit testing. The goal is to confirm that each piece of software operates as intended.

It has been ensured that no faces from static images are matched by the trained dataset using the facial recognition code. The following setup has been made while conducting the experiments.

Camera: Checking to see if the image the camera captures is clear enough to match in various lighting conditions.

Power and touch LCD: The system's power and display are examined for any voltage impacts and are suitably controlled.

Server: By establishing a quick and secure network, it is checked whether the server can be reached. The owner must provide the IP address of his private network to make the module accessible to him while he is away from home. Software testing at the integration testing level involves combining and evaluating many software components together. This level of testing seeks to identify issues with how integrated units interact with one another.

Complementary Elements: Blynk Server Blynk Android App Camera Touch LCD

Power source: Every part of the system is tested to make sure it works well together as a whole. During this stage, any problems that prevent a component from connecting to the server or any errant code that prevents a component from working properly are fixed. After passing all the numerous tests, a secure, quick, and dependable module is produced.

EXPERIMENTAL RESULTS

The final module detects the user's face and takes appropriate action required upon recognition or non-recognition of the face. If the facial recognition validates the face against its existing data the door lock is unlocked or else, the door remains locked as shown in figure 4.1.

Figure 11. Facial recognition

Figure 12. Blynk app unlocked screen

The Blynk app allows the owner to take the appropriate action locally or remotely as shown in figures 4.2 and 4.3. If the owner wants the door to be opened for someone remotely, he just has to press the Unlock button on the Blynk app UI and the door is unlocked. Similarly, the door can be locked automatically when the owner clicks on the 'lock' button. The smart lock system works giving the owner multiple advanced ways to unlock the door and remove the usage of the physical key from his home.

Figure 13. Blynk app locked screen

CONCLUSION AND FUTURE WORK

In this research, a keyless IoT-based method for home security has been suggested. By enabling the user to remotely lock or unlock the door lock, the provided method increases user convenience. The door lock system automatically opens or closes the door when a legitimate user approaches it. This is another effective system function of the proposed work. When compared to existing digital door lock systems, the proposed approach anticipates that if the aforementioned problems are resolved, the system might be more easily marketed as a valuable product, such as a safe security system. Real-time speaking assistants may be used in the future to improve the system's usability and effectiveness. These assistants may also record every activity and credit the owner. A database can be linked to the cloud in case of data loss or power outage. Future security breaches can be prevented by deploying more secure protocols like TLS. There are many opportunities in the field of home automation today, and we can subsequently add GSM and DTMF modules to the proposed approach to increase its functionality. Additionally, on a bigger scale, the live stream may be enabled by utilising x-86 processors rather than ARM processors.

REFERENCES

Abbas, K., Tawalbeh, L. A., Rafiq, A., Muthanna, A., Elgendy, I. A., & Abd El-Latif, A. A. (2021). Convergence of Blockchain and IoT for Secure Transportation Systems in Smart Cities. *Security and Communication Networks*, *2021*, 1–13. Advance online publication. doi:10.1155/2021/5597679

Alamsyah, D. P., Ramdhani, Y., Rhamadhan, S. R., & Susanti, L. (2021). Application of IoT and Cloud Storage in Android-Based Smart Home Technology. *Proceedings of 2021 7th International HCI and UX Conference in Indonesia, CHIuXiD 2021.* 10.1109/CHIuXiD54398.2021.9650605

Alfa, A. A., Alhassan, J. K., Olaniyi, O. M., & Olalere, M. (2021). Blockchain technology in IoT systems: Current trends, methodology, problems, applications, and future directions. *Journal of Reliable Intelligent Environments*, *7*(2), 115–143. Advance online publication. doi:10.100740860-020-00116-z

Bamnote, H., Rakhonde, A., Ghagare, S., Dharpure, S., Kapse, K., & Warjurkar, S. (2021). IoT-based door access control system using face recognition. *Tech-Chronicle, 3*(2).

Deshmukh, Nakrani, Bhuyar, & Shinde. (2019). Face Recognition Using OpenCV Based On IoT for Smart Door. SSRN *Electronic Journal*. doi:10.2139/ssrn.3356332

Gopi Krishna, P., Sai Phani Kumar, P., Sreenivasa Ravi, K., Divya Sravanthi, M., & Likhitha, N. (2019). Smart home authentication and security with IoT using face recognition. *International Journal of Recent Technology and Engineering*, *7*(6).

Hoque, M. A., & Davidson, C. (2019). Design and implementation of an IoT-based smart home security system. *International Journal of Networked and Distributed Computing, 7*(2), 85. Advance online publication. doi:10.2991/ijndc.k.190326.004

Khan, S., Nazir, S., & Ullah Khan, H. (2021). Smart Object Detection and Home Appliances Control System in Smart Cities. *Computers. Materials & Continua, 67*(1), 895–915. Advance online publication. doi:10.32604/cmc.2021.013878

Meenakshi, N., Monish, M., Dikshit, K. J., & Bharath, S. (2019). Arduino Based Smart Fingerprint Authentication System. *Proceedings of 1st International Conference on Innovations in Information and Communication Technology, ICIICT 2019.* 10.1109/ICIICT1.2019.8741459

Neelakandan, S., Beulah, J. R., Prathiba, L., Murthy, G. L. N., Irudaya Raj, E. F., & Arulkumar, N. (2022). Blockchain with deep learning-enabled secure healthcare data transmission and diagnostic model. *International Journal of Modeling, Simulation, and Scientific Computing.* doi:10.1142/S1793962322410069

Ngurah Desnanjaya, I. G. M., & Arsana, I. N. A. (2021). Home security monitoring system with IoT-based Raspberry Pi. *Indonesian Journal of Electrical Engineering and Computer Science, 22*(3), 1295. Advance online publication. doi:10.11591/ijeecs.v22.i3.pp1295-1302

Nur-A-Alam, Ahsan, M., Based, M. A., Haider, J., & Rodrigues, E. M. G. (2021). Smart monitoring and controlling of appliances using lora based iot system. *Designs, 5*(1), 17. Advance online publication. doi:10.3390/designs5010017

Park, Y. T., Sthapit, P., & Pyun, J. Y. (2009). Smart digital door lock for the home automation. *IEEE Region 10 Annual International Conference, Proceedings/ TENCON.* 10.1109/TENCON.2009.5396038

Pawar, S., Kithani, V., Ahuja, S., & Sahu, S. (2018). Smart Home Security Using IoT and Face Recognition. *Proceedings - 2018 4th International Conference on Computing, Communication Control and Automation, ICCUBEA 2018.* 10.1109/ICCUBEA.2018.8697695

R., M., Y., R., R., R., & A., S. (2020). Smart Home Security System using Iot, Face Recognition and Raspberry Pi. *International Journal of Computer Applications, 176*(13). doi:10.5120/ijca2020920105

Radu, L. D. (2020). Disruptive technologies in smart cities: A survey on current trends and challenges. *Smart Cities, 3*(3), 1022–1038. Advance online publication. doi:10.3390martcities3030051

Roy, S., Nasir Uddin, M., Zahirul Haque, M., & Jahidul Kabir, M. (2018). Design and Implementation of the Smart Door Lock System with Face Recognition Method using the Linux Platform Raspberry Pi. *IJCSN-International Journal of Computer Science and Network, 7*(6).

Shamim Hossain, M., Muhammad, G., Rahman, S. M. M., Abdul, W., Alelaiwi, A., & Alamri, A. (2016). Toward end-to-end biomet rics-based security for IoT infrastructure. *IEEE Wireless Communications, 23*(5), 44–51. Advance online publication. doi:10.1109/MWC.2016.7721741

Sheikh, S. (2018). Face Recognition Using Cnn. *International Journal for Research in Applied Science and Engineering Technology, 6*(3), 1411–1414. Advance online publication. doi:10.22214/ijraset.2018.3218

Syed, A. S., Sierra-Sosa, D., Kumar, A., & Elmaghraby, A. (2021). Iot in smart cities: A survey of technologies, practices and challenges. *Smart Cities, 4*(2), 429–475. Advance online publication. doi:10.3390martcities4020024

Touqeer, H., Zaman, S., Amin, R., Hussain, M., Al-Turjman, F., & Bilal, M. (2021). Smart home security: Challenges, issues and solutions at different IoT layers. *The Journal of Supercomputing, 77*(12), 14053–14089. Advance online publication. doi:10.100711227-021-03825-1

Wiyanto, W., & Oktavianti, Y. (2021). Prototype Smart Home Pengendali Lampu Dan Gerbang Otomatis Berbasis IoT Pada Sekolah Islam Pelita Insan Menggunakan Microcontroller Nodemcu V3. *UNISTEK, 8*(1), 68–75. Advance online publication. doi:10.33592/unistek.v8i1.1209

Chapter 13

An Enhanced Method for Running Embedded Applications in a Power-Efficient Manner

N. M. G. Kumar
Sree Vidyanikethan Engineering College, Mohan Babu University, India

S. Lokesh
PSG Institute of Technology and Applied Research, India

Ayaz Ahmad
National Institute of Technology, Mahendru, India

Kirti Rahul Rahul Kadam
Institute of Management Kolhapur, Bharati Vidyapeeth (Deemed), India

Dankan Gowda V.
 https://orcid.org/0000-0003-0724-0333
B.M.S. Institute of Technology and Management, India

ABSTRACT

Many modern items that are in widespread use have embedded systems. Due of embedded processing's ability to provide complex functions and a rich user experience, it has grown commonplace in many types of electronic products during the last 20 years. Power consumption in embedded systems is regarded as a crucial design criterion among other factors like area, testability, and safety. Low power consumption has therefore become a crucial consideration in the design of embedded microprocessors. The proposed new method takes into consideration both the spatial and temporal locality of the accessed data. In the chapter, the new cache replacement is combined with an efficient cache partitioning method to improve the cache hit rate. In this work, a new modification is proposed for the instruction set design to be used in custom made processors.

DOI: 10.4018/978-1-6684-4974-5.ch013

INTRODUCTION

An Embedded system is a combination of an electronic and a computer system. It is a computer or processor-based system designed to perform a dedicated sequence of the task to control or operate a large system (mechanical or electrical) with real-time constraints. It is a system used to perform its functions without human intervention completely or partially. It is

also designed to accomplish a particular task in an efficient method. Mostly, embedded systems are used in operation where timing is very important (Aaron Lindsay and Binoy Ravindran. 2018). In the modern-day embedded systems applications, battery operated devices play an important role. The applications that run on those embedded devices rely more and more on powerful processors and are capable of running real-time applications. Because of the increase in number as well as the complexity of such applications, a significant amount of work has focused on the minimisation of power and energy consumed by the embedded processors (Chen Yang and Leibo Liu. 2018). Apart from giving a result in real-time, these embedded devices are to be designed to satisfy thermal limits as well as battery life limits, thereby directing the research towards low power and low energy enhancements. Along with performance, ease of use, and other such design metrics, power consumption is also a design metric for the present-day embedded systems. Embedded applications are demanding more processing power along with the ever-increasing need for more memory. It calls for active research to satisfy this multiple objective design challenge to develop an application for multicore embedded systems and their memory managing capabilities. The power consumption happening in memory by static and dynamic leakages is calling for a need to work on relevant solutions (Chenjie Yu and Peter Petrov. 2019). The power leakage in the bus and memory also create other side effects on the processor like thermal effects.

The cache contention happening in multicore processors can be addressed by the cache partitioning scheme to maximise the cache space utilisation. It may also increase the execution timing of the task. This work proposes a priority-based cache partitioning approach among the cores to improve both the cache performance and deadline avoidance. An embedded systems benchmark is used to select the set of applications from workloads to work on this cache partitioning problem (Dan, A and Towsley. D. 2020).

This work also presents a method to reduce the power consumed in the instruction fetching data bus during the execution of the instruction. The instruction code fetched from memory is modified so that the bus is loaded less, and therefore the switching capacitance associated with the bus is reduced(Daniel Sanchez and Christos Kozyrakis. 2021). This results in decreasing the power consumed in the data bus during the instruction fetch cycle. To explore a working cache model for a multi-core processor with configurable cache details and to incorporate run time cache optimisation for power minimisation.

The research problem of improving cache design in the context of power and energy optimisation is important because, in today's processors, the cache memory has a share of over 30% of the total processor's power. Cache memory, which is accessed faster and also with a good hit rate, is considered to be more power-efficient (Dimitris Kaseridis and Jeffrey Stuechelix. 2009). By selecting a better cache replacement algorithm which can be implemented either as hardware maintained structure or as a program (Dongwoo Lee and Kiyoung Choi. 2019). This method reduces the bus power of the processor during program execution. Power consumption in embedded systems is one of the important issues. Since a substantial amount of power consumption in a processor happens inside the cache and memory operations. this chapter proposes a technique on that domain to reduce the power consumption in embedded systems

(Dongxing Bao and Xiaoming Liourav. 2019). The key contributions are: A new cache replacement policy is implemented for a set of embedded systems benchmark applications. The standard LRU policy was modified with an extended tracking scheme to get a better cache hit rate, and hence, the power consumption is reduced. This method was applied to data and instruction caches for both L1 and L2 level caches, and a minimum of 7% average hit rate improvement was achieved for cache levels (Fang Juan and Du Wenjuan. 2018). The improved replacement policy was applied along with a priority-based cache partition policy. The compiled simulation architecture was run with two different sets of benchmark applications for validation, which gave a minimum of 4% hit rate improvement for medium-sized L2 cache. A new instruction code was designed specifically for the repeated instructions loop for embedded system applications. This had reduced nearly 70% of bus power inside instruction data bus.

IMPROVED CACHE REPLACEMENT

An enhanced cache replacement policy for using embedded processors is explained in this section. The introductory theory of cache memory and the experimental framework set up for the required simulation is also explained. A new method for cache replacement is explained in this section. The theory of cache and hierarchical memory structure used in the microprocessors are briefly outlined (Fitzgerald, B, Lopez, S. and Sahuquillo, J. 2018). The existing cache replacement policies are given an overview, along with their merits and demerits. The simulation environment to construct the experimental setup is explained. The steps are explained to configure the required simulation environment for a given set of test bench applications are explained (Florin Balasa and Dhiraj K Pradhan. 2012). The individual programs in the test bench applications are given an introductory description. The results are discussed, and the effects of design overheads are discussed.

The cache is a hardware component utilised by one or more CPU cores to lower the typical cost, time, and energy required to retrieve data from main memory. To store copies of the data from frequently utilised main memory locations, a cache must be created as a smaller, quicker memory that is located closer to a processor core. The speed at which instructions and data may be read from and written to the main memory has a significant impact on how long it takes to run a programme (Gan Zhi-Hua and Gu Zhi-min. 2019). Cache memory plays a crucial part in the complicated computing environment of today since memories get slower as they become bigger. Cache memory's primary function is to store data and programme instructions that are often accessed by operating software. These instructions may be accessed quickly, which cuts down on execution time. the computer application. In Figure.1., you can see a basic block diagram of the cache and central processing unit (CPU) placement for a generic single-core CPU.

Figure 1. Basic block diagram of cache [(Dan, A and Towsley. D. 2020)]

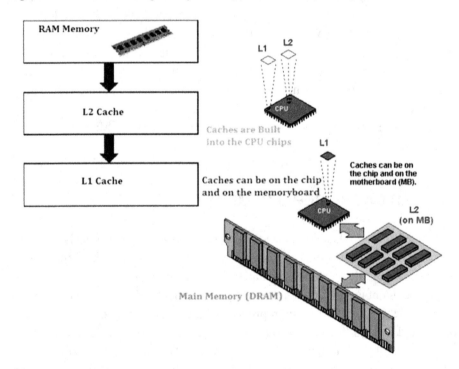

Usually, cache memory contains different layers of architectural placement for the CPU. They are called L1, L2, and L3, which are made up of different memory fabrics (Gang Chen, Biao Hu and Kai Huang. 2019). That is, they are built with different sizes of transistors and metal tracks and designed for various trade-offs between space, power, and speed. The main or internal cache is the level 1 (L1) cache. The processor chip itself contains it. It has a limited storage space of 8 KB to 256 KB. The level 2 (L2) cache is the following level and is less quick than the L1 cache. It is positioned within the CPU chipset packaging but is absent from the core. It typically has a storage size of 64 KB to 2 MB.

The level 3 (L3) cache is slower than L1 and L2 cache, which can be used optionally. If used, it is called as the last level cache (LLC). Last level cache is the highest-level cache that is usually shared by all the functional units on the processor. This LLC is separate from the processor chip on the motherboard (Gang Chen, Kai Huang and Jia Huang. 2019) (Gil Kedar, Avi Mendelson and Israel Cidon. 2019). The size often varies from 2 MB to 8 MB of L3 cache. The L1 cache is unique to each core, while L2 cache can either be separately designed for each core or designed as a common share core. The arrangements can be visualised, as shown below. L3 is usually designed as a shared cache.

Figure 2. Separate L2 cache for 2 cores [(Gang Chen, Kai Huang and Jia Huang. 2019)]

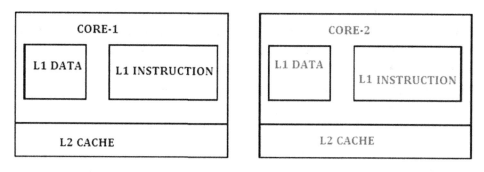

The memory cache is designed to have cache lines, which will hold data from 16 to 128 bytes, depending on the individual design. If we take an example of 256 KB L2 memory cache, to have 64 bytes as line size then, there would be a total of 4096 lines for the total 262144 bytes of cache space. If we have 1 GB main memory, then the main memory sections are divided into 4096 parts of 256 KB each.

The power consumption in the cache is more than that of processing core when considering the chip power consumption, as shown in work done by [(Giovani

Gracioli and Antônio Augusto Fröhlich. 2018)]. Hence it provides the research direction to reduce the power consumption in cache for low power system design. For explaining the effect of cache power reduction because of an increase in hit rate, the method of power measurement used in (Guang Suo and Xue-jun Yang. 2019) is explained along with the equations. When a new data is to be written in a cache segment, space needs to be freed up by eliminating the already existing data in that segment. One cannot randomly select data even though it is less complex to design the control logic involved because it would be detrimental to the execution time and power performance of the processor (Joan J Valls and Mar´ a E Gomez. 2019). A judicial selection of the target segment is important to maximise the utilisation of already stored data inside the cache. The FIFO replacement method for organising and manipulating a cache is analogous to the idea of the FIFO-Queue model. Since the oldest entry point is taken as a factor for removing from the cache, its frequency of operation is not being considered which a disadvantage? Hence this method is no better than the random replacement method. LFU-Here the frequency of use of the same data inside cache is counted and adds a value of the significance of the data. The data that are used least often are discarded first (Jongmoo Choi, Sam H Noh and Sang Lyul Min. 2012). Some of the relevant work done in the cache replacement policy based studies and experiments are explained here. (Juan Antonio Clemente and Rubén Gran. 2019) have given theoretical study results of how the cache hit rate can be increased with the standard least recently used replacement policy by giving extra information in the instruction set coding about whether to kill are keep the concerned data after that instruction. (Juan Fang and Jiang Pu. 2019) briefly explained the available and used replacement policies such as LRU, FIFO, and random in modern cache structures. Then they proposed new, improved replacement policies such as NRU, ENRU, and ENRUNew and evaluated their performance. They have included and calculated the power consumption on cache based on hits. Finally shown that taking the dirty blocks into account in cache policy will reduce the average power consumption by some percentage in mobile devices.

IMPROVED CACHE REPLACEMENT POLICY FOR EMBEDDED SYSTEM APPLICATIONS

In the above referred to previous works, the basic cache replacement policies are considering factors like the recentness of usage or frequency of usage. It needs to be studied with an emphasis on recentness, frequency as well as the total number of occurrences of the particular data in the cache (Juan Fang and Shijian Liu. 2019). Our work has been conceptualised as explained below. In the new proposed Least Recently Used-with Extended Tracking (LRU-ET) scheme, the least recently used replacement policy maintains the list of ranks based on the usage history of the data. If the rank is termed as LRUx, the x-denotes the rank for usage history of the data, which varies from zero to l-1. If x is zero, it means it is the oldest used data, which is going to be removed next and replaced with the new data. If the rank is l, it suggests that it is the latest user data (Junmin Wu, Xiufeng Sui and Yixuan. 2019). The value l depends on the size of the cache used. The other two ranks are known as Total Access Rank TARx and Re Occurrence Rank RORx. Total Access Rank TARx: Here, the total number of times the data is accessed and used is kept tracked, and the rank is given from zero to l-1. The minimum most total access would be one, and the maximum number of access would depend on the data and application (Kai Ma and Xiaorui Wang. 2019). The TAR value zero denotes that it is the least cumulative total among the compared data in cache memory. There can be equal ranks among the available data entries. Re Occurrence Rank RORx: considering the total number of data access is indicating both conditions where the data was reused from the cache as well as it was evicted and came back to the cache, it would help to track the eviction and re-entering for better management of cache replacement policy. The RORx is maintained as a list to rank the elements based on their number of reoccurrence. The x can vary from zero to l-1. There can be equal ranks among the available data entries at that moment based on their reoccurrence trend. Our new algorithm takes into account all the three ranks before evicting a cache line. The rank in all the three contexts is added, and the data with less number of total rank gets evicted. In this work, these rankings have been implemented in an architecture full system simulator and the results obtained to support the proposal.

BENCHMARK APPLICATIONS FOR SIMULATION

To imitate a typical embedded system load is complicated since the variety of domains to be tested is large. The complete diversity of the domains should be reflected in the selected set of benchmark applications. Unlike testing for the performance of desktop computers, the embedded systems should be checked for power consumption with equal importance, like performance. Also, there is a situation where a lot of different operating systems may be used in this domain (Mathew Paul and Peter Petrov. 2018). Hence the set of benchmark applications selected should bring out the individual unique result for embedded systems that are simulated on the above-said aspects. This work has utilised the MiBench benchmark suit readily accessible to academic researchers. In this study, the applications such as basic math, bit count, Qsort, and Susan are taken from Automotive and Industrial Control suit. The programs like jpeg-en, jpeg-dec, and typeset are taken from Consumer Devices suit. Dijkstra and Patricia are applications from Networking, and string search is from Office Automation suit. Adpcm and FFT are from the Telecommunications suit(Matthew R Guthaus and Jeffrey S. 2021). The telecommunications category stresses the importance of explosive growth of the Internet, which is used by many portable consumer devices integrated with wireless communication. A set of MiBench applications were run on the LPC3250 development board, a host machine equipped with Linux OS and a USB serial connector for power supply. A basic study of available data cache and instruction cache memory of the ARM926EJ-S processor was done in this experiment.

IMPLEMENTATION DETAILS FOR SIMULATION FRAMEWORK

The simulation is done in a cycle accurate simulator (Nathan Binkert *et al.* 2011) which can simulate system-level architecture as well as processor microarchitecture modifications. The gem5 event driven simulator consists of the configurable modules of CPU models, Instruction Set Architectures (ISA), I/O devices, wrap up infrastructure, interconnects and memory models. The simulation is done using ARM7 architecture modelled as two cores with individual L1 cache memories for instruction and data and a separate L2 cache memory for instruction and data for both cores. Once the simulation setup is configured, it can be run in two modes. The first one is syscall-emulation mode(Megiddo, N and Modha, DS. 2020). This mode is used for checking the system design and simulates the basic operation status of the target program which would be run on the experimental processor architecture. The second one is the full system mode of operation where it allows the user to load the Linux like operating system on the CPU under study. The down side of the full system emulation mode is that it is much slower than the syscall-emulation mode of operation. In both the modes of operation the individual components like CPU, memory, interconnects and buses are represented as C++ classes for a

behavioural simulation to be done on them. Python code is used for the total system configuration purpose on which any user written program is simulated. The program running on the simulated architecture uses a global clock figuratively known as ticks. Each subsystem which is being simulated refers to that global clock to maintain the synchronous events and update the user defined flags related to time (Musalappa, S and Sundaram, S. 2005). The memory related simulation is given by the m5 classical memory package which is written for multi-level memory hierarchy studies.

The details of execution about memory access timings along with the cache hit or miss statuses are written to a text file in the output directory during simulation.

The configuration of replacement policy is modified to reflect the LRU-ET. The new configuration rules are written as source code and header files by inheriting from the main SimObject file of Gem5 simulator. The new configuration is compiled for setting up fresh framework. The final simulation is functioning as a general execution-driven simulator. The step by step process of setting up the execution driven simulator is given in the Figure 3.

Figure 3. Flow chart for construction of simulation framework

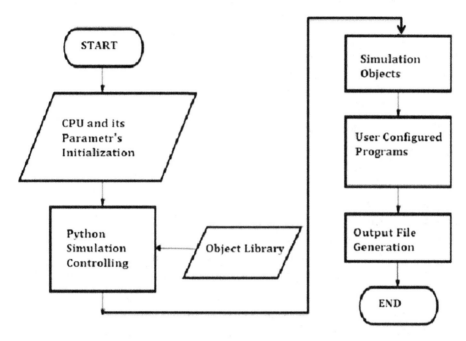

The build configuration can be customised for the required instruction set architecture using following steps. It is followed by process simulation steps. Pseudo code for creating config file in python:

Step 1:   import m5 from m5 objects
Step 2:   instantiate SimObject
Step 3:   Define clock domain
Step 4:   Define CPU type
Step 5:   Define system wide memory bus
Step 6:   Define d-cache
Step 7:   Define i-cache
Step 8:   Define memory controller
Step 9:   Set process command
Step 10:   Create root object
Step 11:   Print simulation details

Step 12:   Exit event

When the Cache SimObject is declared in the cache configuration python file, corresponding SimObject parameters are set one by one. When the SimObject is initiated the parameters are inherited from base Cache Object. The pseudo codes for creating the I-cache and D-cache are given below.

Pseudo code for creating L1 cache:

Step 1:   Define associativity

Step 2:   Define tag latency

Step 3:   Define data latency

Step 4:   Define d-cache size

Step 5:   Define i-cache size

Step 6:   Connect d-cache to cpu.dcache-port

Step 7:   Connect i-cache to cpu.icache-port

Pseudo code for creating L2 cache:

Step 1:   Define associativity

Step 2:   Define tag latency

Step 3:   Define data latency

Step 4:   Define d-cache size

Step 5:   Define i-cache size

Step 6:   Connect L2 cache to memory side bus.slave

Step 7:   Connect L2 cache to CPU side bus.master

Now with the build tools including gcc, the simulation framework can be built to work for studying various cache performance studies. The new SimObjects can be created as per our own definition as shown in the following steps.

Steps for creating new Sim Objects

Step 1:   Define new Sim Object in python class

Step 2:   Implement corresponding SimObject class in C++

Step 3:   Register the C++ and corresponding python files in SCons build system

Step 4:   re-compile the gem5 ISA framework

Step 5:   Modify config scripts to use new SimObject

Thus the framework construction is built for operating the cache with different access rules.

CACHE PARTITIONING

In this section, an effective partitioning method is proposed along with the LRU-ET would be combined to get a better performance of cache in terms of hit rate and hence provide a low power design (Nan Guan and Mingsong Lv. 2019). Embedded processors are many times custom designed by the industry for a particular field of application by a large number of resource people since the customisation may lead to higher enhancements in terms of performance, speed, and power. This section provides a background introduction to cache partitioning schemes, related works, and proposes a priority-based cache partitioning scheme implemented with the LRU-ET replacement policy.

One of the biggest challenges facing developers is to write software applications for MultiCore Processors (MCPs) with unbiased access to shared resources such as cache. MCPs considerably increase cache contention, causing the Worst-Case Execution Time (WCET) to exceed the Average-Case Execution Time (ACET). This becomes a major issue in safety-critical embedded systems programs to meet WCET targets (Nan Jiang and Junmin Wu. 2018). A possible solution is to employ an RTOS that supports cache partitioning, which enables developers to alleviate contention and reduces WCET. That will also maximise available CPU bandwidth without compromising safety criticality. Another solution is cache partitioning, which increases CPU utilisation by reducing WCET. Here the cores are managed by the RTOS so that the cache contention does not happen. Generally, cache partitioning is added with another merit-based enhancement on the memory design to provide highly efficient task handling. Any new policy defined such should be tested with standard workloads to test its validity. The cache partitioning techniques are mostly applied with LRU as the default replacement policy. Since we have the framework constructed for the LRU-ET replacement technique, it may be tested along with the new cache partition method in this work.

In the literature, we find a good number of works to alter the cache arrangement by partitioning techniques to yield speed or power efficiency for the embedded applications. (Giovani Gracioli and Antônio Augusto Fröhlich. 2018) proposed a cache partitioning scheme, named Bankaware, have designed a banking structure based L2 cache configuration and obtained miss rate reductions up to 43%. The design involves multiple complex arrangements of cache banks with each bank having different N-way partitioned arrangement. (Juan Fang and Shijian Liu. 2019) have employed a partitioning technique in which more cache is allotted to slower processes to increase their speed. Thus it decreases the useless waiting time and makes the applications more balanced. The running time was reduced by 7% on average using a balancing algorithm for cache partition. Thus the cache partitioning, when employed with regulated attributes about other factors of cache access, can give a good improvement in hit rate and revel in fair usage of the precious cache space.

CACHE PARTITIONING BASED ON PRIORITY

When multiple cores are involved in a processor, the last level cache is usually designed as a unified cache common to all the cores in the processor. The cache sharing policy can be defined among the processors based on many factors like minimum allotted space requirement, speed balancing based, WCET based, etc. We propose a fair cache partitioning method considering priority

index based allotment of cache space and applying the L RU-ET replacement method on it. Since embedded applications tend to run repetitive tasks based on its priority along with a better cache replacement algorithm is tried. Cache partitioning tuned to the priority of tasks by getting the priority information from the scheduler while running the program. Based on a set value of priority, if the priority of the task is above the cut-off, the task is allotted more space in the cache. For less priority, the cache space allotted is also reduced. This increases the performance of execution along with the hit rate increase in the application.

The two categories of tasks are allotted with different percentages of total cache locations. The high priority tasks were allotted with 70% of the L2 Data cache. The low priority tasks were allotted, with 30% of the L2 Data cache space. The cache partitioning in commercial RTOS uses LRU as the default replacement policy. Since we have the framework constructed for the LRU-ET replacement technique, it is readily available to test the new cache partition method in this work. Along with modification in cache partitioning, the LRU – ET was applied for L2 cache to improve the hit ratio. The Gem5 microarchitecture simulator can be run both in System Emulation mode as well as full system mode for simulating the instruction set architecture. The difference is that in System Emulation mode, the basic memory configurations can be tested without the involvement of operating system simulation. In the full system mode, a kernel can be loaded on to the simulated core, and a patch is to be run on top of the installed kernel to make communication of scheduling information available to the Gem5 simulator. The process of cache partitioning with the help of Linux kernel on the ARM7 instruction set architecture is run on the Gem5 simulator. The flow chart for the cache partitioning is shown in Figure.4.

Embedded processors are many times custom designed by the industry for a particular field of application by a large number of resource people since the customisation may lead to higher enhancements in terms of performance, speed, and accessibility power and heat management. These custom made processors may be upgraded and improved in terms of power consumption to get a better low power designed processor. A new method of coding the instruction set for the embedded system applications is presented in this chapter. Background theory of instruction decoding and instruction set design are given. The improvement in power consumption due to the new method of instruction set design is shown

Figure 4. Cache allotment for cache partitioning

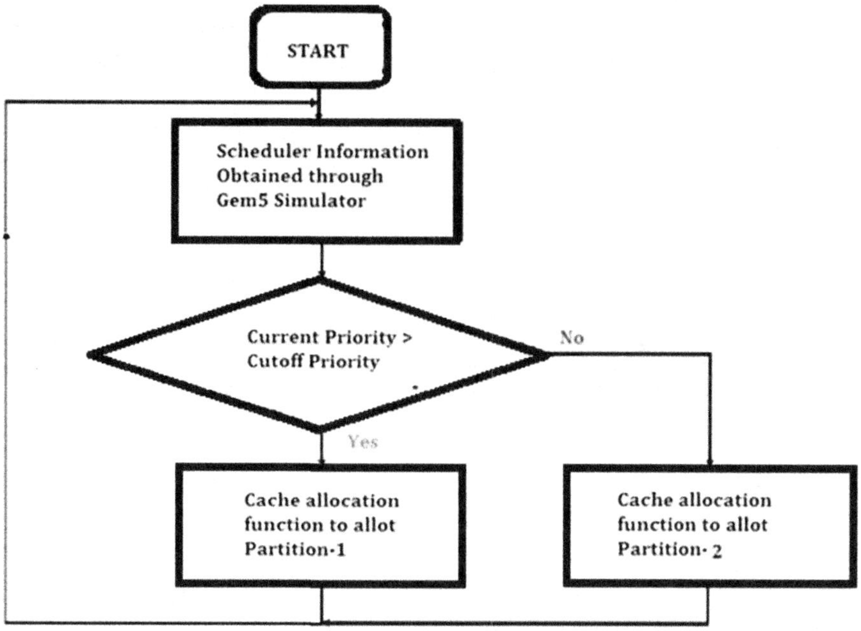

by studying typical embedded application programs on a 32 bit and 64 bit DSP processor architecture. Embedded processors exist in different varieties namely microprocessors, microcontrollers, DSP for signal processing tasks, ASICs which are designed for specific embedded tasks, Application Specific Instruction Set Processors (ASIP) in which the instructions are customised and System on Chip (SoC) in which hardware accelerators and one or many soft cores or hardcore processors are available on single chip. Out of these varieties of processors, the Commercial-Off-The-Shelf (COTS) processors cannot be modified by the end-user since they come with a pre-baked hard processor and all set data paths. But the softcore processors can be modified by the designers to suit the application they are going to serve. Also, in the case of ASIC and ASIP, the tuning of processor features, including instruction set modification is done if it is going to be a cost effective solution.

In any processor architecture, the programmers need a basic set of instructions to execute certain operations on data. The usual categories of instructions are arithmetic, logical, bit manipulation, flow control, on-chip component access, and I/O operations. Any extended components in the future members of the family of that architecture can be derived from the basic set of ISA. For placing the data inside CPU, the named registers may be used, which is encoded in opcode by certain unique combinations of binary numbers for each register. When the opcode is fetched and decoded, the

Figure 5. Lookup table-based instruction compression

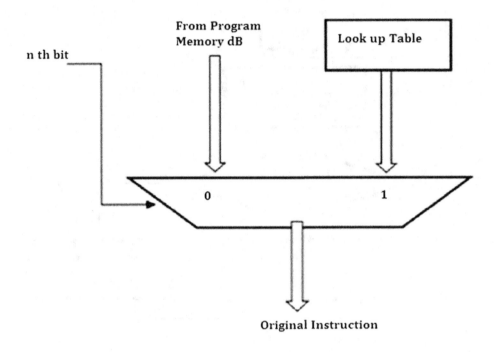

various pipeline stages help the data to be brought, computed, and taken back to result in locations. Since the opcodes are decided at design time, any combination of binary strings can be assigned to the types of operation and the registers. While circuit and VLSI level modifications are done for low power, research has been done at higher levels also. Let us see some of the opcodes, fetch, and bus modifications done in recent research to increase the power efficiency of the processor design.

(Juan Fang and Jiang Pu. 2019) have designed an architectural modification with variable-length instructions to reduce the power and energy consumption of custom processors used in embedded system applications. (Mathew Paul and Peter Petrov. 2018) have designed an instruction re-map table that reduces the flips in the program memory data bus. In work, a set of frequently used loops and a set of instructions in that loop are identified based on program pre-run, and those instructions are compressed to a four-digit code.

These instructions are stored in a lookup table. Whenever the original instructions are needed to be fetched by the ALU, these 4 bits are transferred in the program memory data bus instead of 16-bit regular instructions. An extra 17th bit is also transferred to the CPU for the fetching. cycle, as shown in Figure.5. That 17th bit is used to tag the special instructions, and so the design involves an increase in one digit in normal storage space for instructions. The single-digit tag separates the normal and compressed instructions while being stored in the main memory. During fetching, first, the tag is decoded, and the system decides whether to transfer instruction from the main memory or the lookup table. Because of this arrangement, the author has reduced code size by over 10% and the energy consumed in the program memory data bus by over 40%. One of the demerits of the system is 17 bits for the data bus instead of 16 bits. Since it is a theoretical presentation of design, the design overhead of memory and 17th bit inside CPU and the presence of the lookup table is not calculated. Also, it is not discussed about the increase in execution time for the encoding and decoding process in the instruction fetch and decode phase. (Megiddo, N and Modha, DS. 2020) have worked on a new processor architecture in ASIC. In which a pre-run of the program creates a profile of frequent instruction pairs. At the analysis stage, "frequent instruction pairs" are identified, selected, and encoded as short instructions. By changing the instructions and storing them in the lookup table, an alteration is done in instruction decoding. Decoding is done through lookup tables in the execution time. The decoding unit is added as an extra stage called "Depack" in the pipeline in between fetch and decodes stages. Encoding and decoding stages need extra hardware overhead to realise this method in the instruction decoding pipeline-fetch stage. In this method, they have achieved a 15% reduction in total energy consumption in buses. In this proposed new instruction set design, the basic aim is to reduce switching activity in the program memory data bus. The important instructions are selected based on the frequency of occurrence inside loops and modified using Loop Enhanced Gray code (LEGC) to reduce the bus power.

In the proposed design, the switching activity enhancements mainly target cyclic instructions. Hence the instructions inside loops in any given program is a desirable target to investigate instruction set modification. This formula explains the switching activity reduction and the power reduction in the bus. The Reflected Binary Code (RBC), also known just as Reflected

Binary (RB) or Gray code after Frank Gray, is a binary representation that has been tried successfully for code compression in embedded processors. Apart from low power design, Gray code is popular in applications like electromechanical switches, error correction algorithms, etc. In the Gray code, the ordering of the binary numeral system is done such that two successive values differ in only one bit (binary digit). Thus Gray code is designed to have minimum Hamming distance between successive entries. The resulting binary code reduces the digit flips, and so reduces the bus capacitance induced power loss in the data bus inside the instruction bus set. When a frequently used loop is selected for a particular application, and all the opcodes in that loop are converted into Gray code, the only minimum number of digits are required to express the full set of instructions inside the loop. For example, if the unique number of instructions in a loop is one hundred, instead

Figure 6. Gray code sequence selection using LEGC method

Instruction Code B_2 => Set 1 { 0,1}

Instruction Code B_4 => Set 2 { 0,1,2,3}

Instruction Code B_6=> Set 3 { 0,1,2,5,4,3}

Instruction Code B_8=> Set 4 { 0,1,2,5,6,7,4,3}

Instruction Code B_{10}=> Set 5 { 0,1,2,5,6,9,8,7,4,3}

of representing them with full length 16 or 32-bit instructions, it is enough to use seven digits. This greatly reduces the number of flips happening on the bus. If Gray code is used as it is for loops, then there is a discontinuity of usage of Gray code effect from the last instruction of the loop to the first instruction. It will not have a minimum Hamming Distance due to the jump in the continuity. This is the disadvantage identified and modified in the new code. Notice that the jump between the last instruction opcode 00010 and 00000 is following the Gray code rule. This is an advantage in an embedded systems design where frequently loops are executed for an enormous number of times. For example, in case of a simple loop that is going to be executed one lakh times, during the total execution of all instructions, including the transition from last instruction to the first instruction, for every time repeating the loop will provide only one single flip. This method of finding Gray code equivalent for looped instructions is scalable to any number of the total length of the Gray code. Figure.6. shows the sequence of the Gray code selected using the LEGC method for a loop of total length for 2,4,6,8 and 10.

For any 'n' even number of instructions, while n is even, this can be applied as such. In even number of 'n', we need to add a NOP instruction extra to get an even 'n'. To know about the loop information, the application was to be run in the integrated development environment (IDE) of the particular processor, and the assembly listing should be extracted. From the opcodes obtained, the important loops are selected for selecting the instructions to be modified into the Gray code. The extracted file of opcodes of the full application from the IDE is also fed to a program written to calculate the total number of flips in all the 32 (or 64) digits. The steps of instruction compression analysis are shown in the following Figure.7. and Figure.8.

The comparison of reduction in the number of flips and the total bus energy consumption is calculated using the above steps. Since the LRU-ET cache replacement

Figure 7. Calculation for number of flips in data bus before adopting LEGC

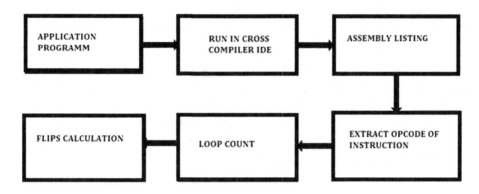

Figure 8. Calculation for number of flips in data bus after adopting LEGC

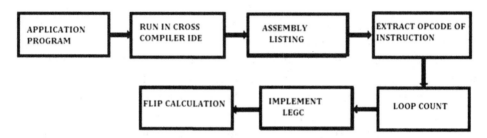

policy is given additional value of cache management based on the priority of a task, the application is tuned more suitable to the embedded systems domain. The simulation of an L2 shared cache applied for two sets of benchmark programs from MiBench set is shown in this section. The Cache Partition with LRU-ET (CP-LRU-ET) gives a hit rate based on the type of application. The Partitioning ratio was validated for two benchmark sets of programs and studied by varying the cache size. With the variation in cache size from 256 KB to 2MB, the improvement in CP-LRU-ET is showing an optimum cache hit rate for medium cache sizes. For smaller and larger cache sizes, the partitioning in this particular ratio is not positive in terms of hit rate improvement. The partitioning ratio among the two categories of tasks on priority can be varied and studied further. The WCET value can be obtained from the scheduling kernel, and the efficiency of tasks with respect to WCET can be improved along with replacement policy implementation.

Since the complexity and size of applications keep increasing in the embedded system applications, it is necessary to do design modifications to accommodate the challenges, including speed and power consumption. The problem of using the full potential of the on-chip components in the multicore environment also calls for improvement in system design. Cache memory optimisation techniques are used to achieve power and speed optimisations in

the embedded system applications. The low power aspect of the working of an embedded processor is nowadays an important metric. One of the approaches to get low power is optimally designing the instruction set of the instruction set architecture. The effect of cache on execution time and power consumption of the processors is also explained. The theory behind cache replacement and the major types of cache replacement policies are also explained.

CONCLUSION

When implementing the new cache replacement policy for a set of embedded systems benchmark applications, there was a clear improvement in the cache hit rate, and hence power consumption is reduced. The hit rate improvement is more pronounced for data cache when applied to data and instruction caches for both L1 and L2 level caches. The improved replacement policy was applied along with a new priority-based cache partition policy. The compiled simulation architecture was run with two different sets of benchmark applications. The results have shown to be optimal for the medium-sized caches in L2 shared cache. A new instruction set code modification was suggested, and the strategy was applied to the DSP processors for 16 bit and 32-bit instruction sets. The results show a tremendous decrease in power for the instruction fetch time data bus as per observations. The cache replacement architecture can be applied to more number of cores, and the effects can be studied. The study can also be done for combinations of cache size, coherency and number of way partitioning modifications. The cache partitioning algorithm can take into account of WCET deadline information about the tasks and can be added to the priority-based partitioning. This can also be tested on a heterogeneous multicore processor. The cache partitioning algorithm can take into account of WCET deadline information about the tasks and can be added to the priority-based partitioning. This can also be tested on a heterogeneous multicore processor.

ACKNOWLEDGMENT

We would like to acknowledge the support given by BMS Institute of Technology and Management, Bangalore, India.

REFERENCES

Balasa & Pradhan. (2012). *Energy-Aware Memory Management for Embedded Multimedia Systems.* CRC Press.

Bao, D., & Liourav, X. (2019). A Cache Scheme Based on LRU-Like Algorithm. *Proceedings of the 2019 IEEE International Conference on Information and Automation*, 2055-2060.

Chen, G., Hu, B., Huang, K., Knoll, A., & Huang, K. (2019). Automatic cache partitioning and timetriggered scheduling for real-time MPSoCs. *International Conference on ReConFigurable Computing and FPGAs (ReConFig14)*, 1-8.

Chen, G., Huang, K., Huang, J., & Knoll, A. (2019). Cache Partitioning and Scheduling for Energy Optimisation of Real-Time MPSoCs. *IEEE 24th International Conference on Application-Specific Systems, Architectures and Processors*, 35-41.

Choi, J. (2012). Design, Implementation, and Performance Evaluation of a Detection-Based Adaptive Block Replacement Scheme. *IEEE Transactions on Computers*, *51*(7), 793–800. doi:10.1109/TC.2002.1017699

Clemente, J. A., Gran, R., Chocano, A., del Prado, C., & Resano, J. (2019). Hardware Architectural Support for Caching Partitioned Recon gurations in Recon gurable System. *IEEE Transactions on Very Large Scale Integration (VLSI) Systems*, *24*(2), 530–543.

Dan, A., & Towsley, D. (2020). An approximate analysis of the LRU and FIFO buffer replacement schemes. *Proc. of the 2020 ACM SIGMETRICS Conf. on Measurement and Modelling of Comp. Systems*, 143-152.

Fang, J., Liu, S., & Zhang, X. (2019). Research on Cache Partitioning and Adaptive Replacement Policy for CPU-GPU Heterogeneous Processors. *16th International Symposium on Distributed Computing and Applications to Business, Engineering and Science (DCABES)*, 19-22.

Fang, J., & Pu, J. (2019). Dynamic Fair Cache Partitioning for Chip Multiprocessor. *Third International Joint Conference on Computational Science and Optimisation*, 2, 283-287.

Fitzgerald, B., Lopez, S., & Sahuquillo, J. (2018). Drowsy Cache Partitioning for Reduced Static and Dynamic Energy in the Cache Hierarchy. *International Green Computing Conference Proceedings*, 1-6.

Gan, Z.-H., & Gu, Z. (2019). WCET-Aware Task assignment and Cache Partitioning for WCRT Minimisation on Multi-core Systems. *Seventh International Symposium on Parallel Architectures, Algorithms and Programming (PAAP)*, 143-148.

Gracioli, G., & Fröhlich, A. A. (2018). An Experimental Evaluation of the Cache Partitioning Impact on Multicore Real-Time Schedulers. *IEEE International Conference on Embedded and Real-Time Computing Systems and Applications*, 72-81.

Guan, N., & Lv, M. (2019). WCET Analysis with MRU Caches: Challenging LRU for Predictability. *IEEE 18th Real-Time and Embedded Technology and Applications Symposium*, 55-64.

Guthaus, Ringenberg, Ernst, Austin, Mudge, & Brown. (2021). MiBench: A free, commercially representative embedded benchmark suite. *Proceedings of the Fourth Annual IEEE International Workshop on Workload Characterization*, 3-14.

Jiang, N., & Wu, J. (2018). Implementation of hardware-assisted virtual machine cache partitioning. *International Conference on Automatic Control and Artificial Intelligence (ACAI 2018)*, 1189-1192.

Juan, F. & Du, W. (2018). A Low-power Oriented Dynamic Hybrid Cache Partitioning for Chip Multi-processor. *2012 International Conference on Industrial Control and Electronics Engineering*, 369-372.

Kaseridis, D., & Stuechelix, J. (2009). Bankaware Dynamic Cache Partitioning for Multicore Architectures. *International Conference on Parallel Processing*, 18-25.

Kedar, G., Mendelson, A., & Cidon, I. (2019). SPACE: Semi-Partitioned Cache for Energy Efficient, Hard Real-Time Systems. *IEEE Transactions on Computers*, 66(4), 717–730. doi:10.1109/TC.2016.2608775

Lee, D., & Choi, K. (2019). Energy-efficient partitioning of hybrid caches in multi-core architecture. *22nd International Conference on Very Large Scale Integration (VLSI-SoC)*, 1-6.

Lindsay, A., & Ravindran, B. (2018). On Cache-Aware Task Partitioning for Multicore Embedded Real-Time Systems. *IEEE International Conference on High Performance Computing and Communications (HPCC)*, 677-684.

Ma, K., Wang, X., & Wang, Y. (2019). DPPC: Dynamic Power Partitioning and Control for Improved Chip Multiprocessor Performance. *IEEE Transactions on Computers*, 63(7), 1736–1750. doi:10.1109/TC.2013.67

Megiddo, N., & Modha, D. S. (2020). *Outperforming LRU with an adaptive replacement cache algorithm*. IEEE Computer.

Musalappa, S., Sundaram, S., & Chu, Y. (2005). Energy savings for data caches: ELRU-SEQ replacement policy. *24th IEEE International Performance, Computing, and Communications Conference*, 641-642. 10.1109/PCCC.2005.1460659

Paul, M., & Petrov, P. (2018). I-cache Configurability for Temperature Reduction through Replicated Cache Partitioning. *IEEE 8th Symposium on Application Specific Processors (SASP)*, 81-86.

Sanchez, D., & Kozyrakis, C. (2021). Scalable and Efficient Fine-Grained Cache Partitioning with Vantage. *IEEE Micro*, *32*(3), 26–37. doi:10.1109/MM.2012.19

Suo, G., & Yang, X. (2019). Balancing Parallel Applications on Multi-core Processors Based on Cache Partitioning. *IEEE International Symposium on Parallel and Distributed Processing with Applications*, 190-195.

Valls, Gomez, Ros, & Sahuquillo. (2019). A Directory Cache with Dynamic Private-Shared Partitioning. *IEEE 23rd International Conference on High-Performance Computing (HiPC)*, 382-391.

Wu, J., Sui, X., Tang, Y., Zhu, X., Wang, J., & Chen, G. (2019). Cache Management with Partitioning-Aware Eviction and Thread-Aware Insertion/Promotion Policy. *International Symposium on Parallel and Distributed Processing with Applications*, 374-381.

Yang, C., Liu, L., Luo, K., Yin, S., & Wei, S. (2018). CIACP: A Correlation- and Iteration- Aware Cache Partitioning Mechanism to Improve Performance of Multiple Coarse-Grained Reconfigurable Arrays. *IEEE Transactions on Parallel and Distributed Systems*, *28*(1), 29–43. doi:10.1109/TPDS.2016.2554278

Yu, C., & Petrov, P. (2019). Off-Chip Memory Bandwidth Minimisation through Cache Partitioning For Multi-Core Platforms. *Design Automation Conference*, 132-137.

Section 3
Low–Power Technologies in Energy Infrastructure

Chapter 14
Production and Use of Electric Vehicle Batteries

Hasan Huseyin Coban

iD https://orcid.org/0000-0002-5284-0568
Ardahan University, Turkey

ABSTRACT

Electromobility is considered the technology of the future due to its ecological and environmental advantages. Modern society relies on the movement of goods and people, but current transport systems have adverse effects on human health and the environment. There are also studies showing that the manufacture of batteries used in electric vehicles can also have a significant environmental impact. How large this environmental impact is affected by which batteries are used and their capacity, among other effects. At the same time, rapid development is taking place in the region, and information on environmental impact risks are rapidly becoming out of date. It is essential for lawmakers to provide up-to-date data on the environmental impact of the manufacture and charging of batteries and how infrastructure design affects the system. For the case study, an electric pickup truck belonging to a chain market in Turkey was used, and emissions from battery production and energy consumption were presented.

INTRODUCTION

On August 27, 1859, the world changed forever. After Edward L. Drake, a miner drilled the first oil well with machinery in Titusville, Pennsylvania, the oil mining industry began to develop rapidly (Christopher L. Liner & T. A. McGilvery, 2019; Ekejiuba, 2020). The oiled city grew rapidly, and everyone wanted to make a

DOI: 10.4018/978-1-6684-4974-5.ch014

fortune in black gold. The first American oil company was founded in 1854 as the Pennsylvania Oil Company. In 1859, when Colonel Drake found oil at 21 meters, many wells were drilled in the same area and production increased rapidly (Ekejiuba, 2020). In this way, the price of oil per barrel fell from $2 in 1859 to 10 $cents in 1862. Petroleum has started to play an important role in determining the economic and political conditions since it was produced for commercial purposes. Oil represented a seemingly inexhaustible source of energy that, along with a host of other innovations, sent man into the industrial revolution at rocket speed.

Day by day, the world has become more dependent on oil than ever before (Yu et al., 2020). Everything from the transportation sector to the energy sector was consuming more and more oil without considering the consequences. But the first data reports on the global climate began to be published and the results were shocking. The burning of more and more oil has resulted in an increase in the amount of greenhouse gases in the atmosphere, which has contributed to the greenhouse effect. The effect is an increase in temperature on the earth's surface. The infrared rays emitted by the sun reach the earth after a journey of millions of kilometers. Some of these rays hit the earth and heat the land and seas, while some of them are reflected to space after hitting the earth. However, greenhouse gases in the air absorb some of the infrared radiation, preventing them from escaping from the atmosphere. The more greenhouse gases in the atmosphere, the more heat is retained. As a result, the average temperature of the Earth rises. In recent years, climate research has made great strides forward, proving at times that the climate issue is something to be taken seriously. Especially with the signing of the Paris agreement (Paris agreement, 2015), most people agree that the world needs to reduce its dependence on oil to reduce greenhouse gas emissions and thus slow global warming.

Transport activities account for approximately 25% of the world's energy usage (Rodrigue, 2020). Emissions from fuels used for transportation activities as a whole account for approximately 20% of global greenhouse gases. Among the modes of transport, road accounts for 70% of global emissions, airlines 12%, sea lines 11%, and railway lines 2% (Civelekoglu & Biyik, 2018). Despite strict regulations on emissions in new vehicles and manufacturers' restrictions on internal combustion engines; the vast majority of vehicle manufacturers agree that the internal combustion engine has reached its maximum potential in terms of carbon dioxide emissions. Therefore, new technologies are needed for the transport industry. One of the technologies that have shown the most promise in recent years is electric vehicles (EVs). In order to drive these electric motors, batteries are thus required that can efficiently store large amounts of electrical energy. As a result, the world's vehicle manufacturers are facing one of the biggest changes ever (Rajaeifar et al., 2022).

Battery production is a technology-intensive and complex process that has seen constant improvements and innovations in recent decades (Nurdiawati & Agrawal,

2022). As more and more vehicle manufacturers choose to conduct research and development within these technologies, batteries will continue to develop. It can be said that the biggest difficulties in battery production are related to battery life, raw materials, chemistry, operating temperature, and energy density (Deng et al., 2020; Houache et al., 2022). It is important to have a good cooling system to prevent areas from overheating and that the lifespan of the battery does not deteriorate due to increased temperature in larger battery systems. This temperature control is also important regarding safety.

Today, the development is moving towards the electrification of vehicles, which can lead to a continued reduced environmental impact globally and locally. At the same time, some studies show that the production of the batteries used in EVs can have a significant environmental impact.

As of 2018, there were 5 million new EVs entering the auto ecosystem, but over the next 20 years, analysts expect this number to hit between 300 to 500 million cars. How will such a staggering number of vehicles be powered; the answer is obvious electric rechargeable batteries. This study will be exposed the global geopolitics of electric car batteries and how the promise of clean and renewable energy is hanging on the precipice of success and failure at the same time. EVs use powertrain batteries that supply energy to all the engine components for the vehicle to function as expected. while the value of powertrain batteries is not in doubt what is in doubt is the supply chain of these batteries and most importantly the raw materials used in producing them. The raw material used in making these batteries is lithium and its variants these include cobalt nickel manganese and several other materials. Currently, the material dominating the market is cobalt manganese material which is used by many electric auto companies. Tesla on the other hand uses a combination of lithium cobalt manganese. Cobalt still represents the most essential mineral without which EV powertrains will be difficult or impossible to make using current technology. Lithium is made in Argentina, Chile, and Bolivia; these three South American nations hold significant deposits of the world's lithium reserves. Cobalt on the other hand is mined mostly in Central Africa in the Democratic Republic of Congo.

The goal of the study: The purpose of this study is to investigate the greenhouse gas emissions as well as the energy usage that occurs during the generation and charging of batteries, depending on which system design is chosen. The aim is to shed light on the manufacturing technical aspects and production methods used in industry today with regard to economy, energy, and materials. The results from the study can be used by traffic authorities in future infrastructure and vehicle procurement.

The method of the study: This study consists of two different parts. In the first part, a literature study focuses on the life cycle and analyzes the production and use of batteries, different charging solutions, and recycling or reuse opportunities have been made. In the second part, greenhouse gas emissions and energy consumption are calculated for various battery and charging infrastructure combinations in Istanbul city for the grocery van.

ELECTRIC VEHICLES IN TURKEY AND GLOBALLY

According to BloombergNEF's 2021 overview, there will be 169 million EV on the road by 2030 (*BloombergNEF Report*, 2021). While getting out of the vehicles as an essential element to achieve the net-zero goal of the countries. Today, in Norway, where roughly 98% of the electricity used is renewable and the share of plug-in vehicles is 84.6%, it is clear that the clean air created by these influences delights drivers and that these developments help protect the planet (Halsnæs et al., 2021). The UK has committed to banning internal combustion car sales by 2030. To replace its 31.5 million vehicles, about 236,000 metric tons of lithium carbonate are needed. Both people and countries from different geographies now want the products they consume and use to be more sustainable and cause lower carbon emissions.

The chain effects of the energy crisis, which emerged in the Covid-19 epidemic and continued with the Russia-Ukraine crisis, are felt all over the world. While 2.3 million EVs were sold in Europe, this figure grew by 66% compared to 2020. Figure 1 shows the number of EVs in different countries (Paoli, 2022; *Turkish Statistical Institute*, 2021). After the sales in 2021, the total number of EVs in the world reached 16 million (TRT Haber, 2022). In the European Union, the share of electric cars in the total market reached 10% in the first quarter of 2022. Accordingly, 36% of the cars sold in EU countries were gasoline, 25.1% hybrid, 16.8% diesel, 10% all-electric (BEV), and 8.9% plug-in hybrid. (PHEV), 3% other fuels and 0.2% natural gas. The highest number of electric vehicles in Europe is in France with 155,000. Germany, in second place, has more than 136,000 EVs. In the Netherlands, the number is 106,508.

The domestic and national EV to be produced by Turkey (TOGG) is expected to be on the roads in 2023. According to Eurostat, the number of EVs in Turkey was 1,176 in 2019. As of 2021, there are 4,869 EVs in Turkey.

PRODUCTION OF THE BATTERIES

William Turlington, Head of Mining, Metals and Industries Finance, Americas for Societe Generale stated that batteries will become a vital component of energy strategy not only for specific industries but for all countries and the world as a whole (Financial Times, 2022). EVs are projected to require 5,000 GWh of capacity by 2040, suggesting a Compound Annual Growth Rate (CAGR) of around 17%

Figure 1. Electric car registrations and sales share, 2016-2021

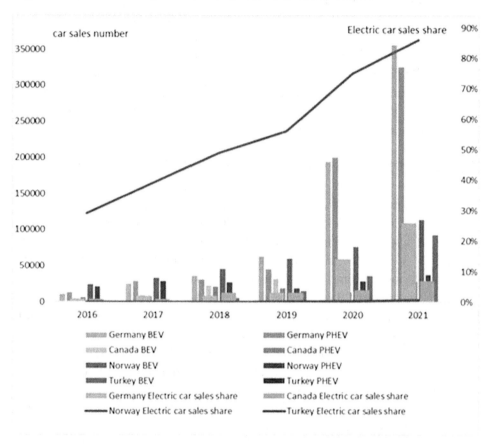

(Ibarra-Gutiérrez et al., 2021). Although energy storage demand is much smaller, it is expected to increase by 11% between 2020 and 2040, from 51 to over 420 GWh.

While the industry and its issues may be complex, the way in which battery-grade lithium is produced is not. Four countries dominate the industry—Argentina, Chile, Australia, and China combined account for 92% of the globe's production. The metal is extracted from the ground at massive sites like the Greenbushes mine in Western Australia, which is the world's largest hard-rock lithium mine. According to the 2021 BP Statistical Review; China, which has only 7.9% of the world's lithium reserves; has spent over $60M to build the lithium industry over the last 10 years. This allowed China to build a solid lithium supply chain and outpace the United States. Considering that EVs will constitute more than 25% of the vehicle market by 2030, it is estimated that many countries will compete for lithium in the near future.

Electric car batteries are most economically based on lithium hydroxide, which is derived from a hard rock called spodumene (Ibarra-Gutiérrez et al., 2021). As demand for this core battery material has increased, the price of Australian spodumene, one of the world's top three mineral resources, has risen from $500 per ton in 2018 to $2,500 per ton in 2021, according to global Benchmark Mineral Intelligence (Australian trade and investment commission, 2018).

As demand increases, the current critical need is to maintain an adequate supply of raw materials to manufacture batteries. This section presents the most common battery types, key terms, energy consumption, and greenhouse gas emissions that may occur during the manufacture of these batteries.

Chemical Structure of The Most Common Battery Types

Battery cells mainly consist of an anode and a cathode as well as a conductive electrolyte (Winter, 2009; Winter & Brodd, 2004). Lithium-ion batteries, the anode consists of graphite, electrically conductive carbon, binder, and additives. The structure of the graphite makes it possible to store lithium ions that move towards the anode during charging. But that's all the graphite does, so it would be possible to reduce this volume and thus increase the energy density in the battery. What many manufacturers do today is to add a small amount of silicon, which makes it possible to store more lithium ions. Silicon is both easier and cheaper to obtain, but there's a reason the anode is only made of silicon (Turcheniuk et al., 2018).

When it comes to the cathode, much of the research goes into developing new variants of its material set. Currently, NMC (lithium-nickel-manganese-cobalt) is the most popular in the EVs market, while Tesla uses NCA (lithium-nickel-cobalt aluminum) instead. Cobalt is by far the most expensive material, which has caused companies to try to develop cathodes without cobalt as a component. NMC typically consists of one-third nickel, one-third manganese, and one-third cobalt (Can, 2021; Chen et al., 2006; Uysal & Gul, 2017).

The Structure of The Lithium-Ion Batteries

It's only been a few decades since lithium-ion batteries reached commercial feasibility but, in that time, they have become the power source of choice for portable electronics thanks to their perfect blend of safety and lightness. However, the latest major tech boom, the latest infatuation of Wall Street and Silicon Valley alike, is centered around the largest consumer electronics product to date: electric vehicles. A lithium-ion battery consists of 3 basic parts cathode, anode, and separator plate (Zhou et al., 2022). When the battery is charged, the electrically charged lithium atoms collect in the carbon layer, and after removing them from the charge, the electrons move towards the cathode. During the realization of this cycle, electromagnetic power is produced. The most common battery type at the moment is so-called Lithium-ion

Table 1. Common battery chemistries

Cathode material	Comment
$LiFePO_4$ Lithium iron phosphate (LFP)	Energy and power-optimized batteries
$Li_4Ti_5O_{12}$ Lithium Titanate (LTO)	Power-optimized batteries
$LiNi_{0.8}Co_{0.15}Al_{0.05}O_2$ Lithium nickel cobalt aluminum oxide (NCA)	Energy-optimized batteries
$LiCoO_2$ Lithium cobalt oxide (LCO)	Mainly in small-scale electronics
$LiNi_{0.33}Mn_{0.33}Co_{0.33}O_2$ Lithium manganese cobalt oxide (NMC)	Energy-optimized batteries
$LiMn_2O_4$ Lithium manganese oxide (LMO)	Historically used in EVS, now less common

batteries (T. Kim et al., 2019; Manthiram, 2020). Compared to other batteries, they have high operational reliability and at the same time a commercial maturity. Li-ion batteries can in turn be divided into different battery chemistries (LCO, LMO, LFP, LTO, NCA, and NMC) which have major differences between them. The different types of battery chemistry affect whether the batteries are most suitable for storing a lot of energy or for being able to be charged quickly with high power, see Table 1. (Jung et al., 2021; Walvekar et al., 2022)

Table 2 shows a typical composition of a battery and shows that the battery cell as such accounts for roughly 60% of the total weight. The corresponding estimate is presented by Romare and Dahllöf who state that the battery cell corresponds to 55 – 60% of the total weight (Romare & Dahllöf, 2017). The active material in the cathode and anode respectively accounts here for 20% and 10% of the total weight. In addition to the active materials in the battery cell, the battery pack also consists of aluminum, copper, plastic and steel, and more. Aluminum, copper, plastic, and steel account for over 60% of the total weight, while manganese, cobalt, nickel, and lithium together account for around 10%.

Several different parameters are used to describe the characteristics of a battery, from how deeply it can be discharged without compromising its performance, or from its storage capacity and lifetime to how fast the battery can be charged. Some of these key concepts are presented below.

Storage Capacity

It is how much energy can be stored in a battery. In automotive applications, it is also relevant to talk about energy density. That is, how much energy can be stored per kg battery pack.

Table 2. Typical weight distribution between different components in a battery pack

Component	Percent Mass (%)
Anode	15 – 24
Cathode	29 – 39
Separator	2 – 3
Cell Casing	3 – 20
Electrolyte	8 – 15
Battery management system	2
Battery Pack Casing/Housing	17 – 23
Passive Cooling System	17 – 20

Power Capacity

It determines the power that can be drawn continuously from a battery in amps. The capacity depends on the surface area of the plates, their number, and the permeability of the separators used.

State of Charge (SOC) and State of Health (SOH)

It is the amount of electrical energy stored in a battery at a given time, defined as the percentage of energy when fully charged. When a new battery is fully charged, the SoC is considered 100% of its rated capacity. On the other hand, when fully discharged, the SoC is considered 0%. SoC information is a criterion to be considered for the charge control method and battery balancing systems of EVs.

Depth of Discharge (DOD)

One of the most important criteria to be considered for the batteries we use to store energy and take it from there when we need it to be long-lasting and trouble-free is the depth of discharge. Each battery has a discharge depth recommended by its manufacturer, that is, "safely usable capacity", depending on the brand, model, and technology used. This capacity is a criterion to be considered for the battery to be the most efficient and long-lasting. The discharge depth is the "useful capacity" of the battery that you can use without damaging it, without losing its capacity and life. You can go below this value, but in this case, the charge-discharge cycle of your battery will be shortened.

Table 3. The cycle life for different battery chemistries

Battery chemistry	Cycle life			
	(Peters et al., 2017)	(Battery University, 2022)	(Ambrose & Kendall, 2016)	(Dai et al., 2019)
LFP	2960	2000	2000-4000	3200
LCO	900	500-1000	500-1000	NA
LMO	1268	300-700	300-700	685
NCA	2200	500	500	1000
NMC	1217	1000-2000	1000-2000	1700
LTO	13850	3000-7000	3000-7000	5000

Lifespan

A battery's cycle life ultimately indicates how many charges cycles the battery can handle under certain conditions before it no longer meets the requirements for a specific application. For batteries used for EVs, it is common for the battery to be used until its storage capacity is less than 80% of the original.

Aging mechanisms can be divided into two categories, degradation caused by mechanical stress and degradation produced by chemical reactions (C-rate (Brückner et al., 2014), temperatere (Ma et al., 2018), région of opération (Egan et al., 2007), calendar aging(McBrayer et al., 2021)). Aging is the phase change process of the lead parts in the battery depending on time. The number of charge cycles (cycle aging) is partly influenced by DOD and how quickly the battery is charged and discharged, but also by parameters such as temperature and humidity.

Different types of batteries are affected to different degrees by cycle aging. Table 3 summarizes data from an extensive literature study as well as the cycle life calculated in the same study for several different battery chemistries. Values for a DOD of 80%.

Solid-State Batteries

As demand for EVs and their batteries increases, getting more cobalt and lithium will be incredibly difficult. But to decarbonize the transport sector, solutions must be found. One option, rather than finding more raw materials, is to need less of them. Of course, the way to do that is by making batteries better. The most promising short-term innovation that could fulfill that mission is solid-state batteries. Whereas traditional EV batteries have a liquidy, viscous lithium-based electrolyte, solid-state batteries rather use a solid, metal composition as their ion transport mechanism.

Solid-state batteries can also be made without cobalt or nickel, which eliminates two problematic and costly necessities in current battery tech. Most significant, however, is solid-state batteries' higher energy density. Traditional lithium-ion compositions used in EV battery packs store about 250Wh of energy per kilogram. That means 1kg of battery could move a Tesla Model-3, 1.3 kilometers. Meanwhile, it's expected that solid-state batteries will be able to store between 400 to 500Wh per kilogram—essentially doubling battery density. That means Tesla could halve the weight of their half-ton battery pack and not only keep the range the same, but increase it as the car would no longer need to carry the rest of the weight of the battery pack. Also, experts believe that, at scale, the production costs of solid-state batteries could be even less than the cheapest current lithium-ion batteries. Most predictions place that enticing end goal more than a decade away. Even if the solid-state battery transition reaches fruition earlier, the world will still need a whole lot more lithium.

Energy Consumption in The Production of Batteries

The production of battery cells is a complex process with strict quality requirements and high energy consumption. In comparison, the final assembly of battery modules, control system, and cooling system into a battery pack is significantly simpler, and not much energy is used here. There are many different studies (Dai et al., 2019; Davidsson Kurland, 2020; Degen & Kratzig, 2022; Degen & Schütte, 2022; Emilsson & Dahllöf, 2019; H. Wu et al., 2021) that try different ways to calculate how much energy is required to produce different types of batteries. However, many studies rely on various forms of simulation and prediction (Thomas et al., 2018). The energy requirement is higher if the factory is located in an area with warm/ moist air compared to an area where the air is cold/dry. The energy consumption is also linked to the air volume rather than the production as such, which means that capacity utilization can be of great importance for the result (Dai et al., 2019).

Table 4 presents the results from three different studies that show how energy consumption varies depending on battery chemistry. Energy consumption for battery manufacturing with current technology is about 350 – 650 MJ/kWh battery (Romare & Dahllöf, 2017). Also, in a recently published study (Benakli, 2022) the energy required to produce NMC batteries in the USA at 520-620 MJ/kWh.

Emissions of Greenhouse Gases During the Production Of Batteries

As the previous Chapter described, the literature shows a large variation in how much energy is used to produce batteries. Correspondingly, there are also large

Table 4. Mean Cumulative Energy Demand for different battery chemistries

Battery Chemistry	Cumulative Energy Demand for battery production (MJ/kWh)		
	(Benakli, 2022)	(Khan, 2021)	(Romare & Dahllöf, 2017)
LFP	860-1090	970	300-2500, average 970
LCO	NA	990	NA
LMO	NA	810	200-1500, average 810
NCA	550-660	1510	NA
NMC	520-620	1030	500-2000, average 1030
LTO	NA	1900	NA

variations in greenhouse gases emitted in relation to the production of batteries. In the literature (Thomas et al., 2018) a range of $38 - 356$ kg CO_2-eq.(equivalent)/kWh battery, (H. C. Kim et al., 2016) 140 kg CO_2-eq./kWh or 11 kg CO_2-eq per kg of battery, (Ellingsen et al., 2014) 172 kg CO_2-eq./kWh is identified. Table 5 reports the results of an extensive literature study. There, too, it appears that there are large variations in the literature. However, the average values stated are very much in line with the $140 - 170$ kg CO_2 eq./kWh stated earlier.

At the same time, the IEA (2020) states that in their analysis they use a value of 65 kg CO_2 eq./kWh and a range of approximately $50 - 150$ kg CO_2 eq./kWh (IEA, 2020). This value also agrees well with the value of 63.4 kg CO_2-eq./kWh LMO battery calculated by (Amarakoon et al., 2014) based on the same calculation model as in IEA (2020).

The Influence of The Current Electricity

The energy required to manufacture the batteries can be affected by the location of the factory, as places with hot and humid weather can increase electricity needs.

Table 5. Emissions of greenhouse gases per kWh battery (Amarakoon et al., 2014)

Battery chemistry	Emissions of greenhouse gases (kg CO_2 eq./kWh battery)
LMO	63.4 (Amarakoon et al., 2014), 55 (Li et al., 2022)
LFP	151 (Amarakoon et al., 2014), 16.11 (Koh et al., 2021)
LTO	180 (Temporelli et al., 2020), 14.19 (Koh et al., 2021)
NCA	66 (Li et al., 2022)
NMC	73 (Li et al., 2022)

Figure 2. Carbon intensity of electricity generation, 2021

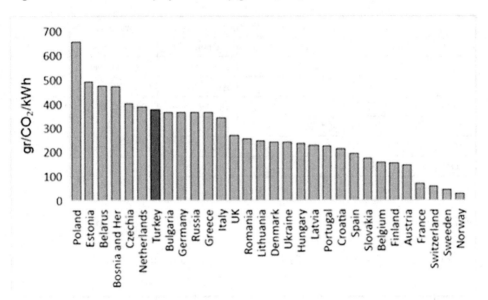

Accordingly, greenhouse gas emissions are highly influenced by how the electricity used is produced, and this is linked to the location of the factory. Figure 2 presents greenhouse gas emissions per kWh of electricity in European countries (Climate Transparency, 2021; Hannah Ritchie et al., 2020). It shows that the emissions vary considerably between 657g CO_2-eq./kWh in Poland to 26g/kWh in Norway and the range is largely linked to the energy used in production. The higher value applies to an electricity generation fuel type with a high proportion of fossil energy and the lower to a production based on renewable energy. The emissions can therefore vary considerably depending on a number of different parameters such as the geographical location of the factories and capacity utilization but also the energy density in the battery and which process energy is used.

Use of Batteries

Each EV may look different depending on electricity consumption. It could be about what capacity the vehicle should have in terms of the number of passengers and whether it should be used in urban traffic or the countryside. Also, parking lots are depots where vehicles can be parked, maintained, and refueled. In cases where the vehicles use liquid or gaseous fuels, refueling normally takes place at the depot in a short time (3-5min). In these cases, the vehicles are deemed to be able to carry enough fuel to cover a day's or week's use without their use is restricted to any great

Table 6. Charging type of EVs

Charging Type		Comments
Plug-in charging		It is flexible and most reminiscent of conventional vehicles and is therefore usually seen as the easiest to introduce in the planning of routes.
Additional charge	Charging at the parking lots	In order to be able to charge at the time of the parking, it is necessary to configure it on the roads, in the parking lots, in the restaurants, and in the shopping centers.
	Charging while traveling	Use electric roads (Coban et al., 2022), which are similar to wire charging but could be used by more actors than just trucks. This could provide the opportunity for smaller energy storage units and more actors who can share the costs. At the same time, it requires a very large expansion of the charging infrastructure.

extent. When it comes to long-haul trucks and buses, the conditions are somewhat different. Even if they are all refueled with the same electricity, there are several different types of charging solutions that affect everything from which batteries can or should be chosen to planning schedules.

Charging EV Batteries

The choice of charging solution has a major impact on the demands placed on the battery's energy storage capacity and charging power. The charging solution can also affect which type of battery chemistry is best suited. As summarized in Table 6 different ways to charge an electric vehicle can be roughly divided into plug-in charging and additional charging.

There are more options for electric vehicle operation that may become relevant in the future, but which are currently not used. One such option is battery replacement. However, this would require more batteries and the size and interface of the batteries to be standardized.

Effects of Battery Pack Weight

As described earlier, different types of batteries can have different energy densities (Wh/kg) (Y. Wu et al., 2020). If a battery is expected to have a certain storage capacity, it will therefore weigh differently depending on the chemistry chosen. Table 7 summarizes the energy density for a number of different types of battery chemistries.

Table 7. Electrical efficiency and energy density of the batteries

Battery Chemistry	Efficiency	Energy density (Wh/kg)
LFP	92%	100
LCO	91%	140
LMO	93%	120
NCA	92%	130
NMC	94%	170
LTO	93%	100

Charging and losses and effects of battery weight

While the batteries are being charged, losses occur partly in the battery and partly in the battery charging infrastructure (MacHiels et al., 2014). When the battery is charged, certain losses occur in the form of heat, which can vary slightly depending on the battery's chemistry. In Table 7, possible losses in the literature are summarized. However, it should be kept in mind that these values are theoretical and losses in practical operations may look different. For comparison, the battery test center (2022) reported that the efficiency of the battery is 95-96% under normal use (Battery test centr, 2022; Peters et al., 2017).

Different types of batteries can have different energy density (Y. Wu et al., 2020). If a battery is expected to have a certain storage capacity, it will therefore weigh differently depending on the chemistry chosen.

When comparing different batteries and charging solutions, it may also be important to consider the weight of the battery itself and how the vehicle affects fuel consumption. How big the effect is can vary between different vehicles and driving. Bi et al (2015) have demonstrated that 10% vehicle mass reduction contributes to about 4.5% energy consumption reduction for battery-powered EVs (Bi et al., 2015).

The energy consumption of EVs can vary greatly depending on different circumstances. Figure 3 gives examples of electric vehicles' energy consumption based on a previous study (Coban et al., 2022). An important factor that affects energy usage is topography, where higher energy use has been measured in hilly cities. Energy consumption is also affected by the driver's driving style.

AN ELECTRIFIED GROCERY VAN

The manufacture and use of batteries require energy inputs and cause greenhouse gas emissions at various parts of their life cycle. Therefore, when comparing different

Figure 3. Energy demand for electric vehicles

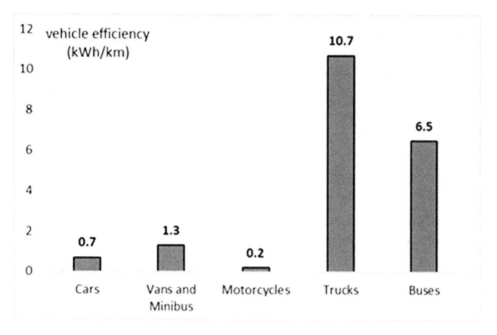

batteries and charging solutions, it is important to consider the direct effects as well as the indirect effects of different solutions over the entire lifecycle. In this study, a general calculation of how large the greenhouse gas emissions are for which different system solutions can be given is presented. As an example, the electric grocery van in Istanbul is used and the calculations are based on data from a number of different studies found in the literature and presented in the previous section. These studies were carried out under different conditions and assumptions, but the data are considered to have sufficient accuracy for the outcome and differences between the different options.

One of the most central concepts in a life cycle analysis is the objective function which is the magnitude to which all emissions and all energy usage are related. Therefore, it is important that the objective function is chosen so that it reflects the benefit to be analyzed. Regarding the battery, there are many studies where the results are reported per kWh of storage capacity. In order to be able to compare batteries with different performances that are used in different ways, it may be more relevant, for example, to report the result per kWh of electricity delivered to or from the battery. The aim is to compare the electrified transportation system with other fuels and systems, and the efficiency and capacity of the vehicles should also be included.

First, a base case is presented based on a grocery chain delivery van in Istanbul city, including battery production and electricity usage. Based on these data, energy

Table 8. Data for the operation of market pickup truck

Battery chemistry	LTO
Battery size	200 kWh
Charging capacity	11 kW
Line length/day	80 km
Annual Mileage (km)	24960 km
Electricity consumption (kWh)	32448 kWh
Electricity consumption	1.3 kWh/km
Greenhouse gas emissions for battery production	120 CO_2-eq./kWh

consumption and greenhouse gas emissions per km are calculated. On the other hand, the importance of battery life on energy consumption and emissions is demonstrated.

Data and Assumptions

Data on battery, mileage, charging solutions, and total electricity consumption are summarized in Table 8.

The van is equipped with LTO batteries with a storage capacity of 200 kWh. Based on the reasoning presented in Chapter 3, it is assumed that 1900 MJ/kWh is required to produce the battery and that greenhouse gas emissions amount to 100 kg/kWh battery. As a sensitivity analysis, it is also shown how the result is affected if the emissions of greenhouse gases vary between 14.19-180 kg CO_2-eq./kWh.

In this study, emissions and energy consumption are calculated based on Turkey's electricity emissions data. The emission intensity of the Turkish energy sector has been determined as 375 grams/kWh (see Figure 2) (Climate Transparency, 2021).

In Chapter 3, the energy consumption in its production per kWh of storage capacity and the emissions that may occur are explained. However, when comparing different types of batteries and charging solutions, it is also important to consider how long the battery meets the quality requirements. However, there appears to be a clear difference between different battery chemistries, just as with energy consumption and emissions. To calculate emissions and energy usage per km of electricity supplied, one must know how much electricity is supplied over the life of the battery. The batteries used in this study are assumed to have a lifespan corresponding to 150,000–250,000 kWh. With an electricity consumption of 32,448 kWh per year, this corresponds to 4 – 5 years. Given a 200-kWh battery with 80% DOD, this corresponds to about 1000 cycles.

However, the pickup truck is never operated to reach 80% DOD. It is assumed that the SOC of the pickup truck battery does not drop below 75% during the summer

Figure 4. Annual greenhouse gas emissions of electricity supplied to the battery.

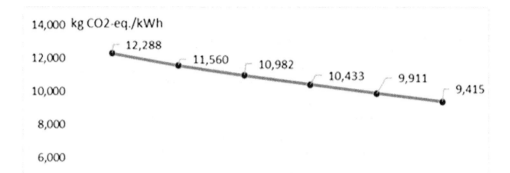

months and only in exceptional cases below 70% during the winter months. The corresponding maximum DOD is thus less than 25% and 30% respectively.

Greenhouse Gas Emissions

In this study, it is assumed that a 200 kWh LTO battery is selected for the chain grocery van. As a base case, it is assumed that battery production produces 120 kg of CO_2-eq./kWh batteries. The sensitivity analysis shows how the result changes when emissions vary between 60 – 180 kg CO_2-eq./kWh. It is assumed that the batteries can be charged with 150,000 – 250,000 kWh during their lifetime. Assuming that the batteries can be charged with 200,000 kWh during their lifetime, the emissions from the production of the battery will be 45 grams of CO_2 eq./kWh of the electricity supplied. The electricity used to run the pickup truck has an emission factor of 375 grams/kWh under Turkish conditions, the total emissions equivalent to 420 grams/ kWh of electricity supplied. With an average electricity consumption of 1.3 kWh per km, it corresponds to 546 grams/km. Figure 4 shows how greenhouse gas emissions vary depending on the lifetime and how large the greenhouse gas emission required by battery manufacturing is.

The calculations presented in this study are based on literature data. However, there are many ways to electrify a vehicle and the solution implemented in Istanbul

Figure 5. Greenhouse gas emissions (grams/km) depending on battery size

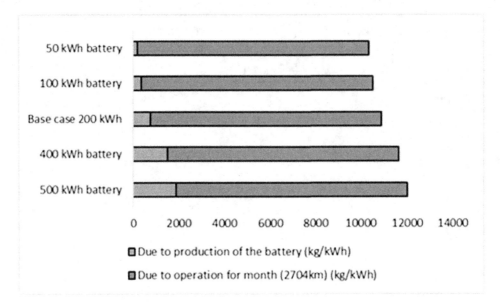

may not be applicable elsewhere. It is even possible that this study was designed elsewhere in a similar way.

An alternative design is to use another vehicle or battery type, which may have other features. However, based on the literature study in this study, it was not possible to detect decisive differences in terms of energy usage and greenhouse gas emissions. Presumably, different battery chemistry could provide a different lifespan. However, a change in battery chemistry was not taken into account in this study, as information on the lifespan is very uncertain. However, it should be noted that in addition to these aspects, other batteries may contain other materials. For example, some batteries contain more or less manganese, nickel, or cobalt.

The most fundamental consideration when comparing different battery and charging solutions is how battery life is affected. It can be stated that the number of charge cycles and the depth of discharge is one of the few factors that affect the life of the battery. In this study, it was not possible to determine a clear relationship between discharge depth and life. Therefore, it is assumed here that the batteries can produce 200,000 kWh over their lifetime. Given this assumption, the results from the calculations to show how greenhouse gas emissions may vary depending on battery size and charging strategy are summarized in Figure 5.

Figure 5 shows how greenhouse gas emissions vary with battery size. As expected, the total emissions appear to be directly related to the size of the battery. If the pickup truck is only plug-in charged, the batteries increase the emissions even more for the winter. Emissions will be significantly reduced when the pickup truck is equipped with a smaller battery and continues to be charged while driving or in short parking. The total emissions therefore largely depend on the size of the battery.

If the pickup truck is equipped with a smaller battery but is still taken to the charging station to be recharged, the extra driving distance must be calculated from a greenhouse gas perspective. This means that the truck goes to the charging station once or several times a day. In the basic case, the calculations are based on here, an electric pickup truck produces 487 grams of CO_2 eq./km. With a driving distance of 80km per day, this corresponds to 39 kg CO_2 eq./day.

DISCUSSION AND CONCLUSION

An electric pickup truck based on EVs can be designed in many different ways. Among other things, it could be about what kind of batteries are used in vehicles, how big the batteries need to be, how and where they need to be charged, and what requirements there are in terms of capacity, and flexibility. Therefore, which solution is most suitable can vary considerably between different uses. When comparing different alternatives, it is also important to clearly define which parameters should be considered and what weight should be given to the different parameters.

The manufacture of batteries requires energy and causes greenhouse gas emissions. Especially when it comes to greenhouse gas emissions, the data described in the literature show that there are great differences between different studies. It is seen that the difference arises from a number of different aspects such as where the production will be made geographically, how high the capacity usage is in the factory, which energy sources it will be made with, and the size of the factories. At the same time, overall, more recent studies seem to report lower emissions data than older ones.

One of the aims of this study has been to compare different battery designs and charging infrastructure to see how long the batteries can last will affect the result. However, it has proven very difficult to find evidence of how different charging strategies and different battery sizes affect battery life. Therefore, it was not possible to make a deeper quantitative comparison.

It can be noted that emissions and energy consumption are directly related to the size of the battery. A smaller battery, therefore, creates less impact, all other things being equal. It would probably be possible to use a much smaller battery if there was more information about how the lifespan of the battery is affected by how the vehicle is used. Charging the batteries more frequently or using the electrified road infrastructure would have made it possible to use smaller batteries (Coban et al., 2022). However, this will require additional infrastructure and associated environmental impact. Such a calculation was not made in this study.

Compared to today's situation, it requires significantly larger batteries, or alternatively, returning the pickup to the charging station during the day probably requires more pickup trucks and drivers. If the batteries are only charged at the charging station, the environmental impact is significantly increased due to the weight of the battery. But it should be noted that this is directly connected with the manufacture of the battery.

When choosing an electric vehicle, it may include issues such as the following in terms of energy consumption and emissions:

- Where the battery cell is manufactured,
- Energy consumption and climate impact per kWh storage capacity and how it is calculated,
- Where the battery pack is installed,
- Long-term data on the State of Charge (SOC) and State of Health (SOH),
- The lifetime in the vehicle, expressed as the number of kWh in which the battery is charged.

Future studies will be directed to consider the recycling of expired batteries. It's not hard to imagine how far batteries are going to go in time which raises the big question; what happens to EV batteries when they die. Demand for EV batteries will increase from around 340 GWh in 2021, to over 3,500 GWh by 2030 (IEA, 2022). The other projection (Zhao et al., 2022) estimates that the transportation sector alone will meet around 90 TWh of this demand in 2030 which raises another question do we have enough materials to support that amount of battery production? The most common materials needed for lithium-ion batteries are cobalt, nickel, lithium, copper, and graphite. The demand for lithium alone is expected to grow from about 300,000 metric tons in 2020 to over 1.7 million metric tons in 2030. Recycling can be an effective solution. Many companies around the world, and dozens of companies in the US and Europe alone, are solving the problem, all with unique techniques and technologies.

REFERENCES

Amarakoon, S., Smith, J., & Segal, B. (2014). *Application of life-cycle assessment to nanoscale technology: Lithium-ion batteries for electric vehicles*. No. EPA 744-R-12-001.

Ambrose, H., & Kendall, A. (2016). Effects of battery chemistry and performance on the life cycle greenhouse gas intensity of electric mobility. *Transportation Research Part D, Transport and Environment*, *47*, 182–194. doi:10.1016/j.trd.2016.05.009

Australian Trade and Investment Commission. (2018). *The Lithium-Ion Battery Value Chain New Economy Opportunities for Australia*. Author.

Battery Test Centre. (2022). *Lithium Ion*. https://batterytestcentre.com.au/project/lithium-ion/

Battery University. (2022). *Types of Lithium-ion*. https://batteryuniversity.com/article/bu-205-types-of-lithium-ion

Benakli, L. (2022). *What are the need for resources of electric cars production, is it sustainable and how?* Université de Liège.

Bi, Z., Song, L., De Kleine, R., Mi, C. C., & Keoleian, G. A. (2015). Plug-in vs. wireless charging: Life cycle energy and greenhouse gas emissions for an electric bus system. *Applied Energy*, *146*, 11–19. doi:10.1016/j.apenergy.2015.02.031

BloombergNEF report. (2021). https://bnef.turtl.co/story/evo-2021/page/1

Brückner, J., Thieme, S., Grossmann, H. T., Dörfler, S., Althues, H., & Kaskel, S. (2014). Lithium-sulfur batteries: Influence of C-rate, amount of electrolyte and sulfur loading on cycle performance. *Journal of Power Sources*, *268*, 82–87. doi:10.1016/j.jpowsour.2014.05.143

Can, S. (2021). *Li22Si5-Grafen Kompozit Anotların Lityum-İyon Pillerdeki Performanslarının İncelenmesi*. http://openaccess.ogu.edu.tr:8080/xmlui/bitstream/handle/11684/4438/311-7706-10232885.pdf?sequence=1&isAllowed=y

Chen, L., Xie, X., Xie, J., Wang, K., & Yang, J. (2006). Binder effect on cycling performance of silicon/carbon composite anodes for lithium ion batteries. *Journal of Applied Electrochemistry*, *36*(10), 1099–1104. doi:10.100710800-006-9191-2

Christopher, L. (2019). *Liner, & T. A. McGilvery*. Historical Overview of Petroleum and Seismology. doi:10.1007/978-3-030-03998-1_2

Civelekoglu, G., & Biyik, Y. (2018). Investigation of Carbon Footprint Change Originated from Transportation Sector. *Bilge International Journal of Science and Technology Research*, *2*(2), 157–166. doi:10.30516/bilgesci.427359

Coban, H. H., Rehman, A., & Mohamed, A. (2022). Analyzing the Societal Cost of Electric Roads Compared to Batteries and Oil for All Forms of Road Transport. *Energies*, *15*(5), 1–20. doi:10.3390/en15051925

Dai, Q., Kelly, J. C., Gaines, L., & Wang, M. (2019). Life Cycle Analysis of Lithium-Ion Batteries for Automotive Applications. *Batteries*, *5*(2), 48. doi:10.3390/batteries5020048

Davidsson Kurland, S. (2020). Energy use for GWh-scale lithium-ion battery production. *Environmental Research Communications*, *2*(1), 012001. doi:10.1088/2515-7620/ab5e1e

Degen, F., & Kratzig, O. (2022). Future in Battery Production: An Extensive Benchmarking of Novel Production Technologies as Guidance for Decision Making in Engineering. *IEEE Transactions on Engineering Management*, 1–19. doi:10.1109/TEM.2022.3144882

Degen, F., & Schütte, M. (2022). Life cycle assessment of the energy consumption and GHG emissions of state-of-the-art automotive battery cell production. *Journal of Cleaner Production*, *330*, 129798. doi:10.1016/j.jclepro.2021.129798

Deng, J., Bae, C., Denlinger, A., & Miller, T. (2020). Electric Vehicles Batteries: Requirements and Challenges. *Joule*, *4*(3), 511–515. doi:10.1016/j.joule.2020.01.013

Egan, M. G., O'Sullivan, D. L., Hayes, J. G., Willers, M. J., & Henze, C. P. (2007). Power-factor-corrected single-stage inductive charger for electric vehicle batteries. *IEEE Transactions on Industrial Electronics*, *54*(2), 1217–1226. doi:10.1109/TIE.2007.892996

Ekejiuba, E. A. I. (2020). *Guide to Petroleum Engineering Career: The Best Practices Petroleum Infographic Cutting Edge Technology Approach*. Dorrance Publishing.

Ellingsen, L. A.-W., Majeau-Bettez, G., Singh, B., Srivastava, A. K., Valøen, L. O., & Strømman, A. H. (2014). Life Cycle Assessment of a Lithium-Ion Battery Vehicle Pack. *Journal of Industrial Ecology*, *18*(1), 113–124. doi:10.1111/jiec.12072

Emilsson, E., & Dahllöf, L. (2019). *Lithium-Ion Vehicle Battery Production - Status 2019 on Energy Use, CO2 Emissions, Use of Metals, Products Environmental Footprint, and Recycling*. IVL Swedish Environmental Research Institute.

Financial Times. (2022). *Full charge ahead: the relentless rise of the battery market*. https://www.ft.com/partnercontent/societe-generale/full-charge-ahead-the-relentless-rise-of-the-battery-market.html

Haber, T. R. T. (2022). *Market share of electric vehicles increased: sales record was broken in 2021*. https://www.trthaber.com/haber/ekonomi/elektrikli-araclarin-pazar-payi-artti-2021de-satis-rekoru-kirildi-673349.html#:~: text=Avrupa'daise 2%2C3,ye göre yüzde 66 büyüdü

Halsnæs, K., Bay, L., Kaspersen, P. S., Drews, M., & Larsen, M. A. D. (2021). Climate services for renewable energy in the nordic electricity market. *Climate (Basel)*, *9*(3), 46. Advance online publication. doi:10.3390/cli9030046

Houache, M. S. E., Yim, C.-H., Karkar, Z., & Abu-Lebdeh, Y. (2022). On the Current and Future Outlook of Battery Chemistries for Electric Vehicles—Mini Review. *Batteries*, *8*(7), 70. doi:10.3390/batteries8070070

Ibarra-Gutiérrez, S., Bouchard, J., Laflamme, M., & Fytas, K. (2021). Perspectives of Lithium Mining in Quebec. *Potential and Advantages of Integration into a Local Battery Production Chain for Electric Vehicles*, *33*, 33. Advance online publication. doi:10.3390/materproc2021005033

IEA. (2020). *Global EV Outlook, Entering the decade of electric drive?* IEA. (2022). *Global Supply Chains of EV Batteries*. https:// iea.blob.core.windows.net/assets/4eb8c252-76b1-4710-8f5e-867e751c8dda/GlobalSupplyChainsofEVBatteries.pdf

Jung, J. C.-Y., Sui, P.-C., & Zhang, J. (2021). A review of recycling spent lithium-ion battery cathode materials using hydrometallurgical treatments. *Journal of Energy Storage*, *35*, 102217. doi:10.1016/j.est.2020.102217

Khan, R. S. (2021). *Life cycle assessment of GHG emissions of light duty vehicles: comparison between internal combustion engine vehicles and battery electric vehicles*. Academic Press.

Kim, H. C., Wallington, T. J., Arsenault, R., Bae, C., Ahn, S., & Lee, J. (2016). Cradle-to-Gate Emissions from a Commercial Electric Vehicle Li-Ion Battery: A Comparative Analysis. *Environmental Science & Technology*, *50*(14), 7715–7722. doi:10.1021/acs.est.6b00830 PMID:27303957

Kim, T., Song, W., Son, D. Y., Ono, L. K., & Qi, Y. (2019). Lithium-ion batteries: Outlook on present, future, and hybridized technologies. *Journal of Materials Chemistry. A, Materials for Energy and Sustainability*, *7*(7), 2942–2964. doi:10.1039/C8TA10513H

Koh, S. C. L., Smith, L., Miah, J., Astudillo, D., Eufrasio, R. M., Gladwin, D., Brown, S., & Stone, D. (2021). Higher 2nd life Lithium Titanate battery content in hybrid energy storage systems lowers environmental-economic impact and balances eco-efficiency. *Renewable & Sustainable Energy Reviews*, *152*, 111704. Advance online publication. doi:10.1016/j.rser.2021.111704

Li, P., Xia, X., & Guo, J. (2022). A review of the life cycle carbon footprint of electric vehicle batteries. *Separation and Purification Technology*, *296*, 121389. Advance online publication. doi:10.1016/j.seppur.2022.121389

Ma, S., Jiang, M., Tao, P., Song, C., Wu, J., Wang, J., Deng, T., & Shang, W. (2018). Temperature effect and thermal impact in lithium-ion batteries: A review. In Progress in Natural Science: Materials International (Vol. 28, Issue 6, pp. 653–666). Elsevier B.V. doi:10.1016/j.pnsc.2018.11.002

MacHiels, N., Leemput, N., Geth, F., Van Roy, J., Buscher, J., & Driesen, J. (2014). Design criteria for electric vehicle fast charge infrastructure based on flemish mobility behavior. *IEEE Transactions on Smart Grid*, *5*(1), 320–327. doi:10.1109/TSG.2013.2278723

Manthiram, A. (2020). A reflection on lithium-ion battery cathode chemistry. In Nature Communications (Vol. 11, Issue 1). Nature Research. doi:10.103841467-020-15355-0

McBrayer, J. D., Rodrigues, M. T. F., Schulze, M. C., Abraham, D. P., Apblett, C. A., Bloom, I., Carroll, G. M., Colclasure, A. M., Fang, C., Harrison, K. L., Liu, G., Minteer, S. D., Neale, N. R., Veith, G. M., Johnson, C. S., Vaughey, J. T., Burrell, A. K., & Cunningham, B. (2021). Calendar aging of silicon-containing batteries. In Nature Energy (Vol. 6, Issue 9, pp. 866–872). Nature Research. doi:10.103841560-021-00883-w

Nurdiawati, A., & Agrawal, T. K. (2022). Creating a circular EV battery value chain: End-of-life strategies and future perspective. *Resources, Conservation and Recycling*, *185*, 106484. Advance online publication. doi:10.1016/j.resconrec.2022.106484

Paoli, L. (2022). *Electric Vehicles*. https://www.iea.org/reports/electric-vehicles

Paris Agreement. (2015). *Paris agreement.* Report of the Conference of the Parties to the United Nations Framework Convention on Climate Change.

Peters, J. F., Baumann, M., Zimmermann, B., Braun, J., & Weil, M. (2017). The environmental impact of Li-Ion batteries and the role of key parameters – A review. In *Renewable and Sustainable Energy Reviews* (Vol. 67, pp. 491–506). Elsevier Ltd., doi:10.1016/j.rser.2016.08.039

Rajaeifar, M. A., Ghadimi, P., Raugei, M., Wu, Y., & Heidrich, O. (2022). Challenges and recent developments in supply and value chains of electric vehicle batteries: A sustainability perspective. In Resources, Conservation and Recycling (Vol. 180). Elsevier B.V. doi:10.1016/j.resconrec.2021.106144

RitchieH.RoserM.RosadoP. (2020). *Energy.* https://ourworldindata.org/energy

Rodrigue, J.-P. (2020). *The Geography of Transport Systems, Transportation and Energy.* Routledge. doi:10.4324/9780429346323

Romare, M., & Dahllöf, L. (2017). *The Life Cycle Energy Consumption and Greenhouse Gas Emissions from Lithium-Ion Batteries.* IVL Swedish Environmental Research Institute.

Temporelli, A., Carvalho, M. L., & Girardi, P. (2020). Life cycle assessment of electric vehicle batteries: An overview of recent literature. In Energies (Vol. 13, Issue 11). MDPI AG. doi:10.3390/en13112864

Thomas, M., Ellingsen, L. A. W., & Hung, C. R. (2018). *Research for TRAN Committee-Battery-powered electric vehicles: market development and lifecycle emissions.* Academic Press.

Transparency, C. (2021). *Climate Transparency Report.* https://www.climate-transparency.org/wp-content/uploads/2021/10/CT2021Turkey.pdf

Turcheniuk, K., Bondarev, D., Singhal, V., & Yushin, G. (2018). Ten years left to redesign lithium-ion batteries. *Nature, 559*(7715), 467–470. doi:10.1038/d41586-018-05752-3 PMID:30046087

Turkish Statistical Institute. (2021). *Vehicles.* https://data.tuik.gov.tr/Bulten/Index?p=Motorlu-Kara-Tasitla ri-Aralik-2021-45703

Uysal, M., & Gul, H. (2017). Lityum İyon Piller İçin Sn-Cu/rGO (İndirgenmiş Grafen Oksit) Anot Malzemelerin, Karakterizasyonu ve Elektrokimyasal Özellikleri. *Academic Platform-Journal of Engineering and Science*, 19–25. doi:10.21541/apjes.336104

Walvekar, H., Beltran, H., Sripad, S., & Pecht, M. (2022). Implications of the Electric Vehicle Manufacturers' Decision to Mass Adopt Lithium-Iron Phosphate Batteries. *IEEE Access: Practical Innovations, Open Solutions, 10*, 63834–63843. doi:10.1109/ACCESS.2022.3182726

Winter, M. (2009). *The Solid Electrolyte Interphase – The Most Important and the Least Understood Solid Electrolyte in Rechargeable Li Batteries.* Academic Press.

Winter, M., & Brodd, R. J. (2004). What are batteries, fuel cells, and supercapacitors? *Chemical Reviews, 104*(10), 4245–4269. doi:10.1021/cr020730k PMID:15669155

Wu, H., Hu, Y., Yu, Y., Huang, K., & Wang, L. (2021). The environmental footprint of electric vehicle battery packs during the production and use phases with different functional units. *The International Journal of Life Cycle Assessment, 26*(1), 97–113. doi:10.100711367-020-01836-3

Wu, Y., Xie, L., Ming, H., Guo, Y., Hwang, J. Y., Wang, W., & Ming, J. (2020). An empirical model for the design of batteries with high energy density. *ACS Energy Letters, 5*(3), 807–816. doi:10.1021/acsenergylett.0c00211

Yu, L., Zha, R., Stafylas, D., He, K., & Liu, J. (2020). Dependences and volatility spillovers between the oil and stock markets: New evidence from the copula and VAR-BEKK-GARCH models. *International Review of Financial Analysis, 68,* 101280. Advance online publication. doi:10.1016/j.irfa.2018.11.007

Zhao, A., McJeon, H., Cui, R., Cyrs, T., Feldmann, J., Iyer, G., Kennedy, K., Kennedy, K., Kennedy, S., O'Keefe, K., Rajpurohit, S., Rowland, L., & Hultman, N. (2022). *An "all-in" Pathway to 2030: Transportation Sector Emissions Reduction Potential.* Academic Press.

Zhou, Z., Zhou, X., Cao, B., Yang, L., & Liew, K. M. (2022). Investigating the relationship between heating temperature and thermal runaway of prismatic lithium-ion battery with LiFePO4 as cathode. *Energy, 256,* 124714. Advance online publication. doi:10.1016/j.energy.2022.124714

Chapter 15
Stochastic Data Envelopment Analysis in Measuring the Efficiency of Electricity Distribution Companies

Zühre Aydın

 https://orcid.org/0000-0002-5992-4983
Energy Market Regulatory Authority, Turkey

Bilal Toklu
Gazi University, Turkey

ABSTRACT

Performance benchmarking of electricity distribution markets is essential for improving industry performance parameters. In a benchmarking study, the most important problem is that regulators often do not have accurate, specific, and sufficient information to determine current input use to achieve expected amount of output. The study combines statistical symmetric error structure with stochastic chance constrained DEA models and compares deterministic data envelopment analysis (DEA) models with stochastic chance-constrained DEA models within random input and output variables. The proposed models were applied on Turkey's electricity distribution units for assessment of energy efficiency. Study revealed that the results obtained with random data softened efficiency frontier. This study contains symmetric error structure and random inputs and outputs for performance benchmarking of electricity distribution markets by stochastic data envelopment analysis within a symmetric error structure in Turkey.

DOI: 10.4018/978-1-6684-4974-5.ch015

INTRODUCTION

Benchmarking the efficiency of electricity distributions is an important case within the energy field and DEA is a widely used optimization method for measuring the relative efficiency of decision-making units (DMUs) (Azadeh et al., 2015).

Since energy is one of the basic concerns of the industrial revolution, improving the efficiency of electricity distribution units is an important issue. The electricity industry plays an important role in the development of countries. Therefore, assessment of the electricity distribution industry plays an important role in improving performance parameters. For these reasons, in the paper, a new approach was applied for assessment of Turkey's distribution units over the period 2011–2016 to evaluate the performance. In this regard, 21 of Turkey's electricity distribution units were considered as the decision-making units in the study. The main objective of the study was efficiency measuring of electricity distribution units with respect to five important indicators, including peak load, transformer capacity, operating cost based on productivity, number of customers and total electricity sales using classical DEA and stochastic chance-constrained data envelopment analysis (CCDEA) approach by using the symmetric error structure. The results were obtained from General Algebraic Modeling System (GAMS) mathematical programming and optimization system.

In this paper, due to the lack of information about some parameters, a stochastic DEA model was applied for evaluating the shaping factor for performance assessment of electricity distribution units. Through the study, the indicators peak load, transformer capacity, operating cost based on productivity, number of customers and total electricity sales were determined as random, and they were frequently applied for the assessment of electricity distribution units in the literature. These indicators were frequently applied for the assessment of electricity distribution (Jamasb et al., 2001).

The aim of the study is to show the deterministic and stochastic approaches on electricity distribution units where uncertainty has a technological structure, especially in the electricity markets. Decision-making problems in electricity markets are debated with uncertainty, which has affected prices, demand, production, equipment availability and similar things. Stochastic programming provides an adequate modelling framework in which problems of decision making under uncertainty are properly formulated (Birge et al., 1997).

This study extended output-oriented deterministic CCR and output-oriented stochastic CCR DEA models within the symmetric error structure to consider the stochastic variations in the input/output data.

To the best of our knowledge, this is the first study on the performance benchmarking of electricity distribution units by using symmetric error structure within output oriented stochastic data envelopment analysis in Turkey. With this study, our aim is to fill the gap of the uncertainty of Turkey's electricity distribution companies in the literature. For the validation of the models, the deterministic output oriented CCR model results were compared to the stochastic output oriented CCR model results.

The rest of the paper is organized as follows: The relevant literature was reviewed in Section 2 as 'Literature review'. The proposed approach is presented in Section 3 as 'Methodology'. The implementation procedure of the approach for the evaluation the performance of electricity distribution units is investigated in Section 4 as 'Case study', and the paper is highlighted in Section 5 as 'Conclusion' and future research directions.

LITERATURE REVIEW

DEA is a technique that is used to evaluate the effectiveness of decision-making units (DMUs), that use multiple inputs to obtain multiple outputs. DEA concepts were first introduced by Farrell (1957)., but later, the approach was pioneered by Charnes et al. (1959) who led the foundations of a literature through operational research and economics. Charnes, Cooper and Rhodes (1978) developed the first DEA model that was called the CCR model. Banker, Charnes and Cooper (1984) extended the DEA to obtain a variable return to scale version of the CCR model called BCC model.

There are many studies on DEA, and many further models were introduced within DEA in the literature. The most commonly used benchmarking models in electricity distribution regulations are: Data Envelopment Analysis (DEA; Charnes et al., 1978), Stochastic Frontier Analysis (SFA; Aigner et al., 1977; Meeusen and van den Broeck, 1977), Corrected Ordinary Least Squares (COLS) (Richmond, 1974) and Stochastic Semi-nonparametric Envelopment of Data (StoNED; Kuosmanen, 2006; Kuosmanen and Kortelainen, 2012).

In addition, DEA is a non-parametric linear programming methodology that creates an efficiency frontier using a convex linear combination of inputs and outputs of DMUs. SFA requires a parametric equation of the efficiency frontier and assumes a compound error, which represents deviations from the frontier. The compound error is the sum of the stochastic inefficiencies and stochastic noise. StoNED is similar to SFA and DEA, with a compound stochastic error and with a non-parametric, piecewise linear frontier (2015). Lopes and Mesquita (2015) showed that these models are very popular among the European electricity distribution regulators for benchmarking.

In various fields, many DEA models have been presented to evaluate DMUs with different kinds of data such as deterministic, stochastic and fuzzy data. However, in many practical problems, managers deal with units with imprecise data (Behzadi et al., 2012). In these situations, analysts may consider imprecise data as random variables. By working with random variables and considering the possibility of uncertain events, different perspectives of the available information can be detected in energy markets. The main advantage of working with random data in DEA is the prediction of efficiencies in future optimization problems (Mirbolouki at al., 2014). Stochastic models where the inputs or outputs are considered to be random data seem to be a reasonable approach for future problems to account for such uncertainties or fluctuations while analyzing such data. Stochastic models where the inputs or outputs are considered to be random variables, seem to be the reasonable approach to account for uncertainties or fluctuations while analyzing such data. In these situations, measurement of input and output values is subject to errors and noise. The noise in data usually leads to mistakes in frontier specification and efficiency scores (Brazdik, 2004).

Land et al. (1993,1994), Li (1998), Cooper et al. (2004), Sengupta (2002), Huang and Li (2001), Olesen (2002), Khodabakhshi (2008,2010) Behzadi et al. (2009) and Jahanshahloo et al. (2010) used stochastic DEA models in their studies.

Recent SDEA approaches led to a change in the constrained optimization problems. Linearization allows solving large-sized chance-constrained optimization problems (Brazdik, 2004).

The objectives of these DEA models include to evaluate efficiencies of homogeneous DMUs that convert a set of inputs into a set of outputs. The results of DEA models divide the DMUs into efficient and inefficient DMUs. The DMU value is evaluated as efficient if it is an element of a production possibility frontier. Since the BCC model's frontier is a piecewise linear set, Banker et al. (1984) defined weak efficiency with nonzero slacks and efficiency with zero slacks.

This section provides a review of other studies which applied the DEA methodology for the electrical distribution industry. Our main purpose was to review the main characteristics and the results of a representative number of DEA models proposed for the electrical distribution markets. The first study of the efficiency of electricity distribution companies by DEA was conducted by Weyman-Jones (1991), and it measured the performance of the electricity distribution sector in the United Kingdom. After this study, the number of applications made in the literature has increased. Data envelopment analysis (DEA) is an ideal approach to evaluate the performance of various systems with multiple inputs and outputs. DEA is a popular optimization method for measuring the relative efficiency of decision-making units (DMUs) (Fallahi et al., 2011). The parametric approaches apply statistical techniques whilst non-parametric approaches use linear and non-linear programming such as data envelopment analysis (DEA) (Fallahi et al., 2011). Azadeh et al. (2009) used integrated DEA, corrected ordinary least squares (COLS), stochastic frontier analysis

(SFA) and PCA methods for efficiency estimation of electricity distribution units. They considered a number of employees, operating costs, network length, transformer capacity, number of customers, size of service area and unit's delivery as the most widely used indicators for evaluation of the electricity distribution units. Azadeh et al. (2009) presented a deterministic approach for the performance assessment of the electricity distribution units in Iran. They used DEA and principle component analysis (PCA) in their proposed approach. Network length, transformer capacity, number of employees, unit's delivery and service area were considered as major indicators in their study.

There are also some studies that analyzed DMUs which are different from distributors, for example, what happens in Amado et al. (2013), which analyzes the medium voltage power lines, and Mullarkey et al. (2015), which looks at electricity distribution counties. They argued that this last variable was better than other contextual variables for the control of the heterogeneity of the 'electricity distribution counties' in the Republic of Ireland. DEA and window analysis (WA) were applied by Sözen et al. (2012) for the efficiency assessment of ten hydro-power plants. They used DEA with respect to the production and energy unit cost in order to optimize hydro-power plants. Additionally, there are also some studies that adopted an international perspective. This is the case, for instance, in the study by Jamasb and Pollitt (Jamasb et al., 2003). which used a sample of 63 companies from different European countries using the data adjusted for the year 1999: Italy, the Netherlands, Norway, Portugal, Spain and the UK.

Gouveia et al. (2015) studied the comparison of results for standard DEA models and the value-based DEA method for benchmarking 40 Portuguese electricity companies. In this study, in order to deal with the underlying uncertainty, the value-based DEA method for performance evaluation was adapted to include the concept of super-efficiency. Gil et al. (2017) studied evaluating 61 Brazilian electricity utilities by the statistical correlation between the DEA efficiency scores and the available environmental variables. Their results showed that major differences between the original and corrected efficiency scores, mainly for utilities were located in harsh environments, which were originally achieved for lower efficiency scores. Arcos-Vargas et al. (2017) used different indicators within the CCR and BCC DEA models for performance evaluation of electricity distributors in Spain. The objective of this study was to give the Regulator a set of models, which provided a theoretical framework to determine the remuneration for distribution.

In the classical DEA method, the input and output are deterministic, which leads to unpredictable measures against future inconsistencies. The disadvantage of classical DEA may be overcome by the inclusion of SDEA in the calculation. SDEA studies are based on Land et al. (1993), where the authors used their models to examine the efficiency of a school program for disabled scholars as in Charnes et al. (1978). In the study by Land et al. (1993), the authors suggested the prospect of stochastic data envelopment analysis and constructed their own model. They introduced the stochastic component of DEA and created chance-constrained problems by introducing the variability to outputs that are conditional on inputs. Outputs were taken as normally distributed random variables. After the stochastic optimization problems were created, Land et al. (1993) transformed these problems into their deterministic equivalents to determine the efficient DMUs.

That is, in the literature, firstly, Charnes et al. (1959) introduced modelling efficiency under uncertainty, by the CCDEA method that was developed by Charnes and Cooper (1959, 1962). This method allowed the constraints to be violated, although not infrequently, so that where the DEA was transformed into a random structure and that seemed to be an approach, accurately and easily incorporate various uncertainties regarding the inputs and outputs of the measured DMUs. The concept of chance constrained the programming approach studied by Khodabakhshi (2010) to develop an output-oriented stochastic super-efficiency model and worked on SDEA, which assessed 17 Iranian power distribution companies by using cost, number of employees, transformer power and line length as inputs, and the amount of energy distributed, the number of customers and the service area as outputs. Mirbolouki et al. (2014) developed a stochastic CCR multiplier model based on the chance-constrained programming approach for measuring the efficiency of 15 Iranian electricity distribution units within the data of 2000 and 2004. In the study, network length, transport capacity and the number of employees were chosen as the inputs, while the number of customers and total electricity sales were chosen as the stochastic outputs. Talluri (2006), Olesen (2002), Behzadi (2012) and Demireli (2013) also used CCDEA in their studies.

Azadeh et al. (2015) measured the performance metrics of 17 different Iranian electricity distribution units by SDEA using the data between 2001 and 2011. The inputs included in the study were line length, transformer power and number of employees, while the outputs were the number of subscribers, total electricity sales are uncertain. According to Talluri (2006), there were different DEA models for inclusion of ambiguities into the DEA in the literature. Sueyoshi (2000) introduced a CCR stochastic model that could be used for future planning. His proposed model could handle stochastic outputs and created accurate results when there was incomplete information. Indeed, the manager could control inputs of their organization whilst they were unable to control outputs because outputs are affected by external factors. An interactive robust data envelopment analysis model was applied by Sadjadi et al. (2011) to determine the input and output values of electricity distribution companies with uncertain data. In their study, network length, transport capacity, number of employees, number of customers and total electricity sales were used. In another study, Sadjadi et al. (2008) developed a new DEA method with the consideration of uncertainty in the output parameters. The method was based on the adaptation of recently developed robust optimization approaches. Khodakaramia et al. (2017) proposed an optimization formulation on the stochastic characterization of electricity markets. Their optimization results showed that the stochastic approach included better modelling of the power system's uncertain behavior. Shiraz et al. (2018), proposed economic efficiency measures for stochastic data with known input and output prices through CCDEA. They transform the stochastic efficiency models into a deterministic equivalent non-linear form. Khodadadipour et al. (2021), proposed a model that employs statistical techniques to evaluate the efficiency of decision-making units (DMUs), 32 thermal power plants, with stochastic data. Based on the proposed model, a SDEA cross-efficiency model is suggested for ranking and discrimination of DMUs. Yenioğlu and Toklu (2021) used deterministic and stochastic DEA models for the evaluation of energy efficiency. They used multiple inputs and outputs as random variables in stochastic models. This study showed that stochastic DEA scores are more flexible than classical DEA and stochastic models give more results closer to the production frontier.

In the symmetric error concept, Brazdik (2004) developed four oriented stochastic DEA models and description of their properties by using the techniques of stochastic problems regarding linearization. The proposed stochastic models were linearized, so the interior point methods for linear problems could be used to solve linear programming problems associated with the models. In the study, the single symmetric error structure that allows transform model from change the constrained problem to the linear programming problem that is introduced. Since DEA is an extreme point technique, noise (even symmetrical noise with zero mean) such as a measurement error can cause significant problems, because the frontier is sensitive to these errors. Therefore, theoretical attempts to incorporate these errors were made. The paper by Gstach (1998) shows that there are research directions in which the future developments on DEA and SDEA could be driven. Gstach (1998) proposed using DEA to estimate a pseudo frontier (nonparametric shape estimation) and then applied a maximum likelihood-technique to the DEA-estimated efficiencies to estimate the scalar value by which this pseudo-frontier must be shifted downward to get the true production frontier (location estimation). He stated that the outcome of a production process might not only deviate from a theoretical maximum due to inefficiency, but it may also deviate due to non-controllable errors. Hence, this raises the case of reliability of DEA in noisy environments. This assumption makes the approach that mixes the parametric and non-parametric approaches to production frontier estimation feasible.

A general remark through efficiency comparisons under noisy conditions seems advisable. This would take care of noise as it has been illustrated well in a recent paper by Brockett and Golany (1996).

By this study, stochastic output oriented CCR DEA was combined with a symmetric error structure. They studied stochastic inputs and outputs through input-oriented CCR CCDEA. They used network length, transport capacity and the number of employees as inputs while the number of customers and total electricity sales were used as outputs. In their paper, they noted that every random variable with normal distribution may be stated as a symmetric error structure.

As a result of the literature review, the difference in the research between assessment performance of electricity distribution units' problem and the information in the literature is presented below with Table1.

As in the Table 1, DEA studies, in which the data were accepted to be random, were carried out on the stochastic performance measurement of electricity distribution companies are few in numbers in the literature. Additionally, this study is supported by symmetric error structure.

As a result, in the literature survey, there is no study for measuring performance of electricity distribution units by stochastic data envelopment methodology in Turkey. Data envelopment analysis has many advantages, such as no requirement for a priori weights or explicit specification of functional relations among the multiple inputs and outputs. However, there is a weakness in DEA models, in fact, deterministic DEA models do not allow random variations in inputs and outputs such as data entry errors. An efficient DMU, may turn inefficient if such random variations are considered.

This study extends output and input oriented chance constrained stochastic DEA within the symmetric error structure to consider stochastic variations in input/output data. In this paper, due to the lack of information about some parameters, a stochastic DEA model is applied for evaluation shaping factor for performance assessment of electricity distribution units with respect to five important indicators OPEX, transport capacity, peak load, the number of customers, and the total electricity sales. These indicators are frequently applied for assessment of electricity distribution.

METHODOLOGY

In order to make a healthy benchmark measurement, the factor level combinations of experimental studies should be random. The main reason for this randomness is making sure that the systematic effects do not affect the results of the experiment. When full randomness is used, it is assumed that experimental observations may be made in homogeneous conditions. However, randomness is often limited because it is difficult to change or control some experimental factors. Another reason for limited randomness is that all observations are not made in homogeneous conditions.

Table 1. Recommended approach and other methods for evaluation of electricity distribution units

	Model	Inputs/Outputs	Stochastic Data
Azadeh et al. [1]	Stochastic output-oriented CCR DEA	network length, transport capacity and the number of employees are chosen as inputs, while number of customers and total electricity sales are chosen as outputs	Outputs are stochastic
Mirbolouki et al. [11]	Stochastic input-oriented CCR CCDEA	network length, transport capacity and the number of employees are chosen as inputs, while number of customers and total electricity sales are chosen as outputs	Inputs and outputs are stochastic
Khodabakhshi [20]	Stochastic output-oriented super-efficiency model BCC CCDEA	network length, transformer capacity, and employee variables are inputs, and delivery of units and service area variables are outputs	Inputs and outputs are stochastic
Khodabakhshi and Kheirollahi [22]	Stochastic input-oriented super-efficiency model BCC DEA	inputs are operating costs, number of employees, transformer capacity and network length and outputs, number of customers and size service area	Inputs and outputs are stochastic
Arcos-Vargas et al. [37]	Used different indicators within the CCR and BCC DEA model	the remuneration, the level of assets, distributed energy; the number of points of supply and a proxy for the variable TIEPI (Time of Interruption Equivalent to the Installed Power in Medium Voltage)	Discretional and non-discretional inputs and outputs
Gouveia et al. [36]	Value-based DEA-BCC DEA	Costs, supply interruptions, number of incidents, complaints per customer are inputs, clients, while network line length are outputs	Deterministic and uncertain inputs and outputs
Sadjadi and Omrani [45]	Robust DEA	inputs; network length (km) transformers capacity, number of employees, outputs; number of customers, total electricity sales	Outputs are stochastic
Omrani, Beiragh, Kaleibari [49]	Principal Component Analysis (PCA) and input-oriented CCR DEA	inputs; transformer capacity, number of transformers, terrestrial network length, aerial network length, number of employees, area, outputs; energy delivery, energy consumption, number of customers, number of street lights	Inputs and outputs are deterministic
Proposed Model	Stochastic output oriented CCR CCDEA within symmetric error structure	operating cost based on productivity, transport capacity and peak load are chosen as inputs, while number of customers and total electricity sales are chosen as outputs	Inputs and outputs are stochastic

In this context, a performance comparison was made by including the symmetric error structure into the chance-constrained CCR DEA. CCR model was used to analyze the set of DMUs that were using the production function with constant returns to scale. The reason for using CCR model was to provide the possibility of separately calculating technical efficiency. Technical efficiency measures the DMU's overall success at utilizing its inputs

The nomenclature for this study is introduced below:

Deterministic Output-Oriented CCR Model

In the efficiency measures of the distribution units, the type of the scale they were producing was also important in determining the type of the DEA model to be used for efficiency measurements.

Mathematical equations of the output-oriented CCR models are given below. It is assumed that there are n homogenous DMUs (DMU_j, $j = 1,....n$) such that all of them use m inputs xij ($i = 1,2,.....m$) to produce s outputs yrj ($r = 1,2,.....s$), $x_j =$ ($x_{1j},... ...,x_{mj}$) and $y_j = (y_{1j},... ...,y_{mj})$ which are nonnegative and nonzero vectors. The CCR production possibility set proposed by Charnes, Cooper, Rhodes in 1978 [6] is as follows:

$$T_{CCR}=\{(X,Y) \mid \sum_{j=1}^{n}\left(X_j \lambda_j\right)\le X \ , \ \sum_{j=1}^{n}\left(Y_j \lambda_j\right)\ge Y \ , \ \lambda j \ge 0, j = 1,....n\}$$

CCR efficiency scores can be obtained by using the envelopment output-oriented Model (1), respectively where x_{i0} and y_{r0} represent the ith input and the rth output vector of DMU_0 under evaluation in model (1) . A DMU is called CCR-efficient if its objective value in model (1) is equal to unity.

$$max\, \theta_0 + \varepsilon \left(\sum_{i=1}^{m} s_i^- + \sum_{r=1}^{s} s_r^+\right)$$

s.t.

$$\sum_{j=1}^{n} x_{ij}\lambda_j + s_i^- = x_{i0}$$

$$\sum_{j=1}^{n} y_{rj}\lambda_j - \theta_0 y_{r0} - s_r^+ = 0 \tag{1}$$

Table 2. Deterministic output-oriented CCR model

j = 1,....n: decision-making units(DMUs)	$\tilde{X}_j = \left(\tilde{x}_{1j}, \ldots\ldots, \tilde{x}_{mj} \right)$: *the random input vector of DMU_j*
r =1,.....s: the set of outputs	$\tilde{Y}_j = \left(\tilde{y}_{1j}, \ldots\ldots, \tilde{y}_{sj} \right)$: *the random output vector of DMU_j*
i =1,.....m: the set of inputs	$x_j = (x_{1j}, \ldots \ldots, x_{mj})$: *the vector corresponding to the expected values of inputs*
DMU_0: the DMU under investigation	$y_j = (y_{1j}, \ldots \ldots, y_{mj})$: *the vector corresponding to the expected values of outputs*
yrj: the amount of output r used by the decision-making unit j	$\tilde{x}_{ij}$: *the means of the input variables*
xij: the amount of input i used by the decision-making unit j th	$\tilde{y}_{rj}$: *the means of the output variables*
y_{r0}: rth output of the DMU_0	$\tilde{\varepsilon}_{ij}$: *input error compared to the average values of the inputs*
x_{i0}: ith input of the DMU_0	$\tilde{\xi}_{rj}$: *output error compared to the average values of the outputs*
θ: efficiency score	*σ: the standard deviation of random inputs and outputs*
α: the level of error between 0 and 1	*λ: density vector, that is weights on the decision units*
Cov: covariance operator	*φ: the cumulative distribution function of the standard normal distribution*
P: means probability measure	*φ⁻¹: inverse of the cumulative distribution function of the standard normal distribution*

$\lambda j_\geq 0, s_i^- \geq 0, \ s_r^+ \geq 0, \ i=1, 2,..,m, \ r= 1,2,\ldots\ldots,s, \ j=1,2,\ldots\ldots,n$

Here the λj represent structural variables, the s_i^-, s_r^+ represent slacks and $\varepsilon > 0$ is a ''non-Archimedean infinitesimal'' defined to be smaller than any positive real number. This means that ε is not a real number.

Stochastic Output Oriented CCR Model Under Chance Constrained

In the stochastic framework, the production possibility set is defined in terms of the mean values of inputs and outputs. In further model development, the transformation from the chance-constrained problem to its deterministic equivalent will be used.

The true production possibility set may be constructed using the true production function. In the nonparametric DEA methodology, this function is not known. Therefore, the general production possibility set of the CCR production function was used by Charnes et al. (1978).

Assuming that the inputs and outputs in the stochastic models are random variables, $\tilde{x}_j = \left(\tilde{x}_{1j},\ldots\ldots,\tilde{x}_{mj} \right)$ and $\tilde{y}_j = \left(\tilde{y}_{1j},\ldots\ldots,\tilde{y}_{sj} \right)$ represent the random input and output vectors of $DMU_j, j=1,\ldots,n$.

It is assumed in the model that all input and output components have normal distribution.

$$\tilde{x}_{ij} \sim N\left(x_{ij}, \sigma_{ij}^2 \right), \quad i = 1, 2, \ldots, m$$

$$\tilde{y}_{rj} \sim N\left(y_{rj}, \sigma_{rj}'^2 \right), \quad r = 1, 2, \ldots, s \tag{2}$$

Using this notation and considering our assumption in which the measurements of the outputs and inputs were probabilistic, in order to carry out a performance analysis of a specific DMU, the chance-constrained CCR model introduced by Cooper et al. in 2004 is as follows:

Max θ

s.t.

$$P\left(\sum_{j=1}^{n} \tilde{x}_{ij} \lambda_j \leq \tilde{x}_{i0} \right) \geq 1 - \alpha, i = 1, \ldots, m$$

$$P\left(\sum_{j=1}^{n} \tilde{y}_{rj} \lambda_j \geq \theta \tilde{y}_{r0} \right) \geq 1 - \alpha, r = 1, \ldots, s \tag{3}$$

$\lambda j \geq 0, j = 1, \ldots. n$

In the model (3), P represents the probability measure, α is the level of error between 0 and 1.

Deterministic Equivalent of the Proposed Model

This subsection presents the deterministic equivalent of the stochastic CCR model (3).

It is assumed that $\tilde{x}_{ij}$ and $\tilde{y}_{rj}$ are the mean values of the input and output variables, which are the values of the inputs and outputs.

The deterministic equivalent of model (3) which is obtained by the method described by Cooper et al. (2004) is as follows:

$$Max \; \theta + \varepsilon \left(\sum_{i=1}^{m} s_i^- + \sum_{r=1}^{s} s_r^+ \right)$$

s.t.

$$\sum_{j=1}^{n} x_{ij}\lambda_j + s_i^- - \phi^{-1}(\alpha)v_i = x_{i0}, \; i=1,\ldots,m$$

$$\sum_{j=1}^{n} y_{rj}\lambda_j - s_r^+ + \phi^{-1}(\alpha)u_r = \theta y_{r0}, \; r=1, \ldots\ldots, s \; (4)$$

$$v_i^2 = \sum_{j \neq 0} \sum_{k \neq 0} \lambda_j \lambda_k cov\left(\bar{x}_{ij}, \bar{x}_{ik}\right) + 2(\lambda_0 - 1)\sum_{j \neq 0}\lambda_j cov\left(\bar{x}_{ij}, \bar{x}_{i0}\right) + (\lambda_0 - 1)^2 var\left(\bar{x}_{i0}\right),$$

$$u_r^2 = \sum_{j \neq 0} \sum_{k \neq 0} \lambda_j \lambda_k cov\left(\bar{y}_{rj}, \bar{y}_{rk}\right) + 2(\lambda_0 - \theta)\sum_{j \neq 0}\lambda_j cov\left(\bar{y}_{rj}, \bar{y}_{r0}\right) + (\lambda_0 - \theta)^2 var\left(\bar{y}_{r0}\right),$$

$\lambda j, \; s_i^-, s_r^+ \geq 0, \; v_i \geq 0, \; u_r \geq 0, \; j = 1,\ldots n, \; r = 1, \ldots\ldots, s, \; i=1,\ldots,m$

It is obvious that the model (4) is nonlinear and quadratic programming model. ϕ is the cumulative distribution function of the standard normal distribution and $\phi-1^{(\alpha)}$ is its inverse in level of α.

DMU0 is defined as a stochastic efficient DMU on the level of α if and only if when $\theta_0^* = 1$ and slack values $s_i^-, s_r^+ \geq 0$ are all zero for all optimal solutions in the models. Here, * is used to define an optimum value (Cooper et al., 2004).

Stochastic Efficiency Based on Symmetric Error Structure

The purpose of the stochastic studies based on the symmetric error structure is to convert the nonlinear chance-constraint models into a deterministic linear model with random inputs and outputs. In the single factor symmetric model, the errors in all variables are driven by a single symmetric shock ε (Brazdik, 2004).

Let us consider the stochastic m inputs and s outputs of the DMUs as follows:

$$\tilde{x}_{ij} = x_{ij} + a_{ij}\tilde{\varepsilon}_{ij}, \; i=1,\ldots,m,$$

$$\tilde{y}_{rj} = y_{rj} + b_{rj}\tilde{\xi}_{rj}, \; r=1,\ldots,s, \tag{5}$$

where a_{ij} and b_{rj} are the positive real values. $\tilde{\varepsilon}_{ij}$ and $\tilde{\xi}_{rj}$ are the random variables with normal distribution.

$$\tilde{\varepsilon}_{ij} \sim N\left(0, \bar{\sigma}^2\right)$$

$$\tilde{\xi}_{rj} \sim N\left(0, \bar{\sigma}^2\right) \tag{6}$$

Therefore, $\tilde{\varepsilon}_{ij}$ and $\tilde{\xi}_{rj}$ input and output errors were compared to the mean values of the inputs and outputs. Since normal distribution is symmetric, then the expression (5) is named a symmetric error structure. Moreover, the following relationships result from the expression (5).

$$\tilde{x}_{ij} \sim N\left(x_{ij}, \bar{\sigma}^2 a^2_{ij}\right)$$

$$\tilde{y}_{rj} \sim N\left(y_{rj}, \bar{\sigma}^2 b^2_{rj}\right) \tag{7}$$

Every random variable with normal distribution may be evaluated within the scope of the symmetric error structure. Assume that the ith input and rth output of every DMU are not correlated. That is, the relation between the same stochastic output and input variables through different DMUs is independent.

$$Cov\left(\tilde{\varepsilon}_{ij}, \tilde{\varepsilon}_{ik}\right) = 0, i = 1,\ldots\ldots,m,$$

$$Cov\left(\tilde{\xi}_{rj}, \tilde{\xi}_{rk}\right) = 0, r = 1,\ldots\ldots, s, \text{ for every } j^l k \tag{8}$$

According to the expressions (5) and (8), one may consider the same error for all DMUs. Thus, the expression becomes $\tilde{\varepsilon}_i = \tilde{\varepsilon}_{ij}$ and $\tilde{\xi}_r = \tilde{\xi}_{rj}$.

Now, consider the ith input for the constraints of the model (3),

$$P\left(\sum_{j=1}^{n}\tilde{x}_{ij}\,\lambda_{j}\leq\tilde{x}_{i0}\right)\geq 1-\alpha,\,i=1,\ldots,m \tag{9}$$

Assume that $\tilde{h}_{i}=\sum_{j=1}^{n}\tilde{x}_{ij}\,\lambda_{j}-\tilde{x}_{i0}$ Then the expressions (5) and (8) become;

$$\tilde{h}_{i}=\left(\sum_{j=1}^{n}x_{ij}\,\lambda_{j}-x_{i0}\right)+\tilde{\varepsilon}_{i}(\sum_{j=1}^{n}a_{ij}\lambda_{j}-a_{i0}) \tag{10}$$

Therefore, we may state;

$$\tilde{h}_{i}\sim N\left(\left(\sum_{j=1}^{n}x_{ij}\lambda_{j}-x_{i0}\right),\bar{\sigma}^{2}\left(\sum_{j=1}^{n}a_{ij}\lambda_{j}-a_{i0}\right)^{2}\right) \tag{11}$$

According to the expression above and the properties of normal distribution, the stochastic constraint (9) may be converted to the following deterministic equivalent:

$$\sum_{j=1}^{n}x_{ij}\lambda_{j}-\phi^{-1}\left(\alpha\right)\bar{\sigma}\left|\sum_{j=1}^{n}a_{ij}\lambda_{j}-a_{i0}\right|\leq x_{i0} \tag{12}$$

Similarly, consider the rth output constraint of model (3) that can be converted to,

$$\sum_{j=1}^{n}y_{rj}\lambda_{j}+\phi^{-1}\left(\alpha\right)\bar{\sigma}\left|\sum_{j=1}^{n}b_{rj}\lambda_{j}-\theta_{0}b_{r0}\right|\geq\theta_{0}y_{r0}. \tag{13}$$

Based on the expressions (12) and (13), the deterministic equivalent of model (3) is accepted as:
Maximize $\theta 0$

$$\sum_{j=1}^{n}x_{ij}\lambda_{j}-\phi^{-1}\left(\alpha\right)\bar{\sigma}\left|\sum_{j=1}^{n}a_{ij}\lambda_{j}-a_{i0}\right|\leq x_{i0},$$

$$\sum_{j=1}^{n}y_{rj}\lambda_{j}+\phi^{-1}\left(\alpha\right)\bar{\sigma}\left|\sum_{j=1}^{n}b_{rj}\lambda_{j}-\theta_{0}b_{r0}\right|\geq\theta_{0}y_{r0}, \tag{14}$$

$\lambda j\geq 0,\,j=1,\ldots n,\,,\,r=1,\,\ldots\ldots,\,s\,,\,i=1,\ldots,m$

Model (14) is a nonlinear programming model. It may be converted into a linear model by using the following transformation formulas.

$$\left| \sum_{j=1}^{n} a_{ij}\lambda_j - a_{i0} \right| = \left(p_i^- + p_i^+ \right), \ i=1,\ldots,m,$$

$$\sum_{j=1}^{n} a_{ij}\lambda_j - a_{i0} = \left(p_i^+ - p_i^- \right), p_i^+ p_i^- = 0, \ , \ i=1,\ldots,m, \tag{15}$$

$$\left| \sum_{j=1}^{n} b_{rj}\lambda_j - \theta_0 b_{r0} \right| = \left(q_r^- + q_r^+ \right), \ r=1,\ldots,s$$

$$\sum_{j=1}^{n} b_{rj}\lambda_j - \theta_0 b_{r0} = \left(q_r^+ - q_r^- \right), q_r^+ q_r^- = 0, \ r=1,\ldots,s$$

$$p_i^+, p_i^-, q_r^+, q_r^- \geq 0$$

By substituting (15) in model (14), we can obtain;
Max θ0

$$\sum_{j=1}^{n} x_{ij}\lambda_j - \phi^{-1}(\alpha)\overline{\sigma}\left(p_i^- + p_i^+ \right) \leq x_{i0},$$

$$\sum_{j=1}^{n} y_{rj}\lambda_j + \phi^{-1}(\alpha)\overline{\sigma}\left(q_r^- + q_r^+ \right) \geq \theta_0 y_{r0}, \tag{16}$$

$$\sum_{j=1}^{n} a_{ij}\lambda_j - a_{i0} = \left(p_i^+ - p_i^- \right)$$

$$\sum_{j=1}^{n} b_{rj}\lambda_j - \theta_0 b_{r0} = \left(q_r^+ - q_r^- \right),$$

$$q_r^+ q_r^- = 0, p_i^+ p_i^- = 0,$$

$$\lambda j \geq 0, j = 1,\ldots n, \, , \, r = 1, \ldots\ldots, s \, , \, i=1,\ldots,m,$$

$$p_i^+, p_i^-, q_r^+, q_r^- \geq 0.$$

Model (16) is nonlinear because of the existence of constraints $q_r^+ q_r^- = 0$ and $p_i^+ p_i^- = 0$. Model (16) without constraints $q_r^+ q_r^- = 0$ and $p_i^+ p_i^- = 0$ is a linear model, while in all its optimal basic solutions, at least one of p_i^+ and p_i^- for every $i=1,\ldots,m$ and at least one of q_r^+ and q_r^- for every $r=1, \ldots\ldots, s$ are zero. Therefore, the constraints $q_r^+ q_r^- = 0$ and $p_i^+ p_i^- = 0$ will be satisfied in the optimal solutions of this linear model. Thus, assuming the use of the optimal basic solution detector algorithms (for example, Simplex), the constraints $q_r^+ q_r^- = 0$ and $p_i^+ p_i^- = 0$ may be removed. So, the linear deterministic equivalent of the model (16) is obtained as follows.

$$\theta_0^*(\alpha) = \max \theta_0$$

$$\sum_{j=1}^{n} x_{ij}\lambda_j - \phi^{-1}(\alpha)\overline{\sigma}\left(p_i^- + p_i^+\right) \leq x_{i0},$$

$$\sum_{j=1}^{n} y_{rj}\lambda_j + \phi^{-1}(\alpha)\overline{\sigma}\left(q_r^- + q_r^+\right) \geq \theta_0 y_{r0}, \tag{17}$$

$$\sum_{j=1}^{n} a_{ij}\lambda_j - a_{i0} = \left(p_i^+ - p_i^-\right)$$

$$\sum_{j=1}^{n} b_{rj}\lambda_j - \theta_0 b_{r0} = \left(q_r^+ - q_r^-\right),$$

$\lambda j \geq 0, j = 1,\ldots.n, , r =1, \ldots\ldots, s , i=1,\ldots,m ,$

$$p_i^+, p_i^-, q_r^+, q_r^- \geq 0$$

CASE STUDY

DEA is a technique that can be best measurement for the effectiveness of units with multiple inputs and multiple outputs. Since it does not require the existence of a functional relationship between the inputs and outputs, it is also free from errors that may arise from the incorrect connection of the functional link. It allows comparison of units which are similar in terms of their qualities. It is an approach that does not require the inputs and outputs those are expressed in the same unit of measurement.

In this section, Turkey's electricity distribution units are considered as the case study. The proposed approach is applied to Turkey's electricity distribution units from 2011 to 2016. 21 distribution units are considered as DMUs. The network length, costs-operational expenditure and peak loads were chosen as the stochastic inputs, while the number of customers and total electricity sales were chosen as the stochastic outputs.

The measurement units for the inputs and outputs were kilometers (km), MVA and MWh. The stochastic and deterministic results of the mathematical models were obtained from the GAMS software.

As discussed widely in the literature, the major sources of uncertainty in the data used in the performance measurement of electricity distribution companies are fuel prices, fuel availability, energy demand and regulatory issues such as unpredictability in the orders and laws.

Uncertainty sources may be calculated mathematically in the first three items, but the fourth item is related to legislation. Deterministic solutions miss versatile decision-making and analysis processes.

The data that were used in the study includes; operating and maintenance costs-operational expenditure (OPEX), personnel, administrative costs, materials, third-party services, insurances, taxes and other uncertainty factors like seasonal peak load, temperature and time slice changes of peak load. So, due to lack of information about these parameters, the theory of uncertainty is imported into the model. In this study, the focus was on the calculated uncertainty for decision-making within the framework of Turkey's electricity market.

Data were added to the model with mean and variance values that are shown in Table 2 and Table 3. For input and output indicators, the standard deviation is computed. We supposed that, $\overline{\sigma} = 1$ in the symmetric error structure, hence $a_{ij} = \sqrt{Var(\overline{x}_{ij})}$ and $b_{rj} = \sqrt{Var(\overline{y}_{rj})}$.

Results of Models

In certain cases, it is necessary to study whether or not the inefficiency of a DMU is caused by its inefficient operation or by the unsuitable conditions under which the DMU is operating.

The results of the deterministic model (1) that were obtained by applying the inputs and outputs are included in Table 4 as follows.

Table 3. Estimated input parameters

DMU	Input1			Input2			Input3		
	Mean	Std_Dev	Variance	Mean	Std_Dev	Variance	Mean	Std_Dev	Variance
DMU1	3.439	2181,20	4757623,31	185911744.00	23656915.40	559649646451814.00	1762.57	170.31	29005.71
DMU2	813	492,26	242319,16	136789334.00	71993787.21	5183105396383440.00	579.61	36.52	1333.80
DMU3	1.439	942,45	888207,47	172211280.00	27054014.79	731919716258406.00	1491.94	64.70	4186.52
DMU4	3.512	2080,61	4328935,72	209053750.00	43986648.32	1934825230770080.00	561.08	30.89	954.07
DMU5	6.707	3914,02	15319582,69	491243784.00	65553798.56	4297300506031670.00	2098.33	1047.49	1097243.93
DMU6	10.769	6383,59	40750180,69	490539910.00	27844715.50	775328181281462.00	4663.59	275.88	76109.68
DMU7	1.065	613,81	376766,50	134277609.00	23107997.60	533979553225536.00	481.94	21.87	478.50
DMU8	1.354	780,40	609026,38	171483054.00	5559505.72	30908103870464.10	561.33	44.71	1999.07
DMU9	3.826	2384,23	5684559,03	399181806.00	56705235.97	3215483786553750.00	3851.67	218.74	47847.87
DMU10	1.278	896,45	803620,28	141993652.00	6350977.12	40334910416893.30	403.49	199.42	39770.31
DMU11	7.216	4332,98	18774709,94	242599902.00	35030340.75	1227124772941810.00	2899.17	251.14	63071.77
DMU12	4.934	2632,18	6928351,83	222105761.00	12516549.35	156664007700391.00	1390.26	1081.61	1169888.22
DMU13	1.217	712,25	507295,62	99594289.00	9771320.33	95478701026719.30	369.33	20.54	421.87
DMU14	3.254	1799,29	3237437,20	314576799.00	47859508.99	2290532600627580.00	2135.99	147.25	21682.18
DMU15	1.961	1119,70	1253725,42	155055496.00	2671017.72	7134335644503.00	1104.09	79.54	6327.33
DMU16	1.976	1167,31	1362617,06	202064914.00	14913923.68	224425119456562.00	1648.04	68.14	4643.19
DMU17	5.695	3494,37	12210641,07	399842323.00	121235920.84	14698148501526300.00	3134.13	236.15	55766.36
DMU18	1.421	883,60	780746,85	110990000.00	22512854.66	506828624971485.00	1325.67	94.42	8914.67
DMU19	4.139	2473,25	6116948,33	278923725.00	41499944.88	1722245424875050.00	1917.46	153.15	23453.68
DMU20	1.393	749,09	561131,20	142461434.00	28650752.49	820865618435018.00	749.67	49.81	2480.67
DMU21	2.344	1463,94	2143124,98	238342910.00	8459558.77	71564134519172.00	1123.84	18.91	357.51

As seen in Table 4, we may state that in both the deterministic and stochastic models the DMUs DMU2, DMU4, DMU8, DMU11, DMU12, DMU15, DMU16, DMU18 are permanently efficient. That is, they are technical efficient under constant returns to scale conditions.

Stochastic DEA model on different levels of α were applied for the assessment of the electricity markets in the study. The SDEA model on different levels of α were ran for the assessment of the electricity markets in this study. The efficiency scores of the stochastic CCR DEA model for each decision-making unit on different levels of α are shown in Tables 5. Figure 1 shows the deviations around $\alpha=0.6$.

In the study, the α levels were considered as 0.005, 0.02, 0.05, 0.1, 0.2, 0.3, 0.4, 0.5, and 0.6 and the program was run separately for each α value. The results of the SDEA model (17) for all α levels are shown in Tables 5. The technical efficiency scores which equal to 1 implies full efficiency. On the other hand, if the score is higher than 1, it means inefficiency. As seen in Table 5, the α level of 0.6 deviated the electricity units' efficiency values.

In stochastic model, we may state that in both the deterministic and the stochastic models the DMUs DMU2, DMU4, DMU8, DMU11, DMU12, DMU15, DMU16, DMU18 are permanently efficient. The results showed that, as the error increased from 0.5 to 0.6, the estimated value of efficiency deviates. Tables 5 shows that

Table 4. Estimated output parameters

DMU	Output1			Output2		
	Mean	Std_Dev	Variance	Mean	Std_Dev	Variance
DMU1	7316471.00	539189.78	290725619211.49	1838326.00	124023.19	15381750618.57
DMU2	3477428.00	161908.62	26214401426.62	598499.00	36932.45	1364006028.67
DMU3	1875126.00	239648.50	57431405142.77	830171.00	113954.85	12985708066.70
DMU4	7112063.00	660832.41	436699478718.61	1687866.00	93922.39	8821415050.57
DMU5	13064862.00	886153.75	785268471865.13	3705568.00	278765.64	77710280227.90
DMU6	21993282.00	1009144.02	1018371650462.02	4477189.00	334546.21	111921167179.90
DMU7	2252326.00	58568.69	3430291016.26	867575.00	37811.66	1429721365.77
DMU8	3092327.00	220742.51	48727255515.56	1183468.00	66552.07	4429177933.37
DMU9	5384669.00	939234.15	882160785559.03	1373547.00	254943.22	64996043781.10
DMU10	2584385.00	255618.14	65340631985.90	811754.00	57714.16	3330923785.07
DMU11	12877336.00	626053.29	391942719688.31	2736291.00	290321.55	84286605111.50
DMU12	10276188.00	575962.05	331732287468.03	2511023.00	183914.14	33824410418.97
DMU13	1968874.00	155113.92	24060326866.06	624615.00	36322.62	1319332871.37
DMU14	7178811.00	851951.20	725820854681.09	1818104.00	101406.04	10283184285.50
DMU15	5490401.00	342759.01	117483737284.82	1542960.00	86365.11	7458931777.47
DMU16	8124983.00	400747.69	160598707255.06	1522795.00	85783.91	7358878670.17
DMU17	12605873.00	1282046.34	1643642823752.60	3339914.00	222674.02	49583716997.60
DMU18	5874359.00	311958.34	97318006749.71	929968.00	51006.94	2601708074.67
DMU19	10009283.00	614333.55	377405715379.90	2770122.00	159424.50	25416170709.87
DMU20	1496625.00	128191.36	16433025011.07	549781.00	53381.78	2849614092.67
DMU21	4743914.00	46426.89	2155456096.42	1794627.00	111635.04	12462381663.47

efficiency of the electricity distribution units deviated while the level of error was higher than a half. Additionally, it is clear that, when the α values increased, the number of efficient units decreased.

According to the results, both in the deterministic and stochastic CCR DEA models, the DMUs DMU1, DMU3, DMU5, DMU6, DMU7, DMU9, DMU10, DMU13, DMU14, DMU17, DMU19, DMU20 and DMU21 are inefficient. In general, we may state that the units DMU2, DMU4, DMU8, DMU11, DMU12, DMU15, DMU16, DMU18 can be selected as benchmarked units for future decisions and planning. The referenced DMUs are efficient on two models. Inefficient DMUs can achieve the same level of effectiveness if they apply the management and organization methods of efficient DMUs.

Table 5. The results of deterministic output-oriented CCR DEA model

	CCR DEA
DMU1	1.0843
DMU2	1.0000
DMU3	1.4929
DMU4	1.0000
DMU5	1.1286
DMU6	1.1178
DMU7	1.0717
DMU8	1.0000
DMU9	2.2532
DMU10	1.1061
DMU11	1.0000
DMU12	1.0000
DMU13	1.2648
DMU14	1.4294
DMU15	1.0000
DMU16	1.0000
DMU17	1.2086
DMU18	1.0000
DMU19	1.0305
DMU20	2.1014
DMU21	1.0724

Validation of the Proposed Approach

The proposed model must be validated by other models. In this study, the stochastic DEA model under the deterministic conditions were compared to the deterministic DEA model to show their consistent performance. For this purpose, the stochastic and deterministic DEA models were compared for validation. In the SDEA model, if $a_{ij} = \sqrt{Var\left(\overline{x}_{ij}\right)} = 0$, $b_{rj} = \sqrt{Var\left(\overline{y}_{rj}\right)} = 0$ and the value of input and output equal to the mean results of SDEA, and deterministic DEA must be similar. Different data envelopes are obtained by assigning different α values between 0 and 1. When $\alpha = 0.5$, the inverse of the distribution function is equal to 0.

When the results of the deterministic and stochastic models are compared with this α value, the results are found to be same. For levels where α is different from

Table 6. The results of stochastic CCR DEA

	α=0.005	α=0.02	α=0.05	α=0.1	α=0.2	α=0.3	α=0.4	α=0.5	α=0.6
DMU1	1.033	1.040	1.043	1.045	1.046	1.052	1.066	1.084	1.544
DMU2	1.000	1.000	1.000	1.000	1.000	1.000	1.000	1.000	2.139
DMU3	1.170	1.203	1.231	1.257	1.319	1.383	1.439	1.493	5.543
DMU4	1.000	1.000	1.000	1.000	1.000	1.000	1.000	1.000	1.567
DMU5	1.000	1.000	1.011	1.036	1.061	1.077	1.100	1.129	1.527
DMU6	1.000	1.000	1.000	1.000	1.002	1.055	1.087	1.118	1.897
DMU7	1.000	1.000	1.000	1.000	1.007	1.036	1.054	1.072	1.815
DMU8	1.000	1.000	1.000	1.000	1.000	1.000	1.000	1.000	1.436
DMU9	1.303	1.441	1.555	1.685	1.845	1.990	2.120	2.253	9.272
DMU10	1.000	1.000	1.000	1.000	1.000	1.000	1.039	1.106	2.205
DMU11	1.000	1.000	1.000	1.000	1.000	1.000	1.000	1.000	2.117
DMU12	1.000	1.000	1.000	1.000	1.000	1.000	1.000	1.000	1.631
DMU13	1.233	1.238	1.241	1.244	1.249	1.252	1.255	1.265	1.669
DMU14	1.284	1.313	1.331	1.348	1.374	1.393	1.411	1.429	3.235
DMU15	1.000	1.000	1.000	1.000	1.000	1.000	1.000	1.000	2.162
DMU16	1.000	1.000	1.000	1.000	1.000	1.000	1.000	1.000	1.364
DMU17	1.000	1.000	1.000	1.000	1.019	1.082	1.140	1.209	1.424
DMU18	1.000	1.000	1.000	1.000	1.000	1.000	1.000	1.000	2.054
DMU19	1.000	1.000	1.000	1.000	1.000	1.000	1.002	1.030	1.245
DMU20	1.772	1.799	1.823	1.844	1.939	2.002	2.053	2.101	2.916
DMU21	1.000	1.000	1.000	1.000	1.014	1.039	1.059	1.072	1.330

0.5, the envelope changes with the effect of the chance constraint. Table 6 shows the results of the deterministic and stochastic models under these conditions. According to Table 6 and Figure 2, the deterministic CCR and stochastic CCR models had the same results at the 0,5 level of α. Thus, the stochastic DEA model were validated and verified by the deterministic DEA model.

CONCLUSION

Stochastic programming is an approach for modelling and solving optimization problems that include stochastic data. The chance-constrained programming approach is one of the most important methods of stochastic programming. Chance-constrained stochastic DEA models create a more flexible efficiency limit than classical DEA

Figure 1. The deviations when α=0

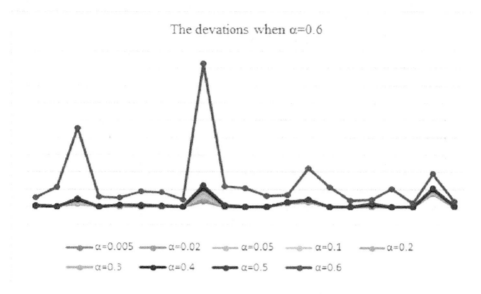

models, since they are based on the assumption of the inputs and/or outputs are randomly determined.

In the study, the deterministic model of nonlinear stochastic CCR model was transformed into linear model by a symmetric error structure. The CCR model evaluates technical efficiencies of DMUs by reference to the efficiency frontier. As it may be seen from the deterministic results a unit that is CCR efficient is also efficient in the stochastic CCR model.

By the results in Table 5 obtained by the chance constrained stochastic CCR method, it was shown that the stochastic efficiency could be deviates with an error rate of more than fifty percent. If a DMU is inefficient on the lowest error level, DMU will be permanently inefficient. These results emphasized that more attention must be paid to the error levels to evaluate the stochastic efficiency. As a result, we may say that Turkey's electricity distribution companies are permanently effective in scale DMU2, DMU4, DMU8, DMU11, DMU12, DMU15, DMU16 and DMU18.

In the performance measurement of the electricity distribution companies, the deterministic and stochastic CCR output-oriented efficiency values within the symmetric error structure showed consistency as shown in the literature. This study demonstrates to us that benchmarking may be used to calculate an indication of performance of the energy markets or utilities, but uncertainty must be handled carefully. Implementation of a symmetric error structure in other DEA models may be recommended for future energy market performance benchmarking studies.

Table 7. Validation-The comparison of the stochastic and deterministic CCR DEA models at the 0,5 level of α.

	Stochastic CCR DEA	CCR DEA
DMU1	1.084	1.084
DMU2	1.000	1.000
DMU3	1.493	1.493
DMU4	1.000	1.000
DMU5	1.129	1.129
DMU6	1.118	1.118
DMU7	1.072	1.072
DMU8	1.000	1.000
DMU9	2.253	2.253
DMU10	1.106	1.106
DMU11	1.000	1.000
DMU12	1.000	1.000
DMU13	1.265	1.265
DMU14	1.429	1.429
DMU15	1.000	1.000
DMU16	1.000	1.000
DMU17	1.209	1.209
DMU18	1.000	1.000
DMU19	1.030	1.030
DMU20	2.101	2.101
DMU21	1.072	1.072

REFERENCES

Amado, C., Santos, S., & Sequeira, J. (2013). Using data envelopment analysis to support the design of process improvement interventions in electricity distribution. *European Journal of Operational Research*, *228*(1), 226–235. doi:10.1016/j.ejor.2013.01.015

Arcos-Vargas, A., Núñez-Hernández, F., & Villa-Caro, G. (2017). A DEA analysis of electricity distribution in Spain: An industrial policy recommendation. *Energy Policy*, *102*, 583–592. doi:10.1016/j.enpol.2017.01.004

Figure 2. Validation of the models at the 0,5 level of α

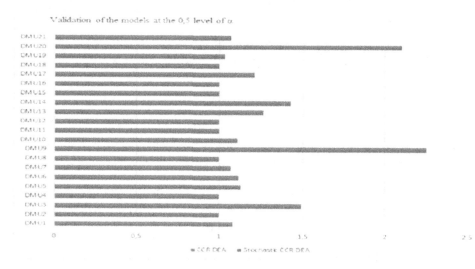

Azadeh, A., Ghaderi, S. F., & Omrani, H. (2009). A Deterministic Approach for Performance Assessment And Optimization Of Power Distribution Units in Iran. *Energy Policy*, *37*(1), 274–280. doi:10.1016/j.enpol.2008.08.027

Azadeh, A., Ghaderi, S. F., Omrani, H., & Eivazy, H. (2009). An integrated DEA–COLS–SFA algorithm for optimization and policy making of electricity distribution units. *Energy Policy*, *37*(7), 2605–2618. doi:10.1016/j.enpol.2009.02.021

Azadeh, A., Motevali Haghighi, S., Zarrin, M., & Khaefi, S. (2015). Performance evaluation of Iranian electricity distribution units by using stochastic data envelopment analysis. *Electrical Power and Energy Systems*, *73*, 919–931. doi:10.1016/j.ijepes.2015.06.002

Banker, R. D., Charnes, A., & Cooper, W. W. (1984). Some Models for Estimating Technical and Scale Inefficiencies in Data Envelopment Analysis. *Management Science*, *30*(9), 1078–1092. doi:10.1287/mnsc.30.9.1078

Behzadi, M. H., & Mirbolouki, M. (2012). Symmetric Error Structure in Stochastic DEA. *Int. J. Industrial Mathematics*, *4*, 335–343.

Behzadi, M. H., Nematollahi, N., & Mirbolouki, M. (2009). Ranking Efficient DMUs with Stochastic Data by Considering Inecient Frontier. *International Journal of Industrial Mathematics*, *1*, 219–226.

Birge, J. R., & Louveaux, F. (1997). *Introduction to stochastic programming.* Springer- Verlag.

Brazdik, F. (2004). Stochastic Data Envelopment Analysis: Oriented and Linearized Models. Joint Workplace of the Center for Economic Research and Graduate Education, Charles University, Prague, and the Economics Institute of the Academy of Sciences of the Czech Republic.

Brockett, P. L., & Golany, B. (1996). Using Rank Statistics for Determining Programmatic Efficiency Differences in Data Envelopment Analysis. *Management Science, 42*(3), 466–472. doi:10.1287/mnsc.42.3.466

Çelen, A. (2013). Efficiency and productivity (TFP) of the Turkish electricity distribution companies: An application of two-stage (DEA&Tobit) analysis. *Energy Policy, 63*, 300–310. doi:10.1016/j.enpol.2013.09.034

Charnes, A., & Cooper, W. W. (1959). Chance-Constrained Programming. *Management Science, 6*(1), 73–79. doi:10.1287/mnsc.6.1.73

Charnes, A., & Cooper, W. W. (1962). Chance Constraints and Normal Deviates. *Journal of the American Statistical Association, 57*(297), 134–148. doi:10.1080/0 1621459.1962.10482155

Charnes, A., Cooper, W. W., & Rhodes, E. (1978). Measuring the efficiency of decision-making units. *European Journal of Operational Research, 2*(6), 429–444. doi:10.1016/0377-2217(78)90138-8

Cooper, W. W., Deng, H., Huang, Z., & Li, S. X. (2002). Chance constrained programming approaches to technical efficiencies and inefficiencies in stochastic data envelopment analysis. *The Journal of the Operational Research Society, 53*(12), 1347–1356. doi:10.1057/palgrave.jors.2601433

Cooper, W. W., Deng, H., Huang, Z., & Li, S. X. (2004). Chance constrained programming approaches to congestion in stochastic data envelopment analysis. *European Journal of Operational Research, 155*(2), 487–501. doi:10.1016/S0377-2217(02)00901-3

Cooper, W. W., Huang, Z., & Li, S. (1996). Satisficing DEA models under chance constraints. *Annals of Operations Research, 66*(4), 259–279. doi:10.1007/BF02187302

Cooper, W. W., Seiford, L. M., & Tone, K. (2007). *Data envelopment analysis: a comprehensive text with models, applications, references and DEA-solver software* (2nd ed.). Springer. doi:10.1007/978-0-387-45283-8

Cooper, W. W., Thompson, R. G., & Thrall, R. M. (1996). Introduction: Extensions and new developments in VZA. *Annals of Operations Research, 66*(1), 3–46. doi:10.1007/BF02125451

Demireli, E., & Özdemir, A. Y. (2013). Seçilmiş Avrupa Ülkelerinde Makroekonomik Performans Ölçümü: Şans Kısıtlı Veri Zarflama Analizi İle Bir Uygulama. *Dumlupınar University Journal of Social Sciences, 37.*

Fallahi, A., Ebrahimi, R., & Ghaderi, S. F. (2011). Measuring efficiency and productivity change in power electric generation management companies by using data envelopment analysis: A case study. *Energy, 36*(11), 6398–6405. doi:10.1016/j. energy.2011.09.034

Farrell, M. J. (1957). The measurement of productive efficiency. J R Stat Soc Ser A. *GEN, 120*, 253–290.

Gil, D. R. G., Costa, M. A., Lopes, A. L. M., & Mayrink, V. D. (2017). Spatial statistical methods applied to the 2015 Brazilian energy distribution benchmarking model: Accounting for unobserved determinants of inefficiencies. *Energy Economics, 64*, 373–383. doi:10.1016/j.eneco.2017.04.009

Gouveia, M. C., Dias, L. C., Antunes, C. H., Boucinha, J., & Inácio, C. F. (2015). Benchmarking of maintenance and outage repair in an electricity distribution company using the value-based DEA method. *Omega, 53*, 104–114. doi:10.1016/j. omega.2014.12.003

Gstach, D. (1998). Another approach to data envelopment analysis in noisy environments: DEA+. *Journal of Productivity Analysis, 9*(2), 161–176. doi:10.1023/A:1018312801700

Huang, Z., & Li, S. X. (2001). Stochastic DEA models with different types of input-output disturbances. *Journal of Productivity Analysis, 15*(2), 95–113. doi:10.1023/A:1007874304917

Jahanshahloo, G. R., Behzadi, M. H., & Mirbolouki, M. (2010). Ranking Stochastic Efficient DMUs based on Reliability. *International Journal of Industrial Mathematics, 2*, 263–270.

Jahanshahloo, G. R., Behzadi, M. H., & Mirbolouki, M. (2010). Ranking Stochastic Efficient DMUs based on Reliability. *International Journal of Industrial Mathematics, 2*, 263–270.

Jamasb, T., & Pollitt, M. (2001). Benchmarking and Regulation: International Electricity Experience. *Utilities Policy*, *9*(3), 107–130. doi:10.1016/S0957-1787(01)00010-8

Jamasb, T., & Pollitt, M. (2003). International Benchmarking and Regulation: An Application to European Electricity Distribution Utilities. *Energy Policy*, *31*(15), 1609–1622. doi:10.1016/S0301-4215(02)00226-4

Khodabakhshi, M. (2010). An Output Oriented Super-Efficiency Measure in Stochastic Data Envelopment Analysis: Considering Iranian Electricity Distribution Companies. *Computers & Industrial Engineering*, *58*(4), 663–671. doi:10.1016/j. cie.2010.01.009

Khodabakhshi, M., & Asgharian, M. (2008). An input relaxation measure of efficiency in stochastic data envelopment analysis. *Applied Mathematical Modelling*, *33*(4), 2010–2023. doi:10.1016/j.apm.2008.05.006

Khodabakhshi, M., & Asgharian, M. (2008). An input relaxation measure of efficiency in stochastic data envelopment analysis. *Applied Mathematical Modelling*, *33*(4), 2010–2023. doi:10.1016/j.apm.2008.05.006

Khodabakhshi, M., & Kheirollahi, H. (2010). Measuring technical efficiency of Iranian electricity distribution units with stochastic data envelopment analysis. *Iranian Conference on Applied Mathematical Modelling.*

Khodadadipour, M., Hadi-Vencheh, A., Behzadi, M. H., & Rostamy-malkhalifeh, M. (2021). Undesirable factors in stochastic DEA cross-efficiency evaluation: An application to thermal power plant energy efficiency. *Economic Analysis and Policy*, *69*, 613–628. doi:10.1016/j.eap.2021.01.013

Khodakaramia, A., Farahanib, F. H., & Aghaeid, J. (2017). Stochastic characterization of electricity energy markets including plug-in electric vehicles. *Renewable & Sustainable Energy Reviews*, *69*, 112–122. doi:10.1016/j.rser.2016.11.094

Land, C. K., Lovell, C. A. K., & Thore, S. (1993). Chance-Constrained Data Envelopment Analysis. *Managerial and Decision Economics*, *14*(6), 541–554. doi:10.1002/mde.4090140607

Land, C. K., Lovell, C. A. K., & Thore, S. (1994). Productive Efficiency under Capitalism and State Socialism: An Empirical Inquiry Using Chance-Constrained Data Envelopment Analysis. *Technological Forecasting and Social Change*, *46*(2), 139–152. doi:10.1016/0040-1625(94)90022-1

Li, S. X. (1998). Stochastic models and variable returns to scales in data envelopment analysis. *European Journal of Operational Research, 104*(3), 532–548. doi:10.1016/S0377-2217(97)00002-7

Lopes, A. L. M., & Mesquita, R. B. (2015). Tariff regulation of electricity distribution: A comparative analysis of regulatory benchmarking models. *Proceedings of the 14th European Workshop on Efficiency and Productivity Analysis*.

Mirbolouki, M., Behzadi, M.H., & Korzaledin, M. (2014). Multiplier, models in stochastic DEA. *Data Envelopment Analysis and Decision Science*.

Mullarkey, S., Caulfield, B., McCormack, S., & Basu, B. (2015). A framework for establishing the technical efficiency of Electricity Distribution Counties (EDCs) using Data Envelopment Analysis. *Energy Conversion and Management, 94*, 112–123. doi:10.1016/j.enconman.2015.01.049

Odyakmaz, N. (2009). *Türkiye'deki Elektrik Dağıtım Şirketlerinin Performansa Dayalı Düzenleme Çerçevesinde Karşılaştırmalı Etkinlik Analizi* [PhD Thesis]. Institute of Social Sciences, Hacettepe University.

Olesen, O. B. (2002). *Comparing and Combining Two Approaches for Chance Constrained DEA. Discussion paper*. The University of Southern Denmark.

Omrani, H., Azadeh, A., Ghaderi, S. F., & Abdollahzadeh, S. (2010). A Consistent Approach for Performance Measurement Of Electricity Distribution Companies. *Int J Energy Sect Manage, 4*(3), 399–416. doi:10.1108/17506221011073879

Omrani, H., Beiragh, R. G., & Kaleibari, S. S. (2015). Performance assessment of Iranian electricity distribution companies by an integrated cooperative game data envelopment analysis principal component analysis approach. *Electrical Power and Energy Systems, 64*, 617–625. doi:10.1016/j.ijepes.2014.07.045

Sadjadi, S. J., & Omrani, H. (2008). Data Envelopment Analysis with Uncertain Data: An Application for Iranian Electricity Distribution Companies. *Energy Policy, 36*(11), 4247–4254. doi:10.1016/j.enpol.2008.08.004

Sadjadi, S. J., Omrani, H., Makui, A., & Shahanaghi, K. (2011). An interactive robust data envelopment analysis model for determining alternative targets in Iranian electricity distribution companies. *Expert Systems with Applications, 38*(8), 9830–9839. doi:10.1016/j.eswa.2011.02.047

Sengupta, J. K. (2002). Efficiency analysis by stochastic data envelopment analysis. *Applied Economics Letters, 7*(6), 379–383. doi:10.1080/135048500351311

Shiraz, R. K., Hatami-Marbini, A., Emrouznejad, A., & Fukuyama, H. (2018). Chance-constrained cost efficiency in data envelopment analysis model with random inputs and outputs. *Operations Research*. Advance online publication. doi:10.1051/ro/2016076

Sözen, A., Alp, I., & Kılınc, C. (2012). Efficiency assessment of the hydro-power plants in Turkey by using data envelopment analysis. *Renewable Energy, 46*, 192–202. doi:10.1016/j.renene.2012.03.021

Sueyoshi, T. (2000). Stochastic DEA for restructure strategy: An application to a Japanese petroleum company. *Omega, 28*(4), 385–398. doi:10.1016/S0305-0483(99)00069-9

Talluri, S., Narasimhan, R., & Nair, A. (2006). Vendor performance with supply risk: A chance-constrained DEA approach. *International Journal of Production Economics, 100*(2), 212–222. doi:10.1016/j.ijpe.2004.11.012

Thakur, T., Deshmukh, S. G., & Kaushik, S. C. (2006). Efficiency evaluation of the state-owned electric utilities in India. *Energy Policy, 34*(17), 2788–2804. doi:10.1016/j.enpol.2005.03.022

Weyman-Jones, G. T. (1991). Productive efficiency in regulated industry: The Area Electricity Boards of England and Wales. *Energy Economics, 3*(2), 116–122. doi:10.1016/0140-9883(91)90043-Y

Yenioğlu, Z. A., & Toklu, B. (2021). Stokastik Veri Zarflama Analizi ile Etkinlik Ölçümü: Türkiye Elektrik Dağıtım Şirketlerinin Karşılaştırmalı Analizi. *Politeknik Dergisi, 24*, 87–101.

Compilation of References

Aassime, A., Johansson, G., Wendin, G., Schoelkopf, R. J., & Delsing, P. (2001). Radio Frequency Single-Electron Transistor as Readout Device for Qubits: Charge Sensitivity and Backaction. *Physical Review Letters*, *86*(15), 86. doi:10.1103/PhysRevLett.86.3376 PMID:11327974

Abbas, K., Tawalbeh, L. A., Rafiq, A., Muthanna, A., Elgendy, I. A., & Abd El-Latif, A. A. (2021). Convergence of Blockchain and IoT for Secure Transportation Systems in Smart Cities. *Security and Communication Networks*, *2021*, 1–13. Advance online publication. doi:10.1155/2021/5597679

Abdelkrim, M. (2019). Modeling and simulation of single-electron transistor (SET) with aluminum island using neural network. *Carpath J Electron Comput Eng*, *12*(1), 23–28. doi:10.2478/cjece-2019-0005

Abroshan, S., & Moghaddam, M. H. Y. (2014, May). SESRT: Score based Event to Sink Reliable Transport in wireless sensor networks. In *Electrical and Computer Engineering (CCECE), 2014 IEEE 27th Canadian Conference on* (pp. 1-6). IEEE.

Achigui, H. F., Sawan, M., & Fayomi, C. J. B. (2007). 1 V fully balanced differential amplifiers: Implementation and experimental results. *Analog Integrated Circuits and Signal Processing*, *53*(1), 19–25. doi:10.100710470-006-9002-z

Adelstein, F., Gupta, S. K., Richard, G., & Schwiebert, L. (2005). *Fundamentals of mobile and pervasive computing*. McGraw-Hill.

Agostini, B., Fabbri, M., Park, J. E., Wojtan, L., Thome, J. R., & Michel, B. (2007). State of the art of high heat flux cooling technologies. *Heat Transfer Engineering*, *28*(4), 258–281. doi:10.1080/01457630601117799

Ahlawat, A., & Malik, V. (2013). An Extended Vice-Cluster Selection Approach to Improve V Leach Protocol in WSN. *2013 Third International Conference on Advanced Computing and Communication Technologies (ACCT)*, 236-240. 10.1109/ACCT.2013.60

Ahsan, M. (2018). Single electron transistor (SET): Operation and application perspectives. *MIST Int J Sci Technol*, *6*, 1.

Akan, Ö. B., & Akyildiz, I. F. (2015). Event-to-sink reliable transport in wireless sensor networks. *IEEE/ACM Transactions on Networking*, *13*(5), 1003–1016. doi:10.1109/TNET.2005.857076

Akyildiz, I. F., & Kasimoglu, I. H. (2014). Wireless sensor and actor networks: Research challenges. *Ad Hoc Networks*, 2(4), 351–367. doi:10.1016/j.adhoc.2004.04.003

Akyildiz, I. F., Su, W., Sankarasubramaniam, Y., & Cayirci, E. (2020). A survey on sensor networks. *Communications Magazine, IEEE*, 40(8), 102–114. doi:10.1109/MCOM.2002.1024422

Alam, M. M., & Hong, C. S. (2019). CRRT: Congestion-aware and ratecontrolled reliable transport in wireless sensor networks. *IEICE Transactions on Communications*, 92(1), 184–199. doi:10.1587/transcom.E92.B.184

Alamsyah, D. P., Ramdhani, Y., Rhamadhan, S. R., & Susanti, L. (2021). Application of IoT and Cloud Storage in Android-Based Smart Home Technology. *Proceedings of 2021 7th International HCI and UX Conference in Indonesia, CHIuXiD 2021*. 10.1109/CHIuXiD54398.2021.9650605

Al-Awami, L., & Hassanein, H. (2018, October). Energy efficient data survivability for WSNs via decentralized erasure codes. *IEEE 37th International Conference on Local Computer Networks*, 577–584.

Alaybeyoglu, E., & Ozenli, D. (2022). Operational Amplifier Design Employing DTMOS Technique with Dual Supply Voltages. *Journal of Circuits, Systems and Computers Vol.*, 31(02), 2250035. doi:10.1142/S0218126622500359

Albert, J. R., Kaliannan, T., Singaram, G., Sehar, F. I. R. E., Periasamy, M., & Kuppusamy, S. (2022). A remote diagnosis using variable fractional order with reinforcement controller for solar-MPPT intelligent system. In *Photovoltaic Systems* (pp. 45–64). CRC Press. doi:10.1201/9781003202288-3

Alfa, A. A., Alhassan, J. K., Olaniyi, O. M., & Olalere, M. (2021). Blockchain technology in IoT systems: Current trends, methodology, problems, applications, and future directions. *Journal of Reliable Intelligent Environments*, 7(2), 115–143. Advance online publication. doi:10.100740860-020-00116-z

Ali, A., Latiff, L. A., Sarijari, M. A., & Fisal, N. (2018, June). Real-time routing in wireless sensor networks. In *Distributed Computing Systems Workshops, 2008. ICDCS '08. 28th International Conference on* (pp. 114-119). IEEE.

Alidina, M., Monteiro, J., Devadas, S., Ghosh, A., & Papaefthymiou, M. (1994). Precomputation-based sequential logic optimization for low power. *IEEE Transactions on Very Large-Scale Integration (VLSI) Systems*, 2(4), 426–436.

Alimohammadi, H., & Jassbi, S. J. (2019). A Fault Tolerant Multipath Routing Protocol for Mobile Ad Hoc Networks. *European Journal of Scientific Research*, 85(2), 317–326.

Ali, N., Leong, K., & Mustapha, A. (2016). Experimental Investigation on Piezoelectric and Electromagnetic Hybrid Micro-Power Generator. *Journal of Engineering and Applied Sciences (Asian Research Publishing Network)*, 11(10).

Alippi, C., Camplani, R., Galperti, C., & Roveri, M. (2010). A Robust, Adaptive, Solar-Powered WSN Framework for Aquatic Environmental Monitoring. *IEEE Sensors Journal*, 11(1), 45–55. doi:10.1109/JSEN.2010.2051539

Ali, S., Fakoorian, A., & Taheri, H. (2017, October). Optimum Reed-Solomon erasure coding in fault tolerant sensor networks. *4th International Symposium on Wireless Communication Systems (ISWCS)*, 6–10.

Almoneef, T. S., & Ramahi, O. M. (2014). Can Split Ring Resonators be viable for electromagnetic energy harvesting? IEEE Antennas and Propagation Society, 424-425.

Almutairi, A. F., Islam, M. S., Samsuzzaman, M., Islam, M. T., Misran, N., & Islam, M. T. (2019). A Complementary Split Ring Resonator Based Metamaterial with Effective Medium Ratio for C-band Microwave Applications. *Results in Physics*, *15*, 102675. doi:10.1016/j.rinp.2019.102675

Alsbouí, T. A. A., Hammoudeh, M., Bandar, Z., & Nisbet, A. (2019). An Overview and Classification of Approaches to Information Extraction in Wireless Sensor Networks. *Proceedings of Fifth International Conference on Sensor Technologies and Applications (SENSORCON)*, 255-260.

Amado, C., Santos, S., & Sequeira, J. (2013). Using data envelopment analysis to support the design of process improvement interventions in electricity distribution. *European Journal of Operational Research*, *228*(1), 226–235. doi:10.1016/j.ejor.2013.01.015

Amarakoon, S., Smith, J., & Segal, B. (2014). *Application of life-cycle assessment to nanoscale technology: Lithium-ion batteries for electric vehicles*. No. EPA 744-R-12-001.

Ambrose, H., & Kendall, A. (2016). Effects of battery chemistry and performance on the life cycle greenhouse gas intensity of electric mobility. *Transportation Research Part D, Transport and Environment*, *47*, 182–194. doi:10.1016/j.trd.2016.05.009

Aminov, P., Jai, P., & Agrawal. (2018). RF Energy Harvesting. *IEEE 64th Electronic Components and Technology Conference (ECTC)*, 1838-1841.

Amiri, M., Tofigh, F., Shariati, N., Lipman, J., & Abolhasan, M. (2020). Wide angle metamaterial absorber with highly insensitive absorption for TE & TM modes. *Scientific Reports*, *10*(1), 2. doi:10.103841598-020-70519-8 PMID:32788706

Anand, Joseph, & Oommen. (2014). Performance Analysis and Implementation of Clock Gating Techniques for Low Power Applications. In *2014 International Conference on Science Engineering and Management Research (ICSEMR)*. IEEE.

Ang, K. H., Chong, G., & Li, Y. (2015). PID control system analysis, design, and technology. *IEEE Transactions on Control Systems Technology*, *13*(4), 559–576.

Anusha & Deepika. (2016). Design and Implementation of Novel 8-Bit Universal Shift Register Using Reversible Logic Gates. *International Journal of Innovation in Engineering and Technology*.

Arcos-Vargas, A., Núñez-Hernández, F., & Villa-Caro, G. (2017). A DEA analysis of electricity distribution in Spain: An industrial policy recommendation. *Energy Policy*, *102*, 583–592. doi:10.1016/j.enpol.2017.01.004

Arrawatia, M., Baghini, M. S., & Kumar, G. (2016). RF energy harvesting system at 2.67 and 5.8GHz. *Asia-Pacific Microwave Conference*, 900-903.

Aslam, A. R., & Altaf, M. A. B. (2019). An 8 Channel patient specific neuromorphic processor for the early screening of autistic children through emotion detection. *Proc. IEEE Int. Symp. Circuits Syst. (ISCAS)*, 1–5.

Asli, A., & Wong, Y. C. (2018). −31 dBm sensitivity high efficiency rectifier for energy scavenging. *International Journal of Electronics and Communication*, *91*, 44–54. doi:10.1016/j.aeue.2018.04.019

Atanasov, A., Kulakov, A., Trajkovic, V., & Davcev, D. (2010). Testbed Environment for Wireless Sensor and Actuator Networks. *Fifth International Conference on Systems and Networks Communications*, 1,6, 22-27. 10.1109/ICSNC.2010.8

Australian Trade and Investment Commission. (2018). *The Lithium-Ion Battery Value Chain New Economy Opportunities for Australia*. Author.

Ayadi, A. (2018). Energy-efficient and reliable transport protocols for wireless sensor networks: State-of-art. *Wireless Sensor Network*, *3*(3), 106–113. doi:10.4236/wsn.2011.33011

Azadeh, A., Ghaderi, S. F., & Omrani, H. (2009). A Deterministic Approach for Performance Assessment And Optimization Of Power Distribution Units in Iran. *Energy Policy*, *37*(1), 274–280. doi:10.1016/j.enpol.2008.08.027

Azadeh, A., Ghaderi, S. F., Omrani, H., & Eivazy, H. (2009). An integrated DEA–COLS–SFA algorithm for optimization and policy making of electricity distribution units. *Energy Policy*, *37*(7), 2605–2618. doi:10.1016/j.enpol.2009.02.021

Azadeh, A., Motevali Haghighi, S., Zarrin, M., & Khaefi, S. (2015). Performance evaluation of Iranian electricity distribution units by using stochastic data envelopment analysis. *Electrical Power and Energy Systems*, *73*, 919–931. doi:10.1016/j.ijepes.2015.06.002

Azghadi, M. R., Eshraghian, J. K., & Linares-Barranco, B. (2020). Hardware Implementation of Deep Network Accelerators Towards Healthcare and Biomedical Applications. *IEEE Transactions on Biomedical Circuits and Systems*.

Baghyalakshmi, D., & Ebenezer, J. & SatyaMurty, S.A.V. (2010, Jan). Low latency and energy efficient routing protocols for wireless sensor networks. *International Conference on Wireless Communication and Sensor Computing*, 1,6. 10.1109/ICWCSC.2010.5415892

Bagmanci, M., Karaaslan, M., Altintaş, O., Karadag, F., Tetik, E., & Bakir, M. (2018). Wideband metamaterial absorber based on CRRs with lumped elements for microwave energy harvesting. *The Journal of Microwave Power and Electromagnetic Energy*, *52*(1), 45–59. doi:10.1080/08327823.2017.1405471

Bahiraei, M., & Heshmatian, S. (2018). Electronics cooling with nanofluids: A critical review. *Energy Conversion and Management*, *172*, 438–456.

Bai, Z., Liu, X., Lian, Z., Zhang, K., Wang, G., Shi, S. F., Pi, X., & Song, F. (2018). A silicon cluster based single electron transistor with potential room-temperature switching. *Chinese Physics Letters*, *35*(3), 037301. doi:10.1088/0256-307X/35/3/037301

Bakır, M., Karaaslan, M., Unal, E., Akgol, O., & Sabah, C. (2017). Microwave metamaterial absorber for sensing applications. *Opto-Electronics Review*, *25*(4), 318–325. doi:10.1016/j. opelre.2017.10.002

Balasa & Pradhan. (2012). *Energy-Aware Memory Management for Embedded Multimedia Systems*. CRC Press.

Bamnote, H., Rakhonde, A., Ghagare, S., Dharpure, S., Kapse, K., & Warjurkar, S. (2021). IoT-based door access control system using face recognition. *Tech-Chronicle, 3*(2).

Banker, R. D., Charnes, A., & Cooper, W. W. (1984). Some Models for Estimating Technical and Scale Inefficiencies in Data Envelopment Analysis. *Management Science*, *30*(9), 1078–1092. doi:10.1287/mnsc.30.9.1078

Bao, D., & Liourav, X. (2019). A Cache Scheme Based on LRU-Like Algorithm. *Proceedings of the 2019 IEEE International Conference on Information and Automation*, 2055-2060.

Baqir, M. A., Ghasemi, M., Choudhury, P. K., & Majlis, B. Y. (2019). Design and analysis of nanostructured subwavelength metamaterial absorber operating in the UV and visible spectral range. *Journal of Electromagnetic Waves and Applications*, *29*(18), 2408–2419. doi:10.1080/0 9205071.2015.1073124

Barbieri, C. (2019). Development of an artificial intelligence model to guide the management of blood pressure, fluid volume, and dialysis dose in end-stage kidney disease patients: Proof of concept and first clinical assessment. *International Journal of Medical Informatics*, *5*(1), 28–33.

Barman, J., & Kumar, V. (2018). Approximate Carry Look Ahead Adder (Cla) for Error Tolerant Applications. In *2018 2nd International Conference on Trends in Electronics and Informatics (ICOEI)*. IEEE. 10.1109/ICOEI.2018.8553739

Baronti, P., Pillai, P., Chook, V. W., Chessa, S., Gotta, A., & Hu, Y. F. (2017). Wireless sensor networks: A survey on the state of the art and the 802.15. 4 and ZigBee standards. *Computer Communications*, *30*(7), 1655–1695. doi:10.1016/j.comcom.2006.12.020

Basak, M. E., & Kaçar, F. (2016). Ultra-Low Voltage VDCC Design by Using DTMOS. *Acta Physica Polonica*, *130*(1), 223–225. doi:10.12693/APhysPolA.130.223

Basak, M. E., & Kaçar, F. (2017). Ultra-Low Voltage VDBA Design By Using PMOS DTMOS Transistors. *Istanbul University-Journal of Electrical & Electronics Engineering*, *17*(2), 3463–3469.

Basanta Singh, N. (2013). Design and Implementation of Hybrid SETCMOS 4-to1 MUX and 2-to-4 Decoder Circuits. *International Journal of Advanced Research in Electrical, Electronics and Instrumentation Engineering, 2*.

Battery Test Centre. (2022). *Lithium Ion*. https://batterytestcentre.com.au/project/lithium-ion/

Battery University. (2022). *Types of Lithium-ion*. https://batteryuniversity.com/article/bu-205-types-of-lithium-ion

Beaumont, A., Dubuc, C., Beauvais, J., & Drouin, D. (2009). Room Temperature Single-Electron Transistor Featuring Gate-Enhanced ON-State Current. *IEEE Electron Device Letters*, *30*(7), 30. doi:10.1109/LED.2009.2021493

Beckett, R. G. (2015). Application and Limitations of Endoscopy in Anthropological and Archaeological Research. *The Anatomical Record*, *298*(6), 1125–1134. doi:10.1002/ar.23145 PMID:25998646

Behi, H., Ghanbarpour, M., & Behi, M. (2017). Investigation of PCM-assisted heat pipe for electronic cooling. *Applied Thermal Engineering*, *127*, 1132–1142.

Behzadi, M. H., & Mirbolouki, M. (2012). Symmetric Error Structure in Stochastic DEA. *Int. J. Industrial Mathematics*, *4*, 335–343.

Behzadi, M. H., Nematollahi, N., & Mirbolouki, M. (2009). Ranking Efficient DMUs with Stochastic Data by Considering Inecient Frontier. *International Journal of Industrial Mathematics*, *1*, 219–226.

Bein, D., Jolly, V., Kumar, B., & Latifi S. (2015). Reliability Modeling in Wireless Sensor Networks. *International Journal of Information Technology*, *11*(2).

Benakli, L. (2022). *What are the need for resources of electric cars production, is it sustainable and how?* Université de Liège.

Benayad, A., & Tellache, M. (2020). A compact energy harvesting multiband rectenna based on metamaterial complementary split ring resonator antenna and modified hybrid junction ring rectifier. *International Journal of RF and Microwave Computer-Aided Engineering*, *30*(2), e22031. doi:10.1002/mmce.22031

Bently, W. F., & Heacock, D. K. (1996). Battery management considerations for multichemistry systems. *IEEE Aerospace and Electronic Systems Magazine*, *11*(5), 23–26. doi:10.1109/62.494184

Bernard, T., Fouchal, H., Linck, S., & Perrin, E. (2018). Impact of routing protocols on packet retransmission over wireless networks. *2013 IEEE International Conference on Communications*, 2979,2983.

Bhattacharjee, P., Majumder, A., & Das, T. D. (2016). A 90 Nm Leakage Control Transistor Based Clock Gating for Low Power Flip Flop Applications. In *2016 IEEE 59th International Midwest Symposium on Circuits and Systems (MWSCAS)*. IEEE. 10.1109/MWSCAS.2016.7870034

Bilen, S. G., McTernan, J. K., Gilchrist, B. E., Bell, I. C., Liaw, D., Voronka, N. R., & Hoyt, R. P. (2012). *Energy Harvesting on Spacecraft Using Electrodynamic Tethers*. Pennsylvania State Univ State College.

Birge, J. R., & Louveaux, F. (1997). *Introduction to stochastic programming*. Springer- Verlag.

Bissengaliyeva, A. (2020). *A Home Automation System Using EnOcean Wireless Technology and Beckhoff Automation*. Academic Press.

Bi, Z., Song, L., De Kleine, R., Mi, C. C., & Keoleian, G. A. (2015). Plug-in vs. wireless charging: Life cycle energy and greenhouse gas emissions for an electric bus system. *Applied Energy, 146,* 11–19. doi:10.1016/j.apenergy.2015.02.031

Black, J. R. (1969). Electro migration - a brief survey and some recent results. *IEEE Transactions on Electron Devices, 16*(4), 338–347. doi:10.1109/T-ED.1969.16754

BloombergNEF report. (2021). https://bnef.turtl.co/story/evo-2021/page/1

Boisseau, S., Despesse, G., & Sylvestre, A. (2010). Optimization of an Electret-Based Energy Harvester. *Smart Materials and Structures, 19*(7), 75015. doi:10.1088/0964-1726/19/7/075015

Bolot, J. C. (2015, September). End-to-end packet delay and loss behavior in the internet. *ACM SIGCOMM Conference on Communications Architectures, Protocols and Applications (SIGCOMM),* 289–298.

Bonamy, R., Chillet, D., Sentieys, O., & Bilavarn, S. (2011). Parallelism level impact on energy consumption in reconfigurable devices. *ACM SIGARCH Computer Architecture News, 39*(4), 104–105. doi:10.1145/2082156.2082186

Borgeson, J. (2012). *Ultra-Low-Power Pioneers: TI Slashes Total MCU Power by 50 Percent with New 'Wolverine' MCU Platform.* Texas Instruments White Paper.

Boukerche, A., Martirosyan, A., & Pazzi, R. (2018). An inter-cluster communication based energy aware and fault tolerant protocol for wireless sensor networks. *Mobile Networks and Applications, 13*(6), 614–626. doi:10.100711036-008-0093-x

Bounouar, M. A., Beaumont, A., El Hajjam, K., Calmon, F., & Drouin, D. (2012). *Room temperature double gate single electron transistor based standard cell library. In 2012 IEEE/ACM international symposium on nanoscale architectures (NANOARCH).* IEEE. doi:10.1145/2765491.2765518

Bourdenas, T., & Sloman, M. (2019, June). Towards self-healing in wireless sensor networks. In *Wearable and Implantable Body Sensor Networks, 2019. BSN 2019. Sixth International Workshop on* (pp. 15-20). IEEE.

Brazdik, F. (2004). Stochastic Data Envelopment Analysis: Oriented and Linearized Models. Joint Workplace of the Center for Economic Research and Graduate Education, Charles University, Prague, and the Economics Institute of the Academy of Sciences of the Czech Republic.

Brockett, P. L., & Golany, B. (1996). Using Rank Statistics for Determining Programmatic Efficiency Differences in Data Envelopment Analysis. *Management Science, 42*(3), 466–472. doi:10.1287/mnsc.42.3.466

Brown, W. C. (1980). The history of the development of the rectenna. *Proc. SPS Microwave Systems Workshop at JSC-NASA,* 271-280.

Brückner, J., Thieme, S., Grossmann, H. T., Dörfler, S., Althues, H., & Kaskel, S. (2014). Lithium-sulfur batteries: Influence of C-rate, amount of electrolyte and sulfur loading on cycle performance. *Journal of Power Sources, 268,* 82–87. doi:10.1016/j.jpowsour.2014.05.143

Buxboim, A., & Schiller, A. (2003). Current characteristics of the single electron transistor at the degeneracy point. *Physical Review. B, 67*(16), 165320. doi:10.1103/PhysRevB.67.165320

Cahill, P., Mathewson, A., & Pakrashi, V. (2018). Experimental Validation of Piezoelectric Energy-Harvesting Device for Built Infrastructure Applications. *Journal of Bridge Engineering, 23*(8), 4018056. doi:10.1061/(ASCE)BE.1943-5592.0001262

Cai, M., & Liao, W.-H. (2020). High-Power Density Inertial Energy Harvester without Additional Proof Mass for Wearables. *IEEE Internet of Things Journal, 8*(1), 297–308. doi:10.1109/JIOT.2020.3003262

Caliò, R., Rongala, U. B., Camboni, D., Milazzo, M., Stefanini, C., De Petris, G., & Oddo, C. M. (2014). Piezoelectric Energy Harvesting Solutions. *Sensors (Basel), 14*(3), 4755–4790. doi:10.3390140304755 PMID:24618725

Camilo, T., Carreto, C., Silva, J. & Boavida, F. (2016). An energy-efficient antbased routing algorithm for wireless sensor networks. *Ant Colony Optimization and Swarm Intelligence*, 49-59.

Can, S. (2021). *Li22Si5-Grafen Kompozit Anotların Lityum-İyon Pillerdeki Performanslarının İncelenmesi.* http://openaccess.ogu.edu.tr:8080/xmlui/bitstream/handle/116 84/4438/311-7706-10232885.pdf?sequence=1&isAllowed=y

Capozzoli, A., & Primiceri, G. (2015). Cooling systems in data centers: State of art and emerging technologies. *Energy Procedia, 83*, 484–493. doi:10.1016/j.egypro.2015.12.168

Cardei, M., Yang, S., & Wu, J. (2018). Algorithms for fault-tolerant topology in heterogeneous wireless sensor networks. *Parallel and Distributed Systems, IEEE Transactions on, 19*(4), 545-558.

Castro-Gonzalez, F., & Sarmiento-Reyes, A. (2014). *Development of a behavioral model of the single-electron transistor for hybrid circuit simulation. In 2014 international Caribbean conference on devices. Circuits and systems (ICCDCS).* IEEE. doi:10.1109/ICCDCS.2014.7016179

Çelen, A. (2013). Efficiency and productivity (TFP) of the Turkish electricity distribution companies: An application of two-stage (DEA&Tobit) analysis. *Energy Policy, 63*, 300–310. doi:10.1016/j.enpol.2013.09.034

Chae, M. J., Yoo, H. S., Kim, J. Y., & Cho, M. Y. (2012). Development of a Wireless Sensor Network System for Suspension Bridge Health Monitoring. *Automation in Construction, 21*, 237–252. doi:10.1016/j.autcon.2011.06.008

Chandrakar, K., & Roy, S. (2019). A SAT-Based Methodology for Effective Clock Gating for Power Minimization. *Journal of Circuits, Systems, and Computers, 28*(01), 1950011. doi:10.1142/S0218126619500117

Chang, Y. W., Cheng, C. H., Wang, J. C., & Chen, S. L. (2008). Heat pipe for cooling of electronic equipment. *Energy Conversion and Management, 49*(11), 3398–3404. doi:10.1016/j.enconman.2008.05.002

Charnes, A., & Cooper, W. W. (1959). Chance-Constrained Programming. *Management Science*, *6*(1), 73–79. doi:10.1287/mnsc.6.1.73

Charnes, A., & Cooper, W. W. (1962). Chance Constraints and Normal Deviates. *Journal of the American Statistical Association*, *57*(297), 134–148. doi:10.1080/01621459.1962.10482155

Charnes, A., Cooper, W. W., & Rhodes, E. (1978). Measuring the efficiency of decision-making units. *European Journal of Operational Research*, *2*(6), 429–444. doi:10.1016/0377-2217(78)90138-8

Chaudhry, A., Niranjan, V., & Kumar, A. (2014). Bandwidth Extension of Analog Multiplier using Dynamic Threshold MOS Transistor. *Proceeding IEEE International Conference on Reliability, Infocom technologies and optimization*. 10.1109/ICRITO.2014.7014682

Chaudhry, A., Niranjan, V., & Kumar, A. (2015). Wideband Analog Multiplier Using DTMOS And Self Cascode Current Mirror. *Proceeding IEEE International Conference on Reliability, Infocom technologies and optimization*.

Chawanda, A., & Luhanga, P. (2012). Piezoelectric Energy Harvesting Devices: An Alternative Energy Source for Wireless Sensors. *Smart Materials Research*, 2012.

Chawla, C., Niranjan, V., & Chopra, V. (2010). Comparative study between Dynamic threshold MOSFET and Conventional MOSFET. *Proceeding National conference on design &. communication technology*.

Chen, G., Hu, B., Huang, K., Knoll, A., & Huang, K. (2019). Automatic cache partitioning and timetriggered scheduling for real-time MPSoCs. *International Conference on ReConFigurable Computing and FPGAs (ReConFig14)*, 1-8.

Chen, G., Huang, K., Huang, J., & Knoll, A. (2019). Cache Partitioning and Scheduling for Energy Optimisation of Real-Time MPSoCs. *IEEE 24th International Conference on Application-Specific Systems, Architectures and Processors*, 35-41.

Chen, J., Cao, X., Cheng, P., Xiao, Y., & Sun, Y. (2018). Distributed collaborative control for industrial automation with wireless sensor and actuator networks. *Industrial Electronics, IEEE Transactions on, 57*(12), 4219-4230.

Chen, X., Dai, Z., Li, W., & Shi, H. (2018). Performance Guaranteed Routing Protocols for Asymmetric Sensor Networks. *Emerging Topics in Computing, IEEE Transactions on, 1*(1), 111-120.

Chen, L., Xie, X., Xie, J., Wang, K., & Yang, J. (2006). Binder effect on cycling performance of silicon/carbon composite anodes for lithium ion batteries. *Journal of Applied Electrochemistry*, *36*(10), 1099–1104. doi:10.100710800-006-9191-2

Chen, T., Li, S. J., Cao, X. Y., Gao, J., & Guo, Z. X. (2019). Ultra-wideband and polarization-insensitive fractal perfect metamaterial absorber based on a three- dimensional fractal tree microstructure with multi-modes. *Applied Physics. A, Materials Science & Processing*, *125*(4), 232. doi:10.100700339-019-2536-6

Chen, W. C., Totachawattana, A., Fan, K., Ponsetto, J. L., Strikwerda, A. C., Zhang, X., Averitt, R. D., & Padilla, W. J. (2018). Single-layer terahertz metamaterials with bulk optical constants. *Physical Review. B, 85*(3), 035112. doi:10.1103/PhysRevB.85.035112

Chen, X., & Tomasz, M. (2017). A Robust method to retrieve the constitutive effective parameters of metamaterials. *Physical Review. E, 70*(1), 016608. doi:10.1103/PhysRevE.70.016608 PMID:15324190

Chen, X., Ye, H., Fan, X., Ren, T., & Zhang, G. (2016). A review of small heat pipes for electronics. *Applied Thermal Engineering, 96*, 1–17. doi:10.1016/j.applthermaleng.2015.11.048

Chi, Y., Sui, B., Yi, X., Fang, L., & Zhou, H. (2010). Advances in the modeling of single electron transistors for the design of integrated circuit. *Journal of Nanoscience and Nanotechnology, 10*(9), 6131–6135. doi:10.1166/jnn.2010.2560 PMID:21133161

Choi, E., Bahadori, M. T., Song, L., Stewart, W. F., & Sun, J. G. R. A. M. (2017). Graph-based Attention Model for Healthcare Representation Learning. *KDD: Proceedings / International Conference on Knowledge Discovery & Data Mining. International Conference on Knowledge Discovery & Data Mining*, 787–795. doi:10.1145/3097983.3098126

Choi, J. (2012). Design, Implementation, and Performance Evaluation of a Detection-Based Adaptive Block Replacement Scheme. *IEEE Transactions on Computers, 51*(7), 793–800. doi:10.1109/TC.2002.1017699

Christopher, L. (2019). *Liner, & T. A. McGilvery*. Historical Overview of Petroleum and Seismology. doi:10.1007/978-3-030-03998-1_2

Civelekoglu, G., & Biyik, Y. (2018). Investigation of Carbon Footprint Change Originated from Transportation Sector. *Bilge International Journal of Science and Technology Research, 2*(2), 157–166. doi:10.30516/bilgesci.427359

Clemente, J. A., Gran, R., Chocano, A., del Prado, C., & Resano, J. (2019). Hardware Architectural Support for Caching Partitioned Recon gurations in Recon gurable System. *IEEE Transactions on Very Large Scale Integration (VLSI) Systems, 24*(2), 530–543.

Coban, H. H., Rehman, A., & Mohamed, A. (2022). Analyzing the Societal Cost of Electric Roads Compared to Batteries and Oil for All Forms of Road Transport. *Energies, 15*(5), 1–20. doi:10.3390/en15051925

Colgan. (2007). A practical implementation of silicon microchannel coolers for high power chips. *IEEE Transactions on Components and Packaging Technologies, 30*(2), 218–225.

Cooper, W. W., Deng, H., Huang, Z., & Li, S. X. (2002). Chance constrained programming approaches to technical efficiencies and inefficiencies in stochastic data envelopment analysis. *The Journal of the Operational Research Society, 53*(12), 1347–1356. doi:10.1057/palgrave. jors.2601433

Cooper, W. W., Deng, H., Huang, Z., & Li, S. X. (2004). Chance constrained programming approaches to congestion in stochastic data envelopment analysis. *European Journal of Operational Research*, *155*(2), 487–501. doi:10.1016/S0377-2217(02)00901-3

Cooper, W. W., Huang, Z., & Li, S. (1996). Satisficing DEA models under chance constraints. *Annals of Operations Research*, *66*(4), 259–279. doi:10.1007/BF02187302

Cooper, W. W., Seiford, L. M., & Tone, K. (2007). *Data envelopment analysis: a comprehensive text with models, applications, references and DEA-solver software* (2nd ed.). Springer. doi:10.1007/978-0-387-45283-8

Cooper, W. W., Thompson, R. G., & Thrall, R. M. (1996). Introduction: Extensions and new developments in VZA. *Annals of Operations Research*, *66*(1), 3–46. doi:10.1007/BF02125451

Daboul, S., Hähnle, N., Held, S., & Schorr, U. (2018). Provably Fast and Near-Optimum Gate Sizing. *IEEE Transactions on Computer-Aided Design of Integrated Circuits and Systems*, *37*(12), 3163–3176. doi:10.1109/TCAD.2018.2801231

Dai, X., Wen, Y., Li, P., Yang, J., & Jiang, X. (2009). A Vibration Energy Harvester Using Magnetostrictive/Piezoelectric Composite Transducer. Sensors. doi:10.1109/ICSENS.2009.5398445

Dai, Q., Kelly, J. C., Gaines, L., & Wang, M. (2019). Life Cycle Analysis of Lithium-Ion Batteries for Automotive Applications. *Batteries*, *5*(2), 48. doi:10.3390/batteries5020048

Dan, A., & Towsley, D. (2020). An approximate analysis of the LRU and FIFO buffer replacement schemes. *Proc. of the 2020 ACM SIGMETRICS Conf. on Measurement and Modelling of Comp. Systems*, 143-152.

Dapino, M. J., Deng, Z., Calkins, F. T., & Flatau, A. B. (1999). Magnetostrictive Devices. Wiley Encyclopedia of Electrical and Electronics Engineering, 1–35. doi:10.1002/047134608X.W4549

Darwin, S., Rani, E. F. I., Raj, E. F. I., Appadurai, M., & Balaji, M. (2022, April). Performance Analysis of Carbon Nanotube Transistors-A Review. In *2022 6th International Conference on Trends in Electronics and Informatics (ICOEI)* (pp. 25-31). IEEE. 10.1109/ICOEI53556.2022.9776858

Das, S., Rossi, D., Martin, K. J. M., Coussy, P., & Benini, L. (2017). A 142mops/Mw Integrated Programmable Array Accelerator for Smart Visual Processing. In *2017 IEEE International Symposium on Circuits and Systems (ISCAS)*. IEEE. 10.1109/ISCAS.2017.8050238

Davidsson Kurland, S. (2020). Energy use for GWh-scale lithium-ion battery production. *Environmental Research Communications*, *2*(1), 012001. doi:10.1088/2515-7620/ab5e1e

Dawar, P., Raghava, N. S., & De, A. (2019). UWB Metamaterial-Loaded Antenna for C-Band Applications. *International Journal of Antennas and Propagation*, *2019*, 1–13. doi:10.1155/2019/6087039

Degen, F., & Kratzig, O. (2022). Future in Battery Production: An Extensive Benchmarking of Novel Production Technologies as Guidance for Decision Making in Engineering. *IEEE Transactions on Engineering Management*, 1–19. doi:10.1109/TEM.2022.3144882

Degen, F., & Schütte, M. (2022). Life cycle assessment of the energy consumption and GHG emissions of state-of-the-art automotive battery cell production. *Journal of Cleaner Production*, *330*, 129798. doi:10.1016/j.jclepro.2021.129798

Deivakani, M., Kumar, S. S., Kumar, N. U., Raj, E. F. I., & Ramakrishna, V. (2021, March). VLSI implementation of discrete cosine transform approximation recursive algorithm. *Journal of Physics: Conference Series*, *1817*(1), 012017. doi:10.1088/1742-6596/1817/1/012017

Demetgul, M., & Guney, I. (2017). Design of the Hybrid Regenerative Shock Absorber and Energy Harvesting from Linear Movement. *Journal of Clean Energy Technologies*, *5*(1), 81–84. doi:10.18178/JOCET.2017.5.1.349

Demireli, E., & Özdemir, A. Y. (2013). Seçilmiş Avrupa Ülkelerinde Makroekonomik Performans Ölçümü: Şans Kısıtlı Veri Zarflama Analizi İle Bir Uygulama. *Dumlupınar University Journal of Social Sciences, 37*.

Deng, Z., Asnani, V. M., & Dapino, M. J. (2015). Magnetostrictive Vibration Damper and Energy Harvester for Rotating Machinery. In Industrial and Commercial Applications of Smart Structures Technologies 2015 (Vol. 9433). SPIE.

Deng, G., & Chen, C. (2012). A SET/MOS hybrid multiplier using frequency synthesis. *IEEE Transactions on Very Large Scale Integration (VLSI) Systems*, *21*(9), 1738–1742. doi:10.1109/TVLSI.2012.2218264

Deng, J., Bae, C., Denlinger, A., & Miller, T. (2020). Electric Vehicles Batteries: Requirements and Challenges. *Joule*, *4*(3), 511–515. doi:10.1016/j.joule.2020.01.013

Deng, Z., & Dapino, M. J. (2017). Review of Magnetostrictive Vibration Energy Harvesters. *Smart Materials and Structures*, *26*(10), 103001. doi:10.1088/1361-665X/aa8347

Deo, A., Niranjan, V., & Kumar, A. (2014). Improving gain of Class-E amplifier using DTMOS for biomedical devices. *Proceeding IEEE International Conference on Medical Imaging, m-health & Emerging Communications Systems*. 10.1109/MedCom.2014.7005564

Derbal, M. C., & Nedil, M. (2020). A High Gain Dual Band Rectenna for RF Energy Harvesting Applications. *Progress in Electromagnetics Research Letters*, *90*, 29–36. doi:10.2528/PIERL19122604

Deshmukh, Nakrani, Bhuyar, & Shinde. (2019). Face Recognition Using OpenCV Based On IoT for Smart Door. SSRN *Electronic Journal*. doi:10.2139/ssrn.3356332

Devi, K. K. A., Sadasivam, S., Din, N. M., & Chakrabarthy, C. K. (2019). Design of a 377 Ω Patch Antenna for Ambient RF Energy Harvesting at Downlink Frequency of GSM 900. *The 17th Asia Pacific Conference on Communications,* 492-495.

Devi, K. K. A., Hau, N. C., Chakrabarty, C. K., & Din, N. M. (2018). Design of Patch Antenna Using Metamaterial at GSM 1800 for RF Energy Scavenging. *IEEE Asia Pacific Conference on Wireless and Mobile*, 157-161.

Deyasi, A., & Sarkar, A. (2019). Effect of temperature on electrical characteristics of single electron transistor. *Microsystem Technologies*, *25*(5), 1875–1880. doi:10.100700542-018-3725-5

Dietterich, T. G. (2000). *Ensemble methods in machine learning*. In *International workshop on multiple classifier systems* (pp. 1-15). Springer.

Ding, F., Dai, J., Chen, Y., Zhu, J., Jin, Y., & Bozhevolnyi, S. I. (2018). Broadband near-infrared metamaterial absorbers utilizing highly lossy metals. *Scientific Reports*, *6*(1), 39445. doi:10.1038rep39445 PMID:28000718

Divakaran, S. K., & Krishna, D. D. & Nasimuddin. (2018). RF energy harvesting systems: An overview and design issues. *International Journal of RF and Microwave Computer-Aided Engineering*, *29*, 1.

Dix, J., Jokar, A., & Martinsen, R. (2008). A microchannel heat exchanger for electronics cooling applications. *Proceedings of 6th International Conference on Nanochannels, Microchannels and Minichannels*.

Dolgov, A., Zane, R., & Popovic, Z. (2019). Power Management System for Online Low Power RF Energy Harvesting Optimization. *IEEE Transactions on Circuits and Systems*, *57*(7), 1802–1811. doi:10.1109/TCSI.2009.2034891

Dubuc, C., Beauvais, J., & Drouin, D. (2008). A nanodamascene process for advanced single-electron transistor fabrication. *IEEE Transactions on Nanotechnology*, *7*(1), 68–73. doi:10.1109/TNANO.2007.913430

Durrani, Z. A. K. (2010). *Single-electron devices and circuits in silicon*. World Scientiðc.

Dutoit, N. E., Wardle, B. L., & Kim, S.-G. (2005). Design Considerations for MEMS-Scale Piezoelectric Mechanical Vibration Energy Harvesters. *Integrated Ferroelectrics*, *71*(1), 121–160. doi:10.1080/10584580590964574

Egan, M. G., O'Sullivan, D. L., Hayes, J. G., Willers, M. J., & Henze, C. P. (2007). Power-factor-corrected single-stage inductive charger for electric vehicle batteries. *IEEE Transactions on Industrial Electronics*, *54*(2), 1217–1226. doi:10.1109/TIE.2007.892996

Ekejiuba, E. A. I. (2020). *Guide to Petroleum Engineering Career: The Best Practices Petroleum Infographic Cutting Edge Technology Approach*. Dorrance Publishing.

Ellingsen, L. A.-W., Majeau-Bettez, G., Singh, B., Srivastava, A. K., Valøen, L. O., & Strømman, A. H. (2014). Life Cycle Assessment of a Lithium-Ion Battery Vehicle Pack. *Journal of Industrial Ecology*, *18*(1), 113–124. doi:10.1111/jiec.12072

Elprocus. (n.d.). https://www.elprocus.com/architecture-of-wireless-sensor-network-and-applications/

Emilsson, E., & Dahllöf, L. (2019). *Lithium-Ion Vehicle Battery Production - Status 2019 on Energy Use, CO2 Emissions, Use of Metals, Products Environmental Footprint, and Recycling.* IVL Swedish Environmental Research Institute.

Escala, O. A. (2019). *Study of the Efficiency of Rectifying Antenna Systems for Energy Harvesting* [Thesis]. UPC-Barcelona, Spain.

Eskandarian, A., Rajeyan, Z., & Ebrahimnezhad, H. (2018). Analysis and simulation of single electron transistor as an analogue frequency doubler. *Microelectronics Journal*, *75*, 52–60. doi:10.1016/j.mejo.2018.02.008

Faghri, A. (2012). Review and advances in heat pipe science and technology. *Journal of Heat Transfer*, *134*(12), 123001. doi:10.1115/1.4007407

Falade, O. P., Rehman, M. U., Gao, Y., Chen, X., & Parini, C. G. (2018). Single Feed Stacked Patch Circular Polarized Antenna for Triple Band GPS Receivers. *IEEE Transactions on Antennas and Propagation*, *60*(10), 4479–4484. doi:10.1109/TAP.2012.2207354

Fallahi, A., Ebrahimi, R., & Ghaderi, S. F. (2011). Measuring efficiency and productivity change in power electric generation management companies by using data envelopment analysis: A case study. *Energy*, *36*(11), 6398–6405. doi:10.1016/j.energy.2011.09.034

Fang, J., Liu, S., & Zhang, X. (2019). Research on Cache Partitioning and Adaptive Replacement Policy for CPU-GPU Heterogeneous Processors. *16th International Symposium on Distributed Computing and Applications to Business, Engineering and Science (DCABES)*, 19-22.

Fang, J., & Pu, J. (2019). Dynamic Fair Cache Partitioning for Chip Multiprocessor. *Third International Joint Conference on Computational Science and Optimisation*, 2, 283-287.

Fante, R. L., & Mccormack, M. T. (1988). Reflection Properties of the Salisbury Screen. *IEEE Transactions on Antennas and Propagation*, *36*(10), 1443–1454. doi:10.1109/8.8632

Fantin Irudaya Raj, E., & Appadurai, M. (2022). Internet of things-based smart transportation system for smart cities. In *Intelligent Systems for Social Good* (pp. 39–50). Springer. doi:10.1007/978-981-19-0770-8_4

Farrell, M. J. (1957). The measurement of productive efficiency. J R Stat Soc Ser A. *GEN*, *120*, 253–290.

Faruque, M. R. I., & Islam, T. (2018). Novel triangular metamaterial design for electromagnetic absorption reduction in human head. *Progress in Electromagnetics Research*, *141*, 463–478. doi:10.2528/PIER13050603

Farzanehnia, A., Khatibi, M., Sardarabadi, M., & Passandideh-Fard, M. (2019). Experimental investigation of multiwall carbon nanotube/paraffin based heat sink for electronic device thermal management. *Energy Conversion and Management*, *179*, 314–325.

Financial Times. (2022). *Full charge ahead: the relentless rise of the battery market.* https://www.ft.com/partnercontent/societe-generale/full-charge-ahead-the-relentless-rise-of-the-battery-market.html

Fitzgerald, B., Lopez, S., & Sahuquillo, J. (2018). Drowsy Cache Partitioning for Reduced Static and Dynamic Energy in the Cache Hierarchy. *International Green Computing Conference Proceedings*, 1-6.

Fonseca, L., Korotkov, A., Likharev, K., & Odintsov, A. (1995). A numerical study of the dynamics and statistics of single electron systems. *Journal of Applied Physics*, *78*(5), 3238–3251. doi:10.1063/1.360752

Fowler, C., & Zhou, J. (2017). *A Highly Efficient Polarization-Independent Metamaterial-Based RF Energy-Harvesting Rectenna for Low-Power Applications.* arXiv, 1705.07717.

Frenkel, J.-D. L., & Bol, D. (2019). MorphIC: A 65-nm 738kSynapse/mm 2 Quad-Core Binary-Weight Digital Neuromorphic Processor With Stochastic Spike-Driven Online Learning. *IEEE Transactions on Biomedical Circuits and Systems*, *13*(5), 999–1010.

Freunek, M., Freunek, M., & Reindl, L. M. (2012). Maximum Efficiencies of Indoor Photovoltaic Devices. *IEEE Journal of Photovoltaics*, *3*(1), 59–64. doi:10.1109/JPHOTOV.2012.2225023

Fromme, Moore, Moore, Rittmann, Torres, & Vermaas. (2014). *Microbial Electrophotosynthesis.* Academic Press.

Fu, C., Jiang, Z., Wei, W. E. I., & Wei, A. (2013). An energy balanced algorithm of LEACH protocol in WSN. *International Journal of Computer Science Issues*, *10*(1), 354.

Fujishima, A., & Honda, K. (1971). Electrochemical Evidence for the Mechanism of the Primary Stage of Photosynthesis. *Bulletin of the Chemical Society of Japan*, *44*(4), 1148–1150. doi:10.1246/bcsj.44.1148

Gajda, I., Greenman, J., Melhuish, C., & Ieropoulos, I. (2015). Self-Sustainable Electricity Production from Algae Grown in a Microbial Fuel Cell System. *Biomass and Bioenergy*, *82*, 87–93. doi:10.1016/j.biombioe.2015.05.017

Galayko, D., Guillemet, R., Dudka, A., & Basset, P. (2011). Comprehensive Dynamic and Stability Analysis of Electrostatic Vibration Energy Harvester (E-VEH). In *2011 16th International Solid-State Sensors, Actuators and Microsystems Conference.* IEEE. 10.1109/TRANSDUCERS.2011.5969592

Gambhir, A., Yadav, D., & Pawar, G. (2017). *Energy Harvesting in Automotive Key Fob Application.* SAE Technical Paper.

Gan, Z.-H., & Gu, Z. (2019). WCET-Aware Task assignment and Cache Partitioning for WCRT Minimisation on Multi-core Systems. *Seventh International Symposium on Parallel Architectures, Algorithms and Programming (PAAP)*, 143-148.

Garg, S., Chaudhary, G., Niranjan, V., & Kumar, A. (2014). Bandwidth Extension of Voltage Follower using DTMOS transistor. *Proceeding IEEE International Conference on Innovative Applications of Computational Intelligence on Power, Energy and Controls with their impact on Humanity.* 10.1109/CIPECH.2014.7018211

Garg, S., & Niranjan, V. (2015). DTMOS transistor with self-cascode subcircuit for achieving high bandwidth in analog applications. *International Journal of Computers and Applications, 127*(11), 19–31. doi:10.5120/ijca2015906538

Garshelis, I. (1974). A Study of the Inverse Wiedemann Effect on Circular Remanence. *IEEE Transactions on Magnetics, 10*(2), 344–358. doi:10.1109/TMAG.1974.1058325

Gatti, R. R. (2013). *Spatially-Varying Multi-Degree-of-Freedom Electromagnetic Energy Harvesting.* Academic Press.

Gatti, R. R. (n.d.). *Novel Multi-Beam Spring Design for Vibration Energy Harvesters.* Academic Press.

Gatti, R. R., Shetty, S. H., & Rao, A. (2022). Building Autonomous IIoT Networks Using Energy Harvesters. In Enterprise Digital Transformation. Auerbach Publications. doi:10.1201/9781003119784-11

Geeks for Geeks. (n.d.). https://www.geeksforgeeks.org/wireless-sensor-network-wsn/

Geng, Y., Mingzhe, J., Wei, O., & Guangchao, J. (2018). AIoT-based remote pain monitoring system: From device to cloud platform. *IEEE Journal of Biomedical and Health Informatics, 22*(6), 1711–1719. doi:10.1109/JBHI.2017.2776351 PMID:29990259

Ghosh, S., Bhattacharyya, S., Chaurasiya, D., & Srivastava, K. V. (2018). An Ultra-wideband Ultrathin Metamaterial Absorber Based on Circular Split Rings. *IEEE Antennas and Wireless Propagation Letters, 14*, 1172–1175. doi:10.1109/LAWP.2015.2396302

Gil, D. R. G., Costa, M. A., Lopes, A. L. M., & Mayrink, V. D. (2017). Spatial statistical methods applied to the 2015 Brazilian energy distribution benchmarking model: Accounting for unobserved determinants of inefficiencies. *Energy Economics, 64*, 373–383. doi:10.1016/j.eneco.2017.04.009

Gil, I., Martin, F., Rottenberg, X., & De Raedt, W. (2017). Tunable stop-band filter at Q-band based on RF-MEMS metamaterials. *Electronics Letters, 43*(21), 1153–1154. doi:10.1049/el:20072164

Gnanasekar, A. K., Deivakani, M., Bathala, N., Raj, E., & Ramakrishna, V. (2021). Novel Low-Noise CMOS Bioamplifier for the Characterization of Neurodegenerative Diseases. In *GeNeDis 2020* (pp. 221–226). Springer. doi:10.1007/978-3-030-78787-5_27

Goldhaber-Gordon, D., Shtrikman, H., Mahalu, D., Abusch-Magder, D., Meirav, U., & Kastner, M.A. (1998). *Kondo effect in a single-electron transistor.* Academic Press.

Gonza'lez, F. J. C., Reyes, A. S., & Saenz, F. J. Z. (2012). Effects of single-electron transistor parameter variations on hybrid circuit design. In *2012 IEEE 3rd Latin American symposium on circuits and systems (LASCAS).* IEEE. 10.1109/LASCAS.2012.6180354

Gopi Krishna, P., Sai Phani Kumar, P., Sreenivasa Ravi, K., Divya Sravanthi, M., & Likhitha, N. (2019). Smart home authentication and security with IoT using face recognition. *International Journal of Recent Technology and Engineering*, *7*(6).

Gorlatova, M., Kinget, P., Kymissis, I., Rubenstein, D., Wang, X., & Zussman, G. (2010). Energy Harvesting Active Networked Tags (EnHANTs) for Ubiquitous Object Networking. *IEEE Wireless Communications*, *17*(6), 18–25. doi:10.1109/MWC.2010.5675774

Gould, P. (2003). Textiles Gain Intelligence. *Materials Today*, *6*(10), 38–43. doi:10.1016/S1369-7021(03)01028-9

Gouveia, M. C., Dias, L. C., Antunes, C. H., Boucinha, J., & Inácio, C. F. (2015). Benchmarking of maintenance and outage repair in an electricity distribution company using the value-based DEA method. *Omega*, *53*, 104–114. doi:10.1016/j.omega.2014.12.003

Goyal, A., Narang, K., & Ahluwalia, G. (2019). Seasonal variation in 24 h blood pressure profile in healthy adults-A prospective observational study. *Journal of Human Hypertension*.

Gozel, M. A., Kahriman, M., & Kasar, O. (2018). Design of an efficiency-enhanced Greinacher rectifier operating in the GSM 1800 band by using rat-race coupler for RF energy harvesting applications. *International Journal of RF and Microwave Computer-Aided Engineering*, *21*(1), e21621. doi:10.1002/mmce.21621

Gracioli, G., & Fröhlich, A. A. (2018). An Experimental Evaluation of the Cache Partitioning Impact on Multicore Real-Time Schedulers. *IEEE International Conference on Embedded and Real- Time Computing Systems and Applications*, 72-81.

Graphics, M. (2009). Low power physical design with Olympus SOC. *Place and Route White Paper*.

Greco, L., Percannella, G., Ritrovato, P., Tortorella, F., & Vento, M. (2020). Trends in IoT-based health care solutions: Moving AI to the edge. *Pattern Recognition Letters*, 346–353. doi:10.1016/j.patrec.2020.05.016

Gstach, D. (1998). Another approach to data envelopment analysis in noisy environments: DEA+. *Journal of Productivity Analysis*, *9*(2), 161–176. doi:10.1023/A:1018312801700

Guan, N., & Lv, M. (2019). WCET Analysis with MRU Caches: Challenging LRU for Predictability. *IEEE 18th Real-Time and Embedded Technology and Applications Symposium*, 55-64.

Guang, N. L. L., Logenthiran, T., & Abidi, K. (2017). Application of Internet of Things (IoT) for home energy management. *IEEE PES Asia-Pacific Power and Energy Engineering Conference (APPEEC)*, 1-6. doi: 10.1109/APPEEC.2017.8308962

Guan, Q., Wang, Y., Ping, B., Li, D., Du, J., Qin, Y., Lu, H., Wan, X., & Xiang, J. (2019). Deep Convolutional Neural Network VGG-16 Model for Differential Diagnosing of Papillary Thyroid Carcinomas in Cytological Images: A Pilot Study. *Journal of Cancer*, *10*(20), 4876.

Guarnieri, M. (2016). The Unreasonable Accuracy of Moore's Law [Historical]. *IEEE Industrial Electronics Magazine*, *10*(1), 40–43. doi:10.1109/MIE.2016.2515045

Guha, D., Biswas, S., & Antar, Y. M. M. (2019). *Microstrip and Printed Antennas: New Trends, Techniques and Applications*. John Wiley & Sons.

Guo, L., Leobandung, E., & Chou, S. Y. (1997). A Silicon Single-Electron Transistor Memory Operating at Room Temperature. *Science*, *275*(5300), 275. doi:10.1126cience.275.5300.649 PMID:9005847

Guo, L., & Lu, Q. (2018). Potentials of piezoelectric and thermoelectric technologies for harvesting energy from pavements. *Renewable & Sustainable Energy Reviews*, *72*, 761–773. doi:10.1016/j.rser.2017.01.090

Guthaus, Ringenberg, Ernst, Austin, Mudge, & Brown. (2021). MiBench: A free, commercially representative embedded benchmark suite. *Proceedings of the Fourth Annual IEEE International Workshop on Workload Characterization*, 3-14.

Gyselinckx, B., Van Hoof, C., Ryckaert, J., Yazicioglu, R. F., Fiorini, P., & Leonov, V. (2005). Human++: Autonomous Wireless Sensors for Body Area Networks. In *Proceedings of the IEEE 2005 Custom Integrated Circuits Conference, 2005*. IEEE. 10.1109/CICC.2005.1568597

Haber, T. R. T. (2022). *Market share of electric vehicles increased: sales record was broken in 2021*. https://www.trthaber.com/haber/ekonomi/elektrikli-araclarin-pazar-payi-artti-2021de-satis-rekoru-kirildi-673349.html#:~: text=Avrupa'daise 2%2C3,ye göre yüzde 66 büyüdü

Halsnæs, K., Bay, L., Kaspersen, P. S., Drews, M., & Larsen, M. A. D. (2021). Climate services for renewable energy in the nordic electricity market. *Climate (Basel)*, *9*(3), 46. Advance online publication. doi:10.3390/cli9030046

Hashemi, S. S., Sawan, M., & Savaria, Y. (2018). A high-efficiency low-voltage CMOS rectifier for harvesting energy in implantable devices. *IEEE Transactions on Biomedical Circuits and Systems*, *6*, 326335. PMID:23853177

Hassan, N., Hisham, A. B., Fareq, A. M. M., Abidin, A. M. Z., Bakar, H., Noor, A. M. S., & Khairy, I. M. (2018). Radio frequency (RF) energy harvesting using metamaterial structure for antenna/rectenna communication network: A review. *Journal of Theoretical and Applied Information Technology*, *96*(6), 1538–1550.

Hergenrother, J. M., Tuominen, M. T., Tighe, T. S., & Tinkham, M. (1993). Fabrication and characterization of Single-Electron tunneling transistors in the superconducting state. *IEEE Transactions on Applied Superconductivity*, *3*(1), 3. doi:10.1109/77.233570

Hirtzlin, T., Bocquet, M., Penkovsky, B., Klein, J.-O., Nowak, E., Vianello, E., Portal, J.-M., & Querlioz, D. (2019). Digital Biologically Plausible Implementation of Binarized Neural Networks With Differential Hafnium Oxide Resistive Memory Arrays. *Frontiers in Neuroscience*, *13*. PMID:31998059

Hoang, D. C., Tan, Y. K., Chng, H. B., & Panda, S. K. (2009). Thermal Energy Harvesting from Human Warmth for Wireless Body Area Network in Medical Healthcare System. *2009 International Conference on Power Electronics and Drive Systems (PEDS)*. 10.1109/PEDS.2009.5385814

Ho, C., Liu, Y. C., Ghalambaz, M., & Yan, W. M. (2020). Forced convection heat transfer of nano-encapsulated phase change material (NEPCM) suspension in a mini-channel heatsink. *International Journal of Heat and Mass Transfer, 155*, 119858.

Hong, H., Cai, X., Shi, X., & Zhu, X. (2018). Demonstration of a highly efficient RF energy harvester for Wi-Fi signals. *International Conference on Microwave and Millimetre Wave Technology (ICMMT)*, 1–4.

Hoque, M. A., & Davidson, C. (2019). Design and implementation of an IoT-based smart home security system. *International Journal of Networked and Distributed Computing, 7*(2), 85. Advance online publication. doi:10.2991/ijndc.k.190326.004

Houache, M. S. E., Yim, C.-H., Karkar, Z., & Abu-Lebdeh, Y. (2022). On the Current and Future Outlook of Battery Chemistries for Electric Vehicles—Mini Review. *Batteries, 8*(7), 70. doi:10.3390/batteries8070070

Hou, L., Tan, S., Zhang, Z., & Bergmann, N. W. (2018). Thermal Energy Harvesting WSNs Node for Temperature Monitoring in IIoT. *IEEE Access: Practical Innovations, Open Solutions, 6*, 35243–35249. doi:10.1109/ACCESS.2018.2851203

Hsieh, J. H., Lee, R. C., Hung, K. C., & Shih, M. J. (2018). Rapid and coding efficient SPIHT algorithm for wavelet-based ECG data compression. *Integration, the VLSI Journal, 60*, 248-256.

Huang, Z., & Li, S. X. (2001). Stochastic DEA models with different types of input-output disturbances. *Journal of Productivity Analysis, 15*(2), 95–113. doi:10.1023/A:1007874304917

Ibarra-Gutiérrez, S., Bouchard, J., Laflamme, M., & Fytas, K. (2021). Perspectives of Lithium Mining in Quebec. *Potential and Advantages of Integration into a Local Battery Production Chain for Electric Vehicles, 33*, 33. Advance online publication. doi:10.3390/materproc2021005033

Ibrahim, R., Chung, T. D., Hassan, S. M., Bingi, K., & Salahuddin, S. K. (2017). Solar Energy Harvester for Industrial Wireless Sensor Nodes. *Procedia Computer Science, 105*, 111–118. doi:10.1016/j.procs.2017.01.184

IEA. (2020). *Global EV Outlook, Entering the decade of electric drive?* IEA. (2022). *Global Supply Chains of EV Batteries.* https://iea.blob.core.windows.net/assets/4eb8c252-76b1-4710-8f5e-867e751c8dda/GlobalSupplyChainsofEVBatteries.pdf

Inokawa, H., Nishimura, T., Singh, A., Satoh, H., & Takahashi, Y. (2018). *Ultrahigh-frequency characteristics of single-electron transistor. In 2018 IEEE international conference on electron devices and solid state circuits (EDSSC)*. IEEE. doi:10.1109/EDSSC.2018.8487153

Inukai, T., Takamiya, M., Nose, K., Kawaguchi, H., Hiramoto, T., & Sakurai, T. (2000, May). Boosted Gate MOS (BGMOS): Device/circuit cooperation scheme to achieve leakage-free giga-scale integration. In *Proceedings of the IEEE 2000 Custom Integrated Circuits Conference (Cat. No. 00CH37044)* (pp. 409-412). IEEE. 10.1109/CICC.2000.852696

Iqbal, N., & Kim, D. H. (2022). IoT task management mechanism based on predictive optimization for efficient energy consumption in smart residential buildings. *Energy and Building, 257*, 111762. doi:10.1016/j.enbuild.2021.111762

Iyer, Y. G., Gandham, S., & Venkatesan, S. (2015). STCP: a generic transport layer protocol for wireless sensor networks. In *Computer Communications and Networks, 2015. ICCCN 2015. Proceedings. 14th International Conference on* (pp. 449-454). IEEE.

Jabbar, H., Song, Y. S., & Jeong, T. T. (2010). RF Energy Harvesting System and Circuits for Charging of Mobile Devices. *IEEE Transactions on Consumer Electronics, 56*(1), 247–253. doi:10.1109/TCE.2010.5439152

Jacob, A.P., Xie, R., Sung, M.G., Liebmann, L., Lee, R.T., & Taylor, B. (2017). Scaling challenges for advanced CMOS devices. *Int J High Speed Electron Syst, 26*(1-2).

Jahanshahloo, G. R., Behzadi, M. H., & Mirbolouki, M. (2010). Ranking Stochastic Efficient DMUs based on Reliability. *International Journal of Industrial Mathematics, 2*, 263–270.

Jain, A., Ghosh, A., Singh, N. B., & Sarkar, S. K. (2015). A new SPICE macro model of single electron transistor for efðcient simulation of single-electronics circuits. *Analog Integr Circ Sig Process, 82*(3), 653–662. doi:10.100710470-015-0491-5

Jamasb, T., & Pollitt, M. (2001). Benchmarking and Regulation: International Electricity Experience. *Utilities Policy, 9*(3), 107–130. doi:10.1016/S0957-1787(01)00010-8

Jamasb, T., & Pollitt, M. (2003). International Benchmarking and Regulation: An Application to European Electricity Distribution Utilities. *Energy Policy, 31*(15), 1609–1622. doi:10.1016/S0301-4215(02)00226-4

Jang, S., Jo, H., Cho, S., Mechitov, K., Rice, J. A., Sim, S.-H., Jung, H.-J., Yun, C.-B., Spencer, B. F. Jr, & Agha, G. (2010). Structural Health Monitoring of a Cable-Stayed Bridge Using Smart Sensor Technology: Deployment and Evaluation. *Smart Structures and Systems, 6*(5–6), 439–459. doi:10.12989ss.2010.6.5_6.439

Jankowski, S., Covello, J., Bellini, H., Ritchie, J., & Costa, D. (2014). *The Internet of Things: Making Sense of the next Mega-Trend.* Goldman Sachs.

Jao, Y.-T., Chang, T.-W., & Lin, Z.-H. (2017). Multifunctional Textile for Energy Harvesting and Self-Powered Sensing Applications. *ECS Transactions, 77*(7), 47–50. doi:10.1149/07707.0047ecst

Jhajharia, H., & Niranjan, V. (2016). Exploiting Body Effect to Improve the Performance of Amplifier. *International Journal of Electronics. Electrical and Computational System, 5*(11), 9–14.

Jia, C., Chaohong, H., Cotofana, S. D., & Jianfei, J. (2004). SPICE implementation of a compact single electron tunneling transistor model. *4th IEEE conference on nanotechnology*, 392–395.

Jiang, N., & Wu, J. (2018). Implementation of hardware-assisted virtual machine cache partitioning. *International Conference on Automatic Control and Artificial Intelligence (ACAI 2018)*, 1189-1192.

Johnson, D. B., & Maltz, D. A. (1996). Dynamic source routing in ad hoc wireless networks. In *Mobile computing* (pp. 153–181). Springer US. doi:10.1007/978-0-585-29603-6_5

Johnsort, D. B. (1994, December). Routing in ad hoc networks of mobile hosts. In *Mobile Computing Systems and Applications, 1994. WMCSA 1994. First Workshop on* (pp. 158-163). IEEE. 10.1109/WMCSA.1994.33

Jooq, M. K. Q., Miralaei, M., & Ramezani, A. (2017). Post-Layout Simulation of an Ultra-Low-Power OTA Using DTMOS Input Differential Pair. *International Journal of Electronics Letters.*, 6(2), 168–180. doi:10.1080/21681724.2017.1335782

Joshi & Jangir. (2019). Design of Low Power and High Speed Shift Register. *IOSR Journal of VLSI and Signal Processing, 9*(1), 28–33.

Joyez, P., & Esteve, D. (1997). Single-electron tunneling at high temperature. *Physical Review, 56*.

Juan, F. & Du, W. (2018). A Low-power Oriented Dynamic Hybrid Cache Partitioning for Chip Multi-processor. *2012 International Conference on Industrial Control and Electronics Engineering*, 369-372.

Jung, J. C.-Y., Sui, P.-C., & Zhang, J. (2021). A review of recycling spent lithium-ion battery cathode materials using hydrometallurgical treatments. *Journal of Energy Storage, 35*, 102217. doi:10.1016/j.est.2020.102217

Juwad, M. F., & Al-Raweshidy, H. S. (2018, May). Experimental Performance Comparisons between SAODV & AODV. In *Modeling & Simulation, 2018. AICMS 08. Second Asia International Conference on* (pp. 247-252). IEEE.

Kafka, T. (2015). Industrial Application of Thermal Energy Harvesting. *IDTex*. http://perpetuapower.com/wp-content/uploads/2015/12/GE_IDTec hEx_Presentation_Berlin_20140402.pdf

Kandasubramanian, B., & Ramdayal, M. (2013). Advancement in Textile Technology for Defence Application. *Defence Science Journal, 63*(3), 331–339. doi:10.14429/dsj.63.2756

Kandilikar, S., & Upadhye, H. (2005). Extending the heat flux limit with enhanced micro-channels in direct single-phase cooling of computer chips. *Proceedings of 21st Semi Therm Symposium,* 8-15.

Kandlikar, S. G., Colin, S., Peles, Y., Garimella, S., Pease, R. F., Brandner, J. J., & Tuckerman, D. B. (2013). Heat transfer in micro-channels, *status and research needs. Journal of Heat Transfer, 135*, 091001.

Kang. (2003). *CMOS Digital Integrated Circuits: Analysis and Design* (3rd ed.). McGraw Hill Pub.

Kanjanop, A., & Kasemsuwan, V. (2011). Low voltage class AB current differencing buffered amplifier (CDBA). *International Symposium on Intelligent Signal Processing and Communications Systems*, 1-5. 10.1109/ISPACS.2011.6146107

Kanjanop, A., Suadet, A., Singhanath, P., Thongleam, T., Kuankid, S., & Kasemsuwan, V. (2011). An ultra low voltage rail-to-rail DTMOS voltage follower. *Proceedings International Conference on Modeling, Simulation and Applied Optimization*, 1-5. 10.1109/ICMSAO.2011.5775534

Karbasian, G., Orlov, A. O., & Snider, G. L. (2015). *Nanodamascene metal insulator-metal single electron transistor prepared by atomic layer deposition of tunnel barrier and subsequent reduction of metal surface oxide. In 2015 silicon nanoelectronics workshop (SNW).* IEEE.

Kargaran, E., Sawan, M., Mafinezhad, K., & Nabovati, H. (2012). Design of 0.4V, 386nW OTA using DTMOS technique for biomedical applications. *Proceedings IEEE International Midwest Symposium on Circuits and Systems*, 270-273. 10.1109/MWSCAS.2012.6292009

Kaseridis, D., & Stuechelix, J. (2009). Bankaware Dynamic Cache Partitioning for Multicore Architectures. *International Conference on Parallel Processing*, 18-25.

Kastner, M. A., & Goldhaber-Gordon, D. (2001). Kondo physics with single electron transistors. *Solid State Communications*, *119*(4-5), 245–252. doi:10.1016/S0038-1098(01)00106-5

Kaur, K., & Noor, A. (2011). Strategies & methodologies for low power VLSI designs: A review. *International Journal of Advances in Engineering and Technology*, *1*(2), 159.

Kedar, G., Mendelson, A., & Cidon, I. (2019). SPACE: Semi-Partitioned Cache for Energy Efficient, Hard Real-Time Systems. *IEEE Transactions on Computers*, *66*(4), 717–730. doi:10.1109/TC.2016.2608775

Keerthi Kiran, R., & Kalpana, A. B. (2015). Low Power 8, 16 & 32 Bit ALU Design Using Clock Gating. *International Journal of Scientific and Engineering Research*, *6*(8).

Keleş, S., & Keleş, F. (2023). Low Voltage-Low Power Wide Range FGMOS Fully Differential Difference Current Conveyor And Application Examples. *International Journal of Electronics*, 1–16. doi:10.1080/00207217.2022.2164079

Keyser, U., Schumacher, H. W., Zeitler, U., Haug, R. J., & Eberl, K. (2000). Fabrication of a single-electron transistor by current-controlled local oxidation of a two-dimensional electron system. *Applied Physics Letters*, *76*(4), 457–459. doi:10.1063/1.125786

Khadem Hosseini, V., Dideban, D., Ahmadi, M. T., & Ismail, R. (2018). An analytical approach to model capacitance and resistance of capped carbon nanotube single electron transistor. *AEÜ. International Journal of Electronics and Communications*, *90*, 97–102. doi:10.1016/j.aeue.2018.04.015

Khan, R. S. (2021). *Life cycle assessment of GHG emissions of light duty vehicles: comparison between internal combustion engine vehicles and battery electric vehicles.* Academic Press.

Khan, F. H. (2014). Chemical hazards of nanoparticles to human and environment (a review). *Oriental Journal of Chemistry*, *29*(4), 1399–1408. doi:10.13005/ojc/290415

Khan, S., Nazir, S., & Ullah Khan, H. (2021). Smart Object Detection and Home Appliances Control System in Smart Cities. *Computers. Materials & Continua*, *67*(1), 895–915. Advance online publication. doi:10.32604/cmc.2021.013878

Khediri, S. E., Nasri, N., Wei, A., & Kachouri, A. (2014). A new approach for clustering in wireless sensor networks based on LEACH. *Procedia Computer Science*, *32*, 1180–1185. doi:10.1016/j.procs.2014.05.551

Khodabakhshi, M. (2010). An Output Oriented Super-Efficiency Measure in Stochastic Data Envelopment Analysis: Considering Iranian Electricity Distribution Companies. *Computers & Industrial Engineering*, *58*(4), 663–671. doi:10.1016/j.cie.2010.01.009

Khodabakhshi, M., & Asgharian, M. (2008). An input relaxation measure of efficiency in stochastic data envelopment analysis. *Applied Mathematical Modelling*, *33*(4), 2010–2023. doi:10.1016/j.apm.2008.05.006

Khodabakhshi, M., & Kheirollahi, H. (2010). Measuring technical efficiency of Iranian electricity distribution units with stochastic data envelopment analysis. *Iranian Conference on Applied Mathematical Modelling*.

Khodadadipour, M., Hadi-Vencheh, A., Behzadi, M. H., & Rostamy-malkhalifeh, M. (2021). Undesirable factors in stochastic DEA cross-efficiency evaluation: An application to thermal power plant energy efficiency. *Economic Analysis and Policy*, *69*, 613–628. doi:10.1016/j.eap.2021.01.013

Khodakaramia, A., Farahanib, F. H., & Aghaeid, J. (2017). Stochastic characterization of electricity energy markets including plug-in electric vehicles. *Renewable & Sustainable Energy Reviews*, *69*, 112–122. doi:10.1016/j.rser.2016.11.094

Khosrojerdi, ARezvani, RPourandoost, A. (2013). 0.8 V 191.9 nW DTMOS Current Mirror OTA in 0.18 μm CMOS Process. *Majlesi Journal of Telecommunication Devices*, *2*(3), 251–254.

Kim, G. W. (2015). Piezoelectric Energy Harvesting from Torsional Vibration in Internal Combustion Engines. *International Journal of Automotive Technology*, *16*(4), 645–651. doi:10.100712239-015-0066-6

Kim, H. C., Wallington, T. J., Arsenault, R., Bae, C., Ahn, S., & Lee, J. (2016). Cradle-to-Gate Emissions from a Commercial Electric Vehicle Li-Ion Battery: A Comparative Analysis. *Environmental Science & Technology*, *50*(14), 7715–7722. doi:10.1021/acs.est.6b00830 PMID:27303957

Kim, N., Hansen, K., Paraoanu, S., & Pekola, J. (2003). Fabrication of Nb based superconducting single electron transistor. *Phys B*, *329*, 1519–1520. doi:10.1016/S0921-4526(02)02419-5

Kim, T., Song, W., Son, D. Y., Ono, L. K., & Qi, Y. (2019). Lithium-ion batteries: Outlook on present, future, and hybridized technologies. *Journal of Materials Chemistry. A, Materials for Energy and Sustainability, 7*(7), 2942–2964. doi:10.1039/C8TA10513H

Kim, Y. J., Joshi, Y. K., & Fedorov, A. G. (2008). An absorption miniature heat pump system for electronics cooling. *International Journal of Refrigeration, 31*(1), 23–33. doi:10.1016/j.ijrefrig.2007.07.003

Klein, M. (2009). Power Consumption at 40 and 45 nm. *White Paper, 298,* 1-21.

Koh, S. C. L., Smith, L., Miah, J., Astudillo, D., Eufrasio, R. M., Gladwin, D., Brown, S., & Stone, D. (2021). Higher 2nd life Lithium Titanate battery content in hybrid energy storage systems lowers environmental-economic impact and balances eco-efficiency. *Renewable & Sustainable Energy Reviews, 152,* 111704. Advance online publication. doi:10.1016/j.rser.2021.111704

Konal, M., & Kacar, F. (2021). DTMOS based low-voltage low-power all-pass filter. *Analog Integr Circ Sig Process, 108*(1), 173–179. doi:10.100710470-021-01878-z

Kononchuk, O., & Nguyen, B. Y. (2014). *Silicon-on-insulator (soi) technology: Manufacture and applications.* Elsevier.

Koppinen, P. J., Stewart, M. D. Jr, & Neil, M. (2013). Fabrication and Electrical Characterization of Fully CMOS-Compatible Si Single-Electron Devices. *IEEE Transactions on Electron Devices, 60*(1), 60. doi:10.1109/TED.2012.2227322

Korotkov, A. N. (1994). Intrinsic noise of the single-electron transistor. *Physical Review, 49*(15), 10381–10392. doi:10.1103/PhysRevB.49.10381 PMID:10009861

Krautschneider, W., Kohlhase, A., & Terletzki, H. (1997). Scaling down and reliability problems of gigabit CMOS circuits. *Microelectronics and Reliability, 37*(1), 19–37. doi:10.1016/0026-2714(96)00236-3

Krestinskaya, O., Salama, K. N., & James, A. P. (2018). Learning in Memristive Neural Network Architectures Using Analog Backpropagation Circuits. *IEEE Transactions on Circuits and Systems. I, Regular Papers, 66*(2), 719–732.

Krishna, J., Kishore, P., & Solomon, A. B. (2017). Heat pipe with nano enhanced-PCM for electronic cooling application. *Experimental Thermal and Fluid Science, 81,* 84–92.

Kuhn, K., Kenyon, C., Kornfeld, A., Liu, M., Maheshwari, A., Shih, W., Sivakumar, S., Taylor, G., VanDerVoorn, P., & Zawadzki, K. (2008). Managing process variation in Intel's 45 nm CMOS technology. *Information Technology Journal, 12,* 2.

Kumar, O., & Kaur, M. (2010). Single Electron Transistor: Applications & Problems. *International Journal of VLSI Design & Communication Systems, 1*(4).

Lageweg, C., Cotofana, S., & Vassiliadis, S. (2002). Static buffered SET based logic gates. In *Proceedings of the 2nd IEEE conference on nanotechnology.* IEEE. 10.1109/NANO.2002.1032295

Lammie, C., Xiang, W., Linares-Barranco, B., & Azghadi, M. R. (2020). *MemTorch: An Open-source Simulation Framework for Memristive Deep Learning Systems.* arXiv preprint arXiv:2004.10971.

Land, C. K., Lovell, C. A. K., & Thore, S. (1993). Chance-Constrained Data Envelopment Analysis. *Managerial and Decision Economics, 14*(6), 541–554. doi:10.1002/mde.4090140607

Land, C. K., Lovell, C. A. K., & Thore, S. (1994). Productive Efficiency under Capitalism and State Socialism: An Empirical Inquiry Using Chance-Constrained Data Envelopment Analysis. *Technological Forecasting and Social Change, 46*(2), 139–152. doi:10.1016/0040-1625(94)90022-1

Lee, D., & Choi, K. (2019). Energy-efficient partitioning of hybrid caches in multi-core architecture. *22nd International Conference on Very Large Scale Integration (VLSI-SoC)*, 1-6.

Lee, D.-Y., Kim, H., Li, H.-M., Jang, A. R., Lim, Y.-D., Cha, S. N., Park, Y. J., Kang, D. J., & Yoo, W. J. (2013). Hybrid Energy Harvester Based on Nanopillar Solar Cells and PVDF Nanogenerator. *Nanotechnology, 24*(17), 175402. doi:10.1088/0957-4484/24/17/175402 PMID:23558434

Lee, Joshi, Orlov, & Snider. (2010). Si single electron transistor fabricated by chemical mechanical polishing. *Journal of Vacuum Science and Technology, 28.*

Liang, C. M., & Terzis, A. (2018). Typhoon: a reliable data dissemination protocol for wireless sensor networks. *5th European conference on Wireless Sensor Networks,* 268-285.

Li-Na, S., Li, L., Xin-Xing, L., Hua, Q., & Xiao-Feng, G. (2015). Fabrication and characterization of a single electron transistor based on a silicon-on-insulator. *Chinese Physics Letters, 32*(4), 047301. doi:10.1088/0256-307X/32/4/047301

Lindsay, A., & Ravindran, B. (2018). On Cache-Aware Task Partitioning for Multicore Embedded Real-Time Systems. *IEEE International Conference on High Performance Computing and Communications (HPCC)*, 677-684.

Lin, J., Yu, W., Zhang, N., Yang, X., Zhang, H., & Zhao, W. (2017). A survey on internet of things: Architecture, enabling technologies, security and privacy, and applications. *IEEE Internet of Things Journal, 4*(5), 1125–1142. doi:10.1109/JIOT.2017.2683200

Li, P., Xia, X., & Guo, J. (2022). A review of the life cycle carbon footprint of electric vehicle batteries. *Separation and Purification Technology, 296*, 121389. Advance online publication. doi:10.1016/j.seppur.2022.121389

Li, S. X. (1998). Stochastic models and variable returns to scales in data envelopment analysis. *European Journal of Operational Research, 104*(3), 532–548. doi:10.1016/S0377-2217(97)00002-7

Liu, J. (2020). Current mirror featuring DTMOS for analog single-event transient mitigation in space application. *Semiconductor Science and Technology, 35*(8).

Liu, H., Zhang, S., Kobayashi, T., Chen, T., & Lee, C. (2014). Flow Sensing and Energy Harvesting Characteristics of a Wind-driven Piezoelectric Pb (Zr0. 52, Ti0. 48) O3 Microcantilever. *Micro & Nano Letters*, *9*(4), 286–289. doi:10.1049/mnl.2013.0750

Liu, J., Han, Y., Xie, L., Wang, Y., & Wen, G. (2014). A 1-V DTMOS-Based fully differential telescopic OTA. *IEEE Asia Pacific Conference on Circuits and Systems (APCCAS)*, 49-52. 10.1109/APCCAS.2014.7032716

Liu, Y., Nishimura, M., Li, L., & Colins, K. (2017). Study on a Low-Cost and Large-Scale Environmentally Adaptive Protocol Stack of Nuclear and Space Wireless Sensor Network Applications under Gamma Radiation. *Nuclear Technology*, *197*(1), 75–87. doi:10.13182/NT16-97

Loh, P. K. K., Hsu, W. J., & Pan, Y. (2017). Reliable and efficient communications in sensor networks. *Journal of Parallel and Distributed Computing*, *67*(8), 922–934. doi:10.1016/j.jpdc.2007.04.008

Lopes, A. L. M., & Mesquita, R. B. (2015). Tariff regulation of electricity distribution: A comparative analysis of regulatory benchmarking models. *Proceedings of the 14th European Workshop on Efficiency and Productivity Analysis*.

Lucia, Balaji, Colin, Maeng, & Ruppel. (2017). Intermittent Computing: Challenges and Opportunities. *2nd Summit on Advances in Programming Languages (SNAPL 2017)*.

Lu, X., & Yang, S.-H. (2010). Thermal Energy Harvesting for WSNs. In *2010 IEEE International Conference on Systems, Man and Cybernetics*. IEEE.

Ma, S., Jiang, M., Tao, P., Song, C., Wu, J., Wang, J., Deng, T., & Shang, W. (2018). Temperature effect and thermal impact in lithium-ion batteries: A review. In Progress in Natural Science: Materials International (Vol. 28, Issue 6, pp. 653–666). Elsevier B.V. doi:10.1016/j.pnsc.2018.11.002

MacHiels, N., Leemput, N., Geth, F., Van Roy, J., Buscher, J., & Driesen, J. (2014). Design criteria for electric vehicle fast charge infrastructure based on flemish mobility behavior. *IEEE Transactions on Smart Grid*, *5*(1), 320–327. doi:10.1109/TSG.2013.2278723

Maeda, K., Okabayashi, N., Kano, S., Takeshita, S., Tanaka, D., Sakamoto, M., Teranishi, T., & Majima, Y. (2012). Logic operations of chemically assembled single-electron transistor. *ACS Nano*, *6*(3), 2798–2803. 10.1021/nn3003086

Mahapatra, R. P., & Yadav, R. K. (2015). Descendant of LEACH based routing protocols in wireless sensor networks. *Procedia Computer Science*, *57*, 1005–1014. doi:10.1016/j.procs.2015.07.505

Mahapatra, S., & Ionescu, A. M. (2005). Realization of multiple valued logic and memory by hybrid SETMOS architecture. *IEEE Transactions on Nanotechnology*, *4*(6), 705–714. doi:10.1109/TNANO.2005.858602

Mahmoud, M. S., & Mohamad, A. A. (2016). *A study of efficient power consumption wireless communication techniques/modules for internet of things (IoT) applications*. Academic Press.

Majid, M. A. (2022, March). Energy-Efficient Adaptive Clustering and Routing Protocol for Expanding the Life Cycle of the IoT-based Wireless Sensor Network. In *2022 6th International Conference on Computing Methodologies and Communication (ICCMC)* (pp. 328-336). IEEE.

Ma, K., Wang, X., & Wang, Y. (2019). DPPC: Dynamic Power Partitioning and Control for Improved Chip Multiprocessor Performance. *IEEE Transactions on Computers, 63*(7), 1736–1750. doi:10.1109/TC.2013.67

Manjula, K., & Bhavana, V.N. (2022). Deep Network Accelerators Towards Healthcare Edge Applications And Systems. *International Research Journal of Modernization in Engineering Technology and Science, 4*(7).

Manthiram, A. (2020). A reflection on lithium-ion battery cathode chemistry. In Nature Communications (Vol. 11, Issue 1). Nature Research. doi:10.103841467-020-15355-0

Masghouni, N., Burton, J., Philen, M. K., & Al-Haik, M. (2015). Investigating the Energy Harvesting Capabilities of a Hybrid ZnO Nanowires/Carbon Fiber Polymer Composite Beam. *Nanotechnology, 26*(9), 95401. doi:10.1088/0957-4484/26/9/095401 PMID:25670370

Matej, R., Viera, S., & Daniel, A. (2018). Design techniques for low-voltage analog integrated circuits. *Journal of Electrical Engineering, 68*(4), 245–255.

Mathews, I., Kelly, G., King, P. J., & Frizzell, R. (2014). GaAs Solar Cells for Indoor Light Harvesting. In *2014 IEEE 40th Photovoltaic Specialist Conference (PVSC)*. IEEE. 10.1109/PVSC.2014.6924971

Matsumoto, K., Ishii, M., Segawa, K., Oka, Y., Vartanian, B.J., & Harris, J.S. (1996). Room temperature operation of a single electron transistor made by the scanning tuneling microscope nanooxidation process for the TiOx/Ti system. *American Institute of Physics, 68*.

Matsumoto, K. (1998). Application of Scanning Tunneling/Atomic Force Microscope Nanooxidation process to room temperature operated Single Electron Transistor and other devices. *Scanning Microscopy, 12*, 6169.

Matsumoto, K. (2000). Room-Temperature Single Electron Devices by Scanning Probe Process. *International Journal of High Speed Electronics and Systems, 10*(01), 83–91. doi:10.1142/S0129156400000118

Maurya, D., Kumar, P., Khaleghian, S., Sriramdas, R., Kang, M. G., Kishore, R. A., Kumar, V., Song, H.-C., Park, J.-M. J., Taheri, S., & Priya, S. (2018). Energy Harvesting and Strain Sensing in Smart Tire for next Generation Autonomous Vehicles. *Applied Energy, 232*, 312–322. doi:10.1016/j.apenergy.2018.09.183

Maymandi-Nejad & Sachdev. (2006). DTMOS technique for low-voltage analog circuits. *IEEE Transactions on VLSI Systems, 14*(10), 1151-1156.

McBrayer, J. D., Rodrigues, M. T. F., Schulze, M. C., Abraham, D. P., Apblett, C. A., Bloom, I., Carroll, G. M., Colclasure, A. M., Fang, C., Harrison, K. L., Liu, G., Minteer, S. D., Neale, N. R., Veith, G. M., Johnson, C. S., Vaughey, J. T., Burrell, A. K., & Cunningham, B. (2021). Calendar aging of silicon-containing batteries. In Nature Energy (Vol. 6, Issue 9, pp. 866–872). Nature Research. doi:10.103841560-021-00883-w

Meeks, S. W., & Timme, R. W. (1977). Rare Earth Iron Magnetostrictive Underwater Sound Transducer. *The Journal of the Acoustical Society of America*, *62*(5), 1158–1164. doi:10.1121/1.381650

Meenakshi, N., Monish, M., Dikshit, K. J., & Bharath, S. (2019). Arduino Based Smart Fingerprint Authentication System. *Proceedings of 1st International Conference on Innovations in Information and Communication Technology, ICIICT 2019.* 10.1109/ICIICT1.2019.8741459

Megiddo, N., & Modha, D. S. (2020). *Outperforming LRU with an adaptive replacement cache algorithm.* IEEE Computer.

Mi, J., Xu, L., Guo, S., Abdelkareem, M. A. A., & Meng, L. (2017). *Suspension Performance and Energy Harvesting Property Study of a Novel Railway Vehicle Bogie with the Hydraulic-Electromagnetic Energy-Regenerative Shock Absorber.* SAE Technical Paper.

Miller, R. (2015). Cheaper Sensors Will Fuel The Age Of Smart Everything. *Tech Crunch.* Retrieved January 31, 2023 (https://techcrunch.com/2015/03/10/cheaper-sensors-will-fuel-the-age-of-smart-everything/)

Miralaie, M., Leilaeioun, M., Abbasian, K., & Hasani, M. (2014). Modeling and analysis of room-temperature silicon quantum dot-based single-electron transistor logic gates. *Journal of Computational and Theoretical Nanoscience*, *11*(1), 15–24. doi:10.1166/jctn.2014.3311

Miralaie, M., & Mir, A. (2016). Performance analysis of single-electron transistor at room-temperature for periodic symmetric functions operation. *Journal of Engineering (Stevenage, England)*, *10*(10), 352–356. doi:10.1049/joe.2016.0139

Mirbolouki, M., Behzadi, M.H., & Korzaledin, M. (2014). Multiplier, models in stochastic DEA. *Data Envelopment Analysis and Decision Science.*

Mishra, B., Kushwah, V. S., & Sharma, R. (2021). Power consumption analysis of MOSFET and Single electron transistor for inverter circuit. *Materials Today: Proceedings*, *47*(Part 19), 6600–6604. doi:10.1016/j.matpr.2021.05.094

Moore, G. E. (1998). Cramming more components onto integrated circuits. *Proceedings of the IEEE*, *86*(1), 82–85. doi:10.1109/JPROC.1998.658762

Morais, R., Fernandes, M. A., Matos, S. G., Serôdio, C., Ferreira, P. J. S. G., & Reis, M. J. C. S. (2008). A ZigBee Multi-Powered Wireless Acquisition Device for Remote Sensing Applications in Precision Viticulture. *Computers and Electronics in Agriculture*, *62*(2), 94–106. doi:10.1016/j.compag.2007.12.004

Mosorov, V., Biedron, S., & Panskyi, T. (2015). Analysis of a new model of low energy adaptive clustering hierarchy protocol in the wireless sensor network. *Восточно-Европейский журнал передовых технологий, 5*(9), 4-8.

Mostafalu, P., & Sonkusale, S. (2014). Flexible and Transparent Gastric Battery: Energy Harvesting from Gastric Acid for Endoscopy Application. *Biosensors & Bioelectronics, 54*, 292–296. doi:10.1016/j.bios.2013.10.040 PMID:24287419

Mowry, T. C. (1994). *Tolerating latency through software-controlled data prefetching.* Stanford University.

Mullarkey, S., Caulfield, B., McCormack, S., & Basu, B. (2015). A framework for establishing the technical efficiency of Electricity Distribution Counties (EDCs) using Data Envelopment Analysis. *Energy Conversion and Management, 94*, 112–123. doi:10.1016/j.enconman.2015.01.049

Müller, Rittenschober, & Springer. (2010). A Wireless Sensor Network Using Energy Harvesting for Agricultural Machinery. *E & I Elektrotechnik Und Informationstechnik, 3*(127), 39–46.

Murshed, S.M.S. (2016). *Electronics Cooling.* Intech Open.

Murshed, S. S., & De Castro, C. N. (2017). A critical review of traditional and emerging techniques and fluids for electronics cooling. *Renewable & Sustainable Energy Reviews, 78*, 821–833. doi:10.1016/j.rser.2017.04.112

Musalappa, S., Sundaram, S., & Chu, Y. (2005). Energy savings for data caches: ELRU-SEQ replacement policy. *24th IEEE International Performance, Computing, and Communications Conference*, 641-642. 10.1109/PCCC.2005.1460659

Nadeau, P., El-Damak, D., Glettig, D., Yong, L. K., Mo, S., Cleveland, C., Booth, L., Roxhed, N., Langer, R., & Chandrakasan, A. P. (2017). Prolonged Energy Harvesting for Ingestible Devices. *Nature Biomedical Engineering, 1*(3), 22. doi:10.103841551-016-0022 PMID:28458955

Neelakandan, S., Beulah, J. R., Prathiba, L., Murthy, G. L. N., Irudaya Raj, E. F., & Arulkumar, N. (2022). Blockchain with deep learning-enabled secure healthcare data transmission and diagnostic model. *International Journal of Modeling, Simulation, and Scientific Computing.* doi:10.1142/S1793962322410069

Neelakandan, S., Rene Beulah, J., Prathiba, L., Murthy, G. L. N., Irudaya Raj, E. F., & Arulkumar, N. (2022). Blockchain with deep learning-enabled secure healthcare data transmission and diagnostic model. *International Journal of Modeling, Simulation, and Scientific Computing, 2241006.*

Ngurah Desnanjaya, I. G. M., & Arsana, I. N. A. (2021). Home security monitoring system with IoT-based Raspberry Pi. *Indonesian Journal of Electrical Engineering and Computer Science, 22*(3), 1295. Advance online publication. doi:10.11591/ijeecs.v22.i3.pp1295-1302

Niranjan, V. (2015). *Performance Improvement Of Low Voltage CMOS Circuits Using Body Bias Approach.* PhD Thesis.

Niranjan, V. (n.d.). *Low power and high performance shift registers using pulsed latch technique*. Academic Press.

Niranjan, V., Kumar, A., & Jain, S.B. (2014b). Composite transistor cell using dynamic body bias for high gain and low-voltage applications. *Journal of Circuits, Systems, and Computers*, *23*(8), 1-18.

Niranjan, V. (2017). Wideband current mirror using transconductance boosting technique. *International Journal of Advance Research in Science and Engineering*, *6*(12), 1159–1171.

Niranjan, V., & Gupta, M. (2009). Low voltage four quadrant analog multiplier using dynamic threshold MOS transistors. *Microelectronics International*, *26*(1), 47–52. doi:10.1108/13565360910923179

Niranjan, V., & Gupta, M. (2011). Body Biasing-A circuit level approach to reduce leakage in Low power CMOS circuits. *Journal of Active and Passive Electronic Devices.*, *6*(1-2), 89–99.

Niranjan, V., Gupta, M., & Jain, S. B. (2012). A Novel 0.5 Volt Analog Multiplier using dynamic body bias technique. *Proceeding National Conference on Advanced VLSI and Embedded Technology*.

Niranjan, V., Kumar, A., & Jain, S. B. (2013a). Low Voltage Flipped Voltage Follower based Current Mirror using DTMOS Technique. *Proceeding IEEE International Conference on Multimedia, Signal Processing and Communication Technologies*, 250-254. 10.1109/MSPCT.2013.6782129

Niranjan, V., Kumar, A., & Jain, S. B. (2013b). Triple Well Subthreshold CMOS Logic Using Body-bias Technique. *Proceeding IEEE International Conference on Signal Processing, Computing and Control*, 1-6. 10.1109/ISPCC.2013.6663447

Niranjan, V., Kumar, A., & Jain, S. B. (2014a). Maximum bandwidth enhancement of current mirror using series-resistor and dynamic body bias technique. *Wuxiandian Gongcheng*, *23*(3), 922–930.

Niranjan, V., Kumar, A., & Jain, S. B. (2014c). Low Voltage Self cascode amplifier using dynamic body bias technique. *Proceeding International Conference on VLSI and Signal Processing*.

Niranjan, V., Kumar, A., & Jain, S. B. (2014d). Low-voltage and High-speed Flipped Voltage Follower Using DTMOS transistor. *Proceeding IEEE International Conference on Signal Propagation and Computer technology*, 145-150. 10.1109/ICSPCT.2014.6884882

Niranjan, V., Kumar, A., & Jain, S. B. (2015). Low-voltage gate and body driven self-biased cascode current mirror with enhanced bandwidth. *International Journal of Circuits and Architecture Design*, *1*(4), 320–342. doi:10.1504/IJCAD.2015.072615

Niranjan, V., Kumar, A., & Jain, S. B. (2017). Improving Bandwidth of Flipped Voltage Follower Using Gate-Body Driven Technique. *Journal of Engineering Science and Technology*, *12*(1), 83–102.

Niranjan, V., Singh, A., & Kumar, A. (2014). Dynamic Threshold MOS transistor for Low Voltage Analog Circuits. *Proceeding International Conference on Recent Trends & Issues in Engineering and Technology.*

Nur-A-Alam, Ahsan, M., Based, M. A., Haider, J., & Rodrigues, E. M. G. (2021). Smart monitoring and controlling of appliances using lora based iot system. *Designs*, *5*(1), 17. Advance online publication. doi:10.3390/designs5010017

Nurdiawati, A., & Agrawal, T. K. (2022). Creating a circular EV battery value chain: End-of-life strategies and future perspective. *Resources, Conservation and Recycling*, *185*, 106484. Advance online publication. doi:10.1016/j.resconrec.2022.106484

Odyakmaz, N. (2009). *Türkiye'deki Elektrik Dağıtım Şirketlerinin Performansa Dayalı Düzenleme Çerçevesinde Karşılaştırmalı Etkinlik Analizi* [PhD Thesis]. Institute of Social Sciences, Hacettepe University.

Olesen, O. B. (2002). *Comparing and Combining Two Approaches for Chance Constrained DEA. Discussion paper.* The University of Southern Denmark.

Omrani, H., Azadeh, A., Ghaderi, S. F., & Abdollahzadeh, S. (2010). A Consistent Approach for Performance Measurement Of Electricity Distribution Companies. *Int J Energy Sect Manage*, *4*(3), 399–416. doi:10.1108/17506221011073879

Omrani, H., Beiragh, R. G., & Kaleibari, S. S. (2015). Performance assessment of Iranian electricity distribution companies by an integrated cooperative game data envelopment analysis principal component analysis approach. *Electrical Power and Energy Systems*, *64*, 617–625. doi:10.1016/j.ijepes.2014.07.045

Ono, Y., Inokawa, H., Takahashi, Y., Nishiguchi, K., & Fujiwara, A. (2010). Single-electron transistor and its logic application. *Nanotechnol Online*, *20*, 45–68.

Orouji, A. A., & Abbasi, A. (2012). Novel partially depleted SOI MOSFET for suppression floating-body effect: An embedded JFET structure. *Superlattices and Microstructures*, *52*(3), 552–559. doi:10.1016/j.spmi.2012.06.006

Paliwal, P., Sharma, J. B., & Nath, V. (2020). Comparative Study of FFA Architectures Using Different Multiplier and Adder Topologies. *Microsystem Technologies*, *26*(5), 1455–1462. doi:10.100700542-019-04678-8

Paoli, L. (2022). *Electric Vehicles.* https://www.iea.org/reports/electric-vehicles

Parekh, R. (2019). Design and simulation of single electron transistor-based SRAM and its memory controller at room temperature. *Int J Integrat Eng*, *11*(6), 186–195.

Paris Agreement. (2015). *Paris agreement.* Report of the Conference of the Parties to the United Nations Framework Convention on Climate Change.

Park, Y. T., Sthapit, P., & Pyun, J. Y. (2009). Smart digital door lock for the home automation. *IEEE Region 10 Annual International Conference, Proceedings/TENCON*. 10.1109/TENCON.2009.5396038

Patel, Agrawal, & Parekh. (2020). Single-electron transistor: review in perspective of theory, modelling, design and fabrication. *Microsystem Technologies-micro-and Nanosystems-information Storage and Processing Systems*, 1-13.

Patel, R., Agrawal, Y., & Parekh, R. (2018). *A vector ðle generation program for simulating single electron transistor based computing system. In 2018 IEEE electron devices Kolkata conference (EDKCON)*. IEEE. doi:10.1109/EDKCON.2018.8770464

Patel, R., Agrawal, Y., & Parekh, R. (2019). Design of prominent set-based high performance computing system. *IET Circuits, Devices & Systems*, *14*(2), 159–167. doi:10.1049/iet-cds.2019.0166

Paul, M., & Petrov, P. (2018). I-cache Configurability for Temperature Reduction through Replicated Cache Partitioning. *IEEE 8th Symposium on Application Specific Processors (SASP)*, 81-86.

Pawar, S., Kithani, V., Ahuja, S., & Sahu, S. (2018). Smart Home Security Using IoT and Face Recognition. *Proceedings - 2018 4th International Conference on Computing, Communication Control and Automation, ICCUBEA 2018*. 10.1109/ICCUBEA.2018.8697695

Payvand, M., Demirag, Y., Dalgaty, T., Vianello, E., & Indiveri, G. (2020). Analog weight updates with compliance current modulation of binary ReRams for on-chip learning. In *Proceedings of the IEEE International Symposium on Circuits and Systems (ISCAS)*. IEEE.

Peng, Z., & Li, X. (2010). The improvement and simulation of LEACH protocol for WSNs. *IEEE International Conference on Software Engineering and Service Sciences*, 500-503. 10.1109/ICSESS.2010.5552317

Peters, J. F., Baumann, M., Zimmermann, B., Braun, J., & Weil, M. (2017). The environmental impact of Li-Ion batteries and the role of key parameters – A review. In *Renewable and Sustainable Energy Reviews* (Vol. 67, pp. 491–506). Elsevier Ltd., doi:10.1016/j.rser.2016.08.039

Preethi, P., Mohan, K. G., Kumar, K. S., & Mahapatra, K. K. (2021, December). Low Power Sorters Using Clock Gating. In *2021 IEEE International Symposium on Smart Electronic Systems (iSES)(Formerly iNiS)* (pp. 6-11). IEEE. 10.1109/iSES52644.2021.00015

Preethi, P., Mohan, K. G., Kumar, K. S., & Mahapatra, K. K. (2022). Sorter Design with Structured Low Power Techniques. *SN Computer Science*, *4*(2), 129. doi:10.100742979-022-01546-7

Prerna & Niranjan, V. (2015). Analog Multiplier Using DTMOS-CCII Suitable for Biomedical Application. *Proceeding IEEE International Conference on Computing, Communication and Automation*.

Priyadarshini, R., Barik, R. K., Dubey, H. C., & Mishra, B. K. (2021). A Survey of Fog Computing Based Healthcare Big Data Analytics and Its Security. *International Journal of Ambient Computing and Intelligence*, *12*(2), 53–72.

Psiborg. (n.d.). https://psiborg.in/wireless-sensor-network-based-on-sub-1-ghz/

Pushparaj, T. L., Raj, E., Rani, E., Darwin, S., & Appadurai, M. (2022). Employing Novel Si-Over-Si Technology to Optimize PV Effect in Solar Array. *Silicon*, 1-13.

Qu, W., & Mudawar, I. (2004). Measurement and correlation of critical heat flux in two-phase microchannel heat sinks. *International Journal of Heat and Mass Transfer*, 47, 2045–2059.

R., M., Y., R., R., R., & A., S. (2020). Smart Home Security System using Iot, Face Recognition and Raspberry Pi. *International Journal of Computer Applications, 176*(13). doi:10.5120/ijca2020920105

Rabaey, J., & Rabaey, J. (2009). Optimizing Power@ Design Time–Circuit-Level Techniques. *Low Power Design Essentials*, 77-111.

Radu, L. D. (2020). Disruptive technologies in smart cities: A survey on current trends and challenges. *Smart Cities*, 3(3), 1022–1038. Advance online publication. doi:10.3390martcities3030051

Rahman, A., Razzak, F., Afroz, R., Akm, M., & Hawlader, M. N. A. (2015). Power Generation from Waste of IC Engines. *Renewable & Sustainable Energy Reviews*, 51, 382–395. doi:10.1016/j.rser.2015.05.077

Rai, C., Khursheed, A., & Haque, F. Z. (2019). Review on Single Electron Transistor (SET): Emerging Device in Nanotechnology, Austin. *Journal of Nanomedicine & Nanotechnology*, 7(1).

Rajaeifar, M. A., Ghadimi, P., Raugei, M., Wu, Y., & Heidrich, O. (2022). Challenges and recent developments in supply and value chains of electric vehicle batteries: A sustainability perspective. In Resources, Conservation and Recycling (Vol. 180). Elsevier B.V. doi:10.1016/j.resconrec.2021.106144

Raj, E. F. I., Appadurai, M., Darwin, S., & Rani, E. F. I. (2022). Internet of Things (IoT) for Sustainable Smart Cities. In *Internet of Things* (pp. 163–188). CRC Press. doi:10.1201/9781003219620-9

Ramasur, D., & Hancke, G. P. (2012). A Wind Energy Harvester for Low Power Wireless Sensor Networks. In *2012 IEEE International Instrumentation and Measurement Technology Conference Proceedings*. IEEE. 10.1109/I2MTC.2012.6229698

Ran, G., Zhang, H., & Gong, S. (2010). Improving on LEACH protocol of wireless sensor networks using fuzzy logic. *Journal of Information and Computational Science*, 7(3), 767–775.

Raut, V., & Dakhole, P. (2016). *Design and implementation of quaternary summation circuit with single electron transistor and MOSFET. In 2016 international conference on electrical, electronics, and optimization techniques (ICEEOT)*. IEEE. doi:10.1109/ICEEOT.2016.7755088

RitchieH.RoserM.RosadoP. (2020). *Energy*. https://ourworldindata.org/energy

Rodrigue, J.-P. (2020). *The Geography of Transport Systems, Transportation and Energy*. Routledge. doi:10.4324/9780429346323

Romare, M., & Dahllöf, L. (2017). *The Life Cycle Energy Consumption and Greenhouse Gas Emissions from Lithium-Ion Batteries*. IVL Swedish Environmental Research Institute.

Roy, S., Nasir Uddin, M., Zahirul Haque, M., & Jahidul Kabir, M. (2018). Design and Implementation of the Smart Door Lock System with Face Recognition Method using the Linux Platform Raspberry Pi. *IJCSN-International Journal of Computer Science and Network, 7*(6).

Sadana, D. K., & Current, M. I. (2006). Fabrication of Silicon-on-insulator (SOI) and strain-Silicon-oninsulator (sSOI) wafers using ion implantation. Ion Implantation: Science and Technology.

Sadhu, A., Das, K., De, D., & Kanjilal, M. R. (2022). Low power design methodology in quantum-dot cellular automata. *Computers & Electrical Engineering, 97*, 107638. doi:10.1016/j.compeleceng.2021.107638

Sadjadi, S. J., & Omrani, H. (2008). Data Envelopment Analysis with Uncertain Data: An Application for Iranian Electricity Distribution Companies. *Energy Policy, 36*(11), 4247–4254. doi:10.1016/j.enpol.2008.08.004

Sadjadi, S. J., Omrani, H., Makui, A., & Shahanaghi, K. (2011). An interactive robust data envelopment analysis model for determining alternative targets in Iranian electricity distribution companies. *Expert Systems with Applications, 38*(8), 9830–9839. doi:10.1016/j.eswa.2011.02.047

Sagar, S., Keke, C., & Amit, S. (2018). Toward practical privacy-preserving analytics for AIoT and cloud-based healthcare systems. *IEEE Internet Computing, 22*(2), 42–51. doi:10.1109/MIC.2018.112102519

Saha, C. R., Huda, M. N., Mumtaz, A., Debnath, A., Thomas, S., & Jinks, R. (2020). Photovoltaic (PV) and Thermo-Electric Energy Harvesters for Charging Applications. *Microelectronics Journal, 96*, 104685. doi:10.1016/j.mejo.2019.104685

Sahað, A., Moaiyeri, M. H., Navi, K., & Hashemipour, O. (2013). Efðcient single-electron transistor inverter-based logic circuits and memory elements. *Journal of Computational and Theoretical Nanoscience, 10*(5), 1171–1178. doi:10.1166/jctn.2013.2824

Sahatiya. (2013). Single Electron Transistor: A Review. *International Journal of Scientific & Engineering Research, 4*.

Sahu & Agrahari. (2021). Low Power Design and Challenges in VLSI with IoT Systems. *International Conference on Advanced Computing and Communication Technology.*

Sahu & Agrahari. (n.d.). Optimization of Parameters in 8-Bit ALU Circuit With Clock Gating Technique. *PRATIBHA: International Journal of Science, Spirituality, Business and Technology, 34.*

Sahu, P., & Agrahari, S. K. (2020). Comparative Analysis of Different Clock Gating Techniques. In *2020 5th IEEE International Conference on Recent Advances and Innovations in Engineering (ICRAIE)*. IEEE. 10.1109/ICRAIE51050.2020.9358375

Sahu, P., & Agrahari, S. K. (2021). Power and Performance Optimization in 16- Bit ALU Using Power Gating. *International Conference on Recent Trends in Electrical, Electronics & Computer Engineering for Environmental and Sustainable Development.*

Sanchez, D., & Kozyrakis, C. (2021). Scalable and Efficient Fine-Grained Cache Partitioning with Vantage. *IEEE Micro, 32*(3), 26–37. doi:10.1109/MM.2012.19

Sandhya, Krishna, & Satamraju. (2015). *A Novel Approach for Auto Clock Gating of Flip-Flops.* IJSER.

Saranya, M., Vijayakumar, V., Ravi, T., & Kannan, V. (2013). Design of Low Power Universal Shift Register. *International Journal of Engineering Research & Technology.*

Sengupta, J. K. (2002). Efficiency analysis by stochastic data envelopment analysis. *Applied Economics Letters, 7*(6), 379–383. doi:10.1080/135048500351311

Shamim Hossain, M., Muhammad, G., Rahman, S. M. M., Abdul, W., Alelaiwi, A., & Alamri, A. (2016). Toward end-to-end biomet rics-based security for IoT infrastructure. *IEEE Wireless Communications, 23*(5), 44–51. Advance online publication. doi:10.1109/MWC.2016.7721741

Shang, L., Kaviani, A. S., & Bathala, K. (2002, February). Dynamic power consumption in Virtex™-II FPGA family. In *Proceedings of the 2002 ACM/SIGDA tenth international symposium on Field-programmable gate arrays* (pp. 157-164). 10.1145/503048.503072

Sharma, D. K. (2012). Effects of Different Clock Gating Techinques on Design (Vol. 3). Academic Press.

Sharma, H., Haque, A., & Jaffery, Z. A. (2019). Maximization of Wireless Sensor Network Lifetime Using Solar Energy Harvesting for Smart Agriculture Monitoring. *Ad Hoc Networks, 94*, 101966. doi:10.1016/j.adhoc.2019.101966

Sharma, R., Jain, G., & Gupta, S. (2015, November). Enhanced Cluster-head selection using round-robin technique in WSN. In *2015 International Conference on Communication Networks (ICCN)* (pp. 37-42). IEEE. 10.1109/ICCN.2015.8

Sheikh, S. (2018). Face Recognition Using Cnn. *International Journal for Research in Applied Science and Engineering Technology, 6*(3), 1411–1414. Advance online publication. doi:10.22214/ijraset.2018.3218

Shelef, Sukenik, & Green. (1984). *Microalgae Harvesting and Processing: A Literature Review.* Academic Press.

Shinde, J., & Salankar, S. S. (2011). Clock Gating—A Power Optimizing Technique for VLSI Circuits. In *2011 annual IEEE India conference.* IEEE. 10.1109/INDCON.2011.6139440

Shiraz, R. K., Hatami-Marbini, A., Emrouznejad, A., & Fukuyama, H. (2018). Chance-constrained cost efficiency in data envelopment analysis model with random inputs and outputs. *Operations Research.* Advance online publication. doi:10.1051/ro/2016076

Shiverware. (n.d.). https://shiverware.com/iot/iot-vs-wsn.html

Singh, V.P., Agrawal, A., & Singh, S.B. (2012). Analytical Discussion of Single Electron Transistor (SET). *International Journal of Soft Computing and Engineering, 2.*

Singh, A., Chaudhry, A., Niranjan, V., & Kumar, A. (2015). Improving Gain Bandwidth Product Using Negative Resistance And DTMOS Technique. *Proceeding 39th National Systems Conference.* 10.1109/NATSYS.2015.7489087

Singh, A., Niranjan, V., & Kumar, A. (2015). A novel technique to achieve high bandwidth at low supply voltage. *Proceeding IEEE International Conference on Computational intelligence and communication technology.* 10.1109/CICT.2015.48

Singh, S. K., Kumar, P., & Singh, J. P. (2017). A Survey on Successors of LEACH Protocol. *IEEE Access: Practical Innovations, Open Solutions, 5,* 4298–4328. doi:10.1109/ACCESS.2017.2666082

Sivakumar, S. A., & Sowmya, R. (2016). Design of Low Power Universal Shift Register Using Pipe Logic Flip Flops. *International Journal of Advanced Research in Computer and Communication Engineering, 5*(5), 55–59.

Soni, S., Niranjan, V., & Kumar, A. (2017). High gain analog cell using biasing technique via gate and body terminals. *Proceeding IEEE International conference on Recent Innovations in signal processing and Embedded Systems.* 10.1109/RISE.2017.8378187

Sözen, A., Alp, I., & Kılınc, C. (2012). Efficiency assessment of the hydro-power plants in Turkey by using data envelopment analysis. *Renewable Energy, 46,* 192–202. doi:10.1016/j.renene.2012.03.021

Srinivasan, N., Prakash, N. S., Shalakha D, Sivaranjani D, Sri Lakshmi G, S., & Sundari, B. B. T. (2015). Power Reduction by Clock Gating Technique. *Procedia Technology, 21,* 631–635. doi:10.1016/j.protcy.2015.10.075

Stark, I. (2006). Invited Talk: Thermal Energy Harvesting with Thermo Life. In *International Workshop on Wearable and Implantable Body Sensor Networks (BSN'06).* IEEE. 10.1109/BSN.2006.37

STMicroelectronics. (2013). *Accurate Power Consumption Estimation For STM32L1 Series of Ultra-Low-Power Microcontrollers.* Retrieved January 31, 2023 https://www.st.com/resource/en/technical_article/dm00024152.pdf

STMicroelectronics. (2017). *STM8L151x6/8 STM8L152x6/8 Datasheet.*

Suadet, A., Thongleam, T., Kanjanop, A., Singhanath, P., Hirunsing, B., Chuenta, W., & Kasemsuwan, V. (2011). A 0.8 V class-AB linear OTA using DTMOS for high-frequency applications. *Proceedings International Conference on Modeling, Simulation and Applied Optimization,* 1-5. 10.1109/ICMSAO.2011.5775478

Suarez, F., Parekh, D. P., Ladd, C., Vashaee, D., Dickey, M. D., & Öztürk, M. C. (2017). Flexible Thermoelectric Generator Using Bulk Legs and Liquid Metal Interconnects for Wearable Electronics. *Applied Energy*, *202*, 736–745. doi:10.1016/j.apenergy.2017.05.181

Subrt, O. (n.d.). *Silicon-On-Insulator-A perspective on low-power, low-voltage supervisory circuits implemented with SOI Technology*. Academic Press.

Sueyoshi, T. (2000). Stochastic DEA for restructure strategy: An application to a Japanese petroleum company. *Omega*, *28*(4), 385–398. doi:10.1016/S0305-0483(99)00069-9

Sui, B., Fang, L., & Zhang, C. (2011). Reconðgurable logic based on tunable periodic characteristics of single-electron transistor. In *2011 24th Canadian conference on electrical and computer engineering (CCECE)*. IEEE. 10.1109/CCECE.2011.6030497

Sun, P., & Reano, R. M. (2010). Submilliwatt thermo-optic switches using free-standing silicon-on-insulator strip waveguides. *Optics Express*, *18*(8), 8406–8411. doi:10.1364/OE.18.008406 PMID:20588686

Suo, G., & Yang, X. (2019). Balancing Parallel Applications on Multi-core Processors Based on Cache Partitioning. *IEEE International Symposium on Parallel and Distributed Processing with Applications*, 190-195.

Syed, A. S., Sierra-Sosa, D., Kumar, A., & Elmaghraby, A. (2021). Iot in smart cities: A survey of technologies, practices and challenges. *Smart Cities*, *4*(2), 429–475. Advance online publication. doi:10.3390martcities4020024

Talluri, S., Narasimhan, R., & Nair, A. (2006). Vendor performance with supply risk: A chance-constrained DEA approach. *International Journal of Production Economics*, *100*(2), 212–222. doi:10.1016/j.ijpe.2004.11.012

Tamil, S. C., & Shanmugasundaram, N. (2018). Clock Gating Techniques: An Overview. *2018 Conference on Emerging Devices and Smart Systems (ICEDSS)*.

Tannu, S., & Sharma, A. (2012). *Low power random number generator using single electron transistor. In 2012 international conference on communication, information and computing technology (ICCICT)*. IEEE. doi:10.1109/ICCICT.2012.6398099

Tan, Y. K., & Panda, S. K. (2010). Energy Harvesting from Hybrid Indoor Ambient Light and Thermal Energy Sources for Enhanced Performance of Wireless Sensor Nodes. *IEEE Transactions on Industrial Electronics*, *58*(9), 4424–4435. doi:10.1109/TIE.2010.2102321

Tao, K., Wu, J., Tang, L., Hu, L., Lye, S. W., & Miao, J. (2017). Enhanced Electrostatic Vibrational Energy Harvesting Using Integrated Opposite-Charged Electrets. *Journal of Micromechanics and Microengineering*, *27*(4), 44002. doi:10.1088/1361-6439/aa5e73

Temporelli, A., Carvalho, M. L., & Girardi, P. (2020). Life cycle assessment of electric vehicle batteries: An overview of recent literature. In Energies (Vol. 13, Issue 11). MDPI AG. doi:10.3390/en13112864

Teubner, J., & Woods, L. (2013). Data processing on FPGAs. *Synthesis Lectures on Data Management, 5*(2), 1–118. doi:10.1007/978-3-031-01849-7

Thakur, T., Deshmukh, S. G., & Kaushik, S. C. (2006). Efficiency evaluation of the state-owned electric utilities in India. *Energy Policy, 34*(17), 2788–2804. doi:10.1016/j.enpol.2005.03.022

Thilakarathne, N. N., Kagita, M. K., & Priyashan, W. D. (2022). Green internet of things: The next generation energy efficient internet of things. In *Applied Information Processing Systems* (pp. 391–402). Springer. doi:10.1007/978-981-16-2008-9_38

Thomas, M., Ellingsen, L. A. W., & Hung, C. R. (2018). *Research for TRAN Committee-Battery-powered electric vehicles: market development and lifecycle emissions.* Academic Press.

Tibu, M., Chiriac, H., Ovari, T., & Lupu, N. (2015). Efficient Electromagnetic Energy Harvesting Devices. In *2015 IEEE International Magnetics Conference (INTERMAG).* IEEE.

Tiemann, J. J. (1996). *Apparatus for Converting Vibratory Motion to Electrical Energy.* Academic Press.

Tiwari, V., Malik, S., & Wolfe, A. (1994, October). Compilation techniques for low energy: An overview. In *Proceedings of 1994 IEEE Symposium on Low Power Electronics* (pp. 38-39). IEEE. 10.1109/LPE.1994.573195

Tong, H.-M., Lai, Y.-S., & Wong, C. (2013). *Advanced Flip Chip Packaging.* Springer. doi:10.1007/978-1-4419-5768-9

Touqeer, H., Zaman, S., Amin, R., Hussain, M., Al-Turjman, F., & Bilal, M. (2021). Smart home security: Challenges, issues and solutions at different IoT layers. *The Journal of Supercomputing, 77*(12), 14053–14089. Advance online publication. doi:10.100711227-021-03825-1

Transparency, C. (2021). *Climate Transparency Report.* https://www.climate-transparency.org/wp-content/uploads/2021/10/CT2021Turkey.pdf

Tsiolakis, T., Alexiou, G. P., & Konofaos, N. (2010). Low power single electron OR/NOR gate operating at 10GHz. *2010 IEEE computer society annual symposium on VLSI*, 273–276.

Tsividis, Y. (1999). *Operation and Modelling of the MOS Transistor* (2nd ed.). McGraw-Hill.

Tucker, J. (1992). Complementary digital logic based on the Coulomb blockade. *Journal of Applied Physics, 72*(9), 4399–4413. doi:10.1063/1.352206

Tuckerman, D., & Pease, R. (1981). High-performance heat sinking for VLSI. *IEEE Electron Device Letters, 2*(5), 126–129. doi:10.1109/EDL.1981.25367

Tuli, S., Basumatary, N., Gill, S. S., Kahani, M., Arya, R. C., Wander, G. S., & Buyya, R. (2020). HealthFog: An Ensemble Deep Learning based Smart Healthcare System for Automatic Diagnosis of Heart Diseases in Integrated IoT and Fog Computing Environments. *Future Generation Computer Systems, Elsevier, 104*, 187–200.

Turcheniuk, K., Bondarev, D., Singhal, V., & Yushin, G. (2018). Ten years left to redesign lithium-ion batteries. *Nature, 559*(7715), 467–470. doi:10.1038/d41586-018-05752-3 PMID:30046087

Turkish Statistical Institute. (2021). *Vehicles.* https://data.tuik.gov.tr/Bulten/Index?p=Motorlu-Kara-Tasitla ri-Aralik-2021-45703

Tyagi, S., & Kumar, N. (2013). A systematic review on clustering and routing techniques based upon LEACH protocol for wireless sensor networks. *Journal of Network and Computer Applications, 36*(2), 623–645. doi:10.1016/j.jnca.2012.12.001

Tzannes, N. S. (1966). Joule and Wiedemann Effects-The Simultaneous Generation of Longitudinal and Torsional Stress Pulses in Magnetostrictive Materials. *IEEE Transactions on Sonics and Ultrasonics, 13*(2), 33–40. doi:10.1109/T-SU.1966.29373

Ubbelohde, N., Fricke, C., Flindt, C., Hohls, F., & Haug, R. J. (2012). Measurement of finite-frequency current statistics in a single-electron transistor. *Nature Communications, 3*(1), 612. doi:10.1038/ncomms1620 PMID:22215087

Uygur, A., & Kuntman, H. (2014). A very compact, 0.4 V DTMOS CCII employed in an audio-frequency filter. *Analog Integrated Circuits and Signal Processing, 81*(1), 89–98. doi:10.100710470-014-0365-2

Uysal, M., & Gul, H. (2017). Lityum İyon Piller İçin Sn-Cu/rGO (İndirgenmiş Grafen Oksit) Anot Malzemelerin, Karakterizasyonu ve Elektrokimyasal Özellikleri. *Academic Platform-Journal of Engineering and Science*, 19–25. doi:10.21541/apjes.336104

Valentian, A., Rummens, F., & Vianello, E. (2019). Fully Integrated Spiking Neural Network with Analog Neurons and RRAM Synapses. *Proceedings of the IEEE International Electron Devices Meeting (IEDM)*, 14.13.1–14.13.4.

Valls, Gomez, Ros, & Sahuquillo. (2019). A Directory Cache with Dynamic Private-Shared Partitioning. *IEEE 23rd International Conference on High-Performance Computing (HiPC)*, 382-391.

Varadharajan, S. K., & Nallasamy, V. (2017, March). Low power VLSI circuits design strategies and methodologies: A literature review. In *2017 Conference on Emerging Devices and Smart Systems (ICEDSS)* (pp. 245-251). IEEE. 10.1109/ICEDSS.2017.8073688

Venkataratnam, A., & Goel, A. K. (2008). Design and simulation of logic circuits with hybrid architectures of single-electron transistors and conventional MOS devices at room temperature. *Microelectronics Journal, 39*(12), 1461–1468. doi:10.1016/j.mejo.2008.08.002

Viera, S., Matej, R., Martin, K., Daniel, A., Lukas, N., Michal, S., & Miroslav, P. (2018). Ultra-Low Voltage Analog IC Design: Challenges, Methods and Examples. *Wuxiandian Gongcheng, 27*(1), 171–185.

von Büren, T., & Tröster, G. (2007). Design and Optimization of a Linear Vibration-Driven Electromagnetic Micro-Power Generator. *Sensors and Actuators. A, Physical, 135*(2), 765–775. doi:10.1016/j.sna.2006.08.009

Walvekar, H., Beltran, H., Sripad, S., & Pecht, M. (2022). Implications of the Electric Vehicle Manufacturers' Decision to Mass Adopt Lithium-Iron Phosphate Batteries. *IEEE Access: Practical Innovations, Open Solutions, 10*, 63834–63843. doi:10.1109/ACCESS.2022.3182726

Wang, C.-C. (2017). A quick overview of compact air-cooled heat sinks applicable for electronic cooling - recent progress. *Inventions (Basel, Switzerland), 2*(1), 5. doi:10.3390/inventions2010005

Wang, L., & Yuan, F. G. (2008). Vibration Energy Harvesting by Magnetostrictive Material. *Smart Materials and Structures, 17*(4), 45009. doi:10.1088/0964-1726/17/4/045009

Wang, W. S., O'Donnell, T., Wang, N., Hayes, M., O'Flynn, B., & O'Mathuna, C. (2008). Design Considerations of Sub-MW Indoor Light Energy Harvesting for Wireless Sensor Systems. *ACM Journal on Emerging Technologies in Computing Systems, 6*(2), 1–26. doi:10.1145/1773814.1773817

Wasshuber, C. (2001). *Computational single-electronics.* Springer.

Weber, Weber, Weber, & Weber. (2010). Internet of Things as Tool of Global Welfare. *Internet of Things: Legal Perspectives,* 101–25.

Wei, Y., & Joshi, Y.K. (2004). Stacked microchannel heat sinks for liquid cooling of microelectronic components. *J Elect Pack, 126*(6).

Wei, X., & Liu, J. (2008). Power Sources and Electrical Recharging Strategies for Implantable Medical Devices. *Frontiers of Energy and Power Engineering in China, 2*(1), 1–13. doi:10.100711708-008-0016-3

Wen, Y., Ares, N., Schupp, F., Pei, T., Briggs, G., & Laird, E. (2020). A coherent nano mechanical oscillator driven by single-electron tunnelling. *Nature Physics, 16*(1), 75–82. doi:10.103841567-019-0683-5 PMID:31915459

Weyman-Jones, G. T. (1991). Productive efficiency in regulated industry: The Area Electricity Boards of England and Wales. *Energy Economics, 3*(2), 116–122. doi:10.1016/0140-9883(91)90043-Y

Williams, C. (2015). *MGMT9.* 4LTR Press.

Winter, M. (2009). *The Solid Electrolyte Interphase – The Most Important and the Least Understood Solid Electrolyte in Rechargeable Li Batteries.* Academic Press.

Winter, M., & Brodd, R. J. (2004). What are batteries, fuel cells, and supercapacitors? *Chemical Reviews, 104*(10), 4245–4269. doi:10.1021/cr020730k PMID:15669155

Wiyanto, W., & Oktavianti, Y. (2021). Prototype Smart Home Pengendali Lampu Dan Gerbang Otomatis Berbasis IoT Pada Sekolah Islam Pelita Insan Menggunakan Microcontroller Nodemcu V3. *UNISTEK*, *8*(1), 68–75. Advance online publication. doi:10.33592/unistek.v8i1.1209

Wu, H., Hu, Y., Yu, Y., Huang, K., & Wang, L. (2021). The environmental footprint of electric vehicle battery packs during the production and use phases with different functional units. *The International Journal of Life Cycle Assessment*, *26*(1), 97–113. doi:10.100711367-020-01836-3

Wu, J., Sui, X., Tang, Y., Zhu, X., Wang, J., & Chen, G. (2019). Cache Management with Partitioning-Aware Eviction and Thread-Aware Insertion/Promotion Policy. *International Symposium on Parallel and Distributed Processing with Applications*, 374-381.

Wu, Y., Xie, L., Ming, H., Guo, Y., Hwang, J. Y., Wang, W., & Ming, J. (2020). An empirical model for the design of batteries with high energy density. *ACS Energy Letters*, *5*(3), 807–816. doi:10.1021/acsenergylett.0c00211

Xia, Q., Xie, K., Liu, X., Wu, Z., & Wang, N. (2016). Influence and Efficiency of Energy Harvesting on the Process of De-Orbiting Using Bare Electrodynamic Tether System. *52nd AIAA/SAE/ASEE Joint Propulsion Conference*. 10.2514/6.2016-5110

Xu, J., Jin, N., Lou, X., Peng, T., Zhou, Q., & Chen, Y. (2012). Improvement of LEACH protocol for WSN. *2012 9th International Conference on Fuzzy Systems and Knowledge Discovery*, 2174-2177. 10.1109/FSKD.2012.6233907

Yadav, L., & Sunitha, C. (2014). Low energy adaptive clustering hierarchy in wireless sensor network (LEACH). *International Journal of Computer Science and Information Technologies*, *5*(3), 4661–4664.

Yang, C., Liu, L., Luo, K., Yin, S., & Wei, S. (2018). CIACP: A Correlation- and Iteration-Aware Cache Partitioning Mechanism to Improve Performance of Multiple Coarse-Grained Reconfigurable Arrays. *IEEE Transactions on Parallel and Distributed Systems*, *28*(1), 29–43. doi:10.1109/TPDS.2016.2554278

Yang, X. H., Tan, S. C., He, Z. Z., & Liu, J. (2018). Finned heat pipe assisted low melting point metal PCM heat sink against extremely high power thermal shock. *Energy Conversion and Management*, *160*, 467–476.

Yang, Y., Zhang, H., Zhu, G., Lee, S., Lin, Z.-H., & Zhong, L. W. (2013). Flexible Hybrid Energy Cell for Simultaneously Harvesting Thermal, Mechanical, and Solar Energies. *ACS Nano*, *7*(1), 785–790. doi:10.1021/nn305247x PMID:23199138

Yan, W. M., Ho, C., Tseng, Y. T., Qin, C., & Rashidi, S. (2020). Numerical study on convective heat transfer of nanofluid in a minichannel heat sink with micro-encapsulated PCM-cooled ceiling. *International Journal of Heat and Mass Transfer*, *153*, 119589.

Yao, S., Lu, Y. W., & Kandlikar, S. G. (2012). Pool boiling heat transfer enhancement through nanostructures on silicon microchannels. *ASME J Nanotechnol Eng Med*, *3*, 031002.

Yenioğlu, Z. A., & Toklu, B. (2021). Stokastik Veri Zarflama Analizi ile Etkinlik Ölçümü: Türkiye Elektrik Dağıtım Şirketlerinin Karşılaştırmalı Analizi. *Politeknik Dergisi*, *24*, 87–101.

Yeo, Y. C., Lu, Q., Lee, W. C., King, T. J., Hu, C., Wang, X., ... Ma, T. P. (2000). Direct tunneling gate leakage current in transistors with ultrathin silicon nitride gate dielectric. *IEEE Electron Device Letters*, *21*(11), 540–542. doi:10.1109/55.877204

Yildirim, M. (2021). Design of Low-Voltage and Low-Power DTMOS Based Analog Multiplier Utilizing Current Squarer. *International Journal of Electronics Letters*, *9*(1), 1–13. doi:10.108 0/21681724.2021.1889041

Yuan, X., Gao, X., Yang, J., Shen, X., Li, Z., You, S., Wang, Z., & Dong, S. (2020). The Large Piezoelectricity and High Power Density of a 3D-Printed Multilayer Copolymer in a Rugby Ball-Structured Mechanical Energy Harvester. *Energy & Environmental Science*, *13*(1), 152–161. doi:10.1039/C9EE01785B

Yu, C., & Petrov, P. (2019). Off-Chip Memory Bandwidth Minimisation through Cache Partitioning For Multi-Core Platforms. *Design Automation Conference*, 132-137.

Yu, L., Zha, R., Stafylas, D., He, K., & Liu, J. (2020). Dependences and volatility spillovers between the oil and stock markets: New evidence from the copula and VAR-BEKK-GARCH models. *International Review of Financial Analysis*, *68*, 101280. Advance online publication. doi:10.1016/j.irfa.2018.11.007

Yunus Cengel, A. (2003). *Heat Transfer-A Practical Approach*. McGraw-Hill.

Yu, Y., Oh, J. H., & Hwang, S. (2000) Implementation of single electron circuit simulation by SPICE: KOSEC-SPICE. *Proc Asia Pac Workshop Fundam Appl Adv Semicond Device, 100*(150), 85–90.

Zareei, S., & Deng, J. D. (2016). Energy Management Policy for Fitness Gadgets: A Case Study of Human Daily Routines. In *2016 26th International Telecommunication Networks and Applications Conference (ITNAC)*. IEEE. 10.1109/ATNAC.2016.7878774

Zhang, J., Liu, S., Kong, L., Nshimiyimana, J. P., Hu, X., Chi, X., Wu, P., Liu, J., Chu, W., & Sun, L. (2018, May). Room Temperature Carbon Nanotube Single-Electron Transistors with Mechanical Buckling–Defined Quantum Dots. *Advanced Electronic Materials*, *4*(5), 1700628. doi:10.1002/aelm.201700628

Zhang, S., Feng, D., Shi, L., Wang, L., Jin, Y., Tian, L., Li, Z., Wang, G., Zhao, L., & Yan, Y. (2021). A review of phase change heat transfer in shape-stabilized phase change materials (ss-PCMs) based on porous supports for thermal energy storage. *Renewable & Sustainable Energy Reviews*, *135*, 110127.

Zhang, X., & Wang, D. L. (2017). Deep learning based binaural speech separation in reverberant environments. *IEEE/ACM Transactions on Audio, Speech, and Language Processing*, *25*(5), 1075–1084.

Compilation of References

Zhao, A., McJeon, H., Cui, R., Cyrs, T., Feldmann, J., Iyer, G., Kennedy, K., Kennedy, K., Kennedy, S., O'Keefe, K., Rajpurohit, S., Rowland, L., & Hultman, N. (2022). *An "all-in" Pathway to 2030: Transportation Sector Emissions Reduction Potential.* Academic Press.

Zhou, Z., Zhou, X., Cao, B., Yang, L., & Liew, K. M. (2022). Investigating the relationship between heating temperature and thermal runaway of prismatic lithium-ion battery with LiFePO4 as cathode. *Energy*, *256*, 124714. Advance online publication. doi:10.1016/j.energy.2022.124714

Zhu, Gu, Dick, Shang, & Knobel. (2009). Characterization of Single-Electron Tunneling Transistors for Designing Low-Power Embedded Systems. *IEEE Transactions on Very Large Scale Integration Systems, 17*.

Zhu, G., Bai, P., Chen, J., & Zhong, L. W. (2013). Power-Generating Shoe Insole Based on Triboelectric Nanogenerators for Self-Powered Consumer Electronics. *Nano Energy*, *2*(5), 688–692. doi:10.1016/j.nanoen.2013.08.002

Zurbuchen, A., Pfenniger, A., Stahel, A., Stoeck, C. T., Vandenberghe, S., Koch, V. M., & Vogel, R. (2013). Energy Harvesting from the Beating Heart by a Mass Imbalance Oscillation Generator. *Annals of Biomedical Engineering*, *41*(1), 131–141. doi:10.100710439-012-0623-3 PMID:22805983

About the Contributors

Ayaz Ahmad is Assistant Professor of department of Mathematics, National Institute of Technology, Patna. A Gold medalist from Aligarh Muslim University. He has rich and illustrious teaching and research experience of 13 years at NITP and 6 years at Maulana Azad College of Engineering and Technology, Neora, Patna. He has published several papers and served as a reviewer in renowned international and national journal of repute. Dr Ayaz Ahmad is the adaption author of the several text books also.

M. Appadurai holds a B.E. degree in Mechanical Engineering from the Anna University, Chennai and a M.E. degree in Engineering Design from the Anna University, Tirunelveli and a Ph.D. degree in Mechanical Engineering from the Anna University, Chennai. He worked as an Assistant Professor in the Department of Mechanical Engineering at Infant Jesus College of Engineering and Technology for five years. He published several book chapters, books, conference papers and journals in the national and international level. He is presently working as an Assistant Professor at Dr. Sivanthi Aditanar College of Engineering for seven years. His Professional interests include Solar Energy, Finite Element Analysis and Hybrid Electric Vehicles.

Zühre Aydın received her PhD degree on Management Information Systems from the Gazi University, Ankara. She worked in IT and Energy Transition departments as an engineer in the fields of Digital Transformation, IT Project Management, Information Security, Data Analysis and Database Administration between 2007-2022. She continues as a lecturer of Database Management Systems and Information Retrieval courses. She writes and presents widely on issues of database management systems, data security, energy data analysis, energy efficiency, operational research, cybersecurity in critical infrastructures, energy forecasting within machine learning. Aydın, has been working at Digital Transformation Group of Energy Transition Department in Energy Market Regulatory Authority, Türkiye.

A. Yasmine Begum is working as Associate Professor In the Department of Electronics and Instrumentation Engineering. Her Research interests include Wireless sensor Networks and its applications, Internet of Things, Process Control and Machine learning Applications.

M. Chithambara Thanu born on 28th August 1980 is a mechanical engineer. He obtained his Doctoral degree under Mechanical engineering in the expertise of composite material from Anna University, Chennai. He got his Master's degree in CAD from Sathyabama University. He is also Graduate Engineer in Mechanical Engineering from Arulmigu Kalasalingam College of Engineering, Srivilliputhur. He has more than 17 years of teaching experience in various institutions. He has ISTE membership and published/presented technical papers in various international and national conferences and journals. Now, he is presently working as an Assistant Professor in the Department of Mechanical Engineering, Dr. Sivanthi Aditanar college of Engineering for 16 years. His Professional interests include Composite materials, Localized Hybridization, Automobile Engineering and Robotics.

Hasan Huseyin Coban, born in 1985, obtained his BSc degree in electronics engineering from Vilnius Gediminas Technical University in 2008. He received the MSc. diploma in Electrical Engineering from Klaipeda University in 2010 and PhD degree from Riga Technical University in 2016. He worked as a project engineer in the industry between 2010 and 2018. His main research interests include Sustainable Energy Technologies, Renewable Energy Systems, Energy Policy, Energy Access, ANN, Electric Vehicles, and Flexibility & Optimization in electric power systems. Since February 2018, he has been an Assistant Professor at Ardahan University Electrical & Electronics Engineering department.

Palanikkumar D. has completed his Bachelor of Engineering Degree in CSE from Sri Krishna College of Engineering and Technology, Bharathiar University, Coimbatore; he did his Master of Engineering in Computer Science and Engineering from Government College of Engineering, Tirunelveli, Anna University. During his Masters programme, he was awarded stipend being a GATE qualifier. He received his Doctor of philosophy in Computer Science and Engineering from Anna University, Chennai with Web Services and SOA as his research realm. He has guided more than 100 Post and Undergraduate students. He has published as many as 60 papers in reputed International Journals, International and National Conferences. He also has published 5 patents in Indian patent journal. He has chaired and organized several Conferences/Workshops/Seminars and served as Guest Speaker in many events in India and abroad. He is a reviewer of various referred journals including Springer, Elsevier, IGI Global and so on. He has more than 15 years of experience in industry and academia. He has served in various portfolios such as BI/DW solutions architect, Onsite Project Coordinator, Deputy Director Sports, Research Coordinator, Chief Warden, Academic Coordinator and Coordinator for first Year students during his tenure in industry and academics. He is and was been a member in various academic and professional committees and chapters like IEEE, ISTE, IAENG, ICES and ISRD. He is serving as a member in 2 industry boards which focuses on automation. He works in India and travelled to countries like Singapore, Maldives, Srilanka, Mauritius and Thailand for off shore projects. He was awarded a sanction of 40 Lakh INR towards a project from DST under Young Scientist Scheme. He served as a member board of studies for SKCET, the alma matter for the year 2012 and also in few other institutions. Being in teaching, he enjoys tutoring to the school kids in villages by conducting lecture sessions to them during weekends. He was awarded for his proficiency in studies for his higher secondary mark. He is also a recipient of Nehru Yuva-Kendra award for helping differently able people. He is been working with Dr NGP ITECH for the past four and half years and currently as Professor & Head in the Department of CSE in Dr NGP Institute of Technology.

S. Darwin received the Ph.D. degree in Information and Communication Engineering from Anna University, Chennai, India in 2021. He received the Master Degree in the department of VLSI Design from Anna University, Chennai, India in 2011. He is currently working as an Associate Professor in the Department of Electronics and Communication Engineering, Tamilnadu, India. His current research interests include modeling and simulation of multigate junctionless transistors, TFETs, negative capacitance FET, IoT in healthcare and artificial intelligence applications in healthcare and semiconductor field. He is a Member of MISTE.

E. Fantin Irudaya Raj completed his BE degree in Electrical and Electronics Engineering, ME degree in Power Electronics and Drives, and currently pursuing his Ph.D. degree from Anna University, Chennai. Presently, He is working as an Assistant Professor at Dr. Sivanthi Aditanar College of Engineering, Tamilnadu, India, and has more than ten years of teaching experience with various institutions. His area of research includes Power Electronic Drives, the Internet of Things, Image Processing, and Artificial Intelligence Techniques. He participated and presented his research ideas at more than 50 national and international conferences, and published more than 25 research articles in various internationally reputed journals. He also published one book in the engineering series and contributed so many book chapters with various international publishers like Springer, Taylor & Francis, IEEE, Wiley, and Elsevier. He is also acting as a reviewer for more than 18 international journals. He is an active and lifetime member of the Indian Society for Technical Education (ISTE), the Institution of Engineers (India), the International Association of Engineers, the Institute of Research Engineers and Doctors, and the Institute of Scholars. Furthermore, he received the Emerging Scientist Award, Research Excellence Award, and Young Researcher Award from reputed international organizations.

Dankan Gowda V. is Assistant Professor of Electronics and Communication Engineering, B M S Institute of Technology and Management, Bangalore. He has rich and illustrious teaching and research experience of 12 years at various institutions and industries. He has published several papers and served as a reviewer in renowned international and national journal of repute. Dr Dankan Gowda V is the adaption author of the several text books also.

Priyanka H. completed her PHD in computer science and engineering, currently she is working as an assistant professor in PES University. Her area of interest is cloud computing, cloud forensics, artificial intelligence, neural networks, computer networks intelligence systems. She has published many paper in national and international conference and journals.

Ramakrishna N. Hegde is the Professor and Head of the Department of Automobile and Aeronautical Engineering, Srinivas Institute of Technology, Mangaluru. He is an authorized Boiler Operation Engineer and he has served in the capacity of Board of Examiners for Mechanical, Aeronautical and Automobile Engineering Boards of VTU, Belagavi. His research interests include, Heat transfer, Nano-fluids, Compact heat exchangers, I.C. Engine Combustion, Non-Conventional Energy and CFD. He is an active reviewer for reputed international journals like, Experimental heat transfer (Taylor & Francis), International Journal of Heat Transfer (Elsevier), International Journal of Mechanical Science and Technology (Springer Link), International Journal of Green Energy and Heat Transfer Asian Research (Wiley International) etc. He has authored several books in Engineering and Management and published more than 50 papers in reputed international journals, National and International Conferences.

Sandeep Kumar Hegde completed PhD from VTU University. He has published research papers in various international journals and conferences. Presently he is working as assistant professor in the Department of Computer Science and Engg at NMAMIT, Nitte. His area of interest includes big data analytics, machine learning and computational intelligence. He is a life member of ISTE.

Jothimani K. is currently working as a Professor of Computer Science and Engineering at Graphic Era Deemed to be University, Dehradun. She has completed her Ph.D. in Computer Science from Bharathiyar University, Coimbatore. She has published several papers in reputed journals and received the best paper award at conferences. She has also played a multitude of roles like resource person, conference chair, convenor of Ph.D. defense, technical program committee member, and reviewer. She is associated with the Editor of the International Journal on Data Science and Technology part of the Science Publishing Group. Her area of interest includes Data Mining, Machine Learning and IoT. Presently she is guiding students in the Deep Learning domain. Orcid: 0000-0002-4179-3348.

Bhagya Jyothi K. L. has completed Master's degree in Computer Applications at Visvesvaraya Technological University, Belagavi. She is working as Associate Professor in Computer Science and Engineering at K.V.G College of Engineering. She is pursuing her PhD in the area of Video Processing under VTU. She has published many papers in reputed journal and presented her work in international conferences. Her area of research are Image and Video processing, Pattern Recognition, Machine Learning, and Deep Learning.

Kirti Rahul Kadam is renowned faculty who has rich experience in teaching in the selected subjects. She has published Research Papers and International Books and Patents .She is working as an assistant professor in Bharati Vidyapeeth Deemed to be University Institute of Management, Kolhapur, Maharashtra, India.

N. M. G. Kumar is currently working as professor at Department of EEE, Sree Vidyanikethan Engineering College, Tirupati, AP, India and obtained his B.E in Electrical and Electronics Engineering from SMVIT, Bangalore, M.Tech and Ph.D from SVU College Engineering, SV University Tirupati. He has 22 years of teaching and 3 years of Industrial experience. He published 21 international journals and 50 national and international conferences with 5 patents. Areas of interest are power system planning, security, and state estimation, application of FACTS devices in Transmission and distribution systems, power system reliability studies, real time application of power system studies with soft computing techniques.

S. Lokesh did B.E. in Computer Science and Engineering, M.E.,in Computer Science and Engineering and Ph.D., in Information and Communication Engineering from Anna University. He is having 15 years of Teaching experience and currently working at PSG Institute of Technology and Applied Research, Coimbatore. His research areas are Human Computer Interaction, Speech Recognition, Cluster Computing, Data Analytics and Machine Learning.

Shankara Murthy H. M. is the Assistant Professor in the Department of Mechanical Engineering, Sahyadri College of Engineering and Management, Adyar,

Mangaluru, Karnataka, India. His research interests include, Heat transfer, Heat Exchangers, Nano-fluids, Renewable energy harnessing, Alternative Fuels for Engines, and CFD. He is an active reviewer for reputed international journals like, Heat Transfer Engineering (Taylor and Francis), International Journal of Mechanical Sciences (Elsevier), etc. He has published more than 10 papers in reputed international journals, National and International Conferences.

Vandana Niranjan is working as Professor in the Department of Electronics and Communication Engineering at Indira Gandhi Delhi Technical University Delhi, India. She graduated in the year 2000 and received her B.E. degree in Electronics and Communication Engineering from Government Engineering College (now University Institute of Technology of Rajiv Gandhi Proudyogiki Vishwavidyalaya) Bhopal. In the year 2002, she received her M.Tech degree from the Department of Electronics and Communication Engineering at Indian Institute of Technology (I.I.T) Roorkee with VLSI Design as specialization. In the year 2015, she was awarded her Ph.D degree in the area of Low Voltage VLSI Design from University School of Engineering & Technology, GGSIP University Delhi. She has a teaching and research experience of approximately 20 years at Indira Gandhi Delhi Technical University Delhi. Her areas of interest includes MOSFET body bias techniques and Low-voltage Low-Power Analog CMOS circuit design. She has several publications to her credit in various international journals and conferences and Book chapters.

C. Padmavathy is currently working as Assistant Professor (Selection Grade) in the Department of Computer Science and Engineering, Sri Ramakrishna Engineering College, Coimbatore She has completed her Master's degree in computer science in 2005 and a Bachelor engineering degree in Computer science and Engineering in 1999. Currently she is pursuing her Ph.D in Anna University .She has 20 years of teaching experience. Her areas of interest are wireless network, IOT, Artificial Intelligence, Data structures, mobile computing, and compiler design. She has guided more than 40 UG projects and 10 PG projects. She had published 18 papers in international journals, presented 11 papers in National and International Conferences. She is an active member in various Professional bodies like ACM, IEEE, ISTE, CSI, IAENG, and ISOC.

Sumathi Pawar completed her PhD under the Guidance of Dr Niranjan N Chiplunkar through Vesweswaria Technological University in 2017. She completed her M-Tech in NITK Suratkal and has got 20 years of teaching experience. She was working as Head of the Department of ISE in Canara Engineering College and has published more than 20 papers in journals and conferences. She was active trainer under coding skill developments and was successful in achieving 100% placement in 4 years under her leadership.

Rajalaxmi Prabhu B. obtained her B.E and MTech degree in Computer Science and Engg from VTU University, India. Currently she is pursuing PhD from VTU University. She has published research papers in various international journals and conferences. Presently she is working as assistant professor in the Department of Computer Science and Engg at NMAMIT, Nitte. Her area of interest includes big data analytics, machine learning and computational intelligence. He is a life member of ISTE.

Sahana Prasad is currently working as an IT Architect in the field of technical and cloud consulting at T-Systems International GmbH, part of Deutsche Telekom AG, the largest telecommunications sector in Europe. She graduated from Frankfurt University of Applied Sciences in June 2022 with a master's degree in Information Technology. As part of her curriculum, she has worked on cutting edge technologies such as Machine Learning Algorithms, Development of Autonomous Pickup Vehicle and Cloud Computing.

Preethi received B.E. and masters from VTU. Total 14 years of Teaching experience. Currently working as Assistant professor at Presidency University, Bangalore. Area of interest: Computer Architecture and Internet of Things.

Wilma Puthran is currently working as Cloud Network Engineer at Microsoft for Azure Team. Previously she worked for Intel Corporation as Network Engineer for 3 years which also includes 1 year duration working as Graduate Technical Intern. She completed her Master's in computer Networking from Manipal Institute of Technology and Bachelors in Electronics and Communication from Sahyadri College of Engineering. Her research interests include Wireless and Wired network, Cloud Computing, Artificial Intelligence, Network Security, Data Analytics and Machine Learning.

Sapna R. received B.E. and M.E. from VTU. Total 5 years of Teaching experience. Currently working as Assistant professor at Presidency University, Bangalore and pursuing her PhD under VTU. Her areas of interest: IoT, Semantic Web, Machine Learning.

Niranjana Rai is the Assistant Professor in the Department of Mechanical Engineering, Canara Engineering College, Mangaluru, Karnataka, India. His research interests include, Heat transfer, Heat Exchangers, Nano-fluids, IC Engines, and CFD.

He has published more than 08 papers in reputed international journals, National and International Conferences.

Manjula Rao did her PhD in computer science and engineering, currently she is working as an Associate professor in NMAMIT, Nitte (Deemed to be University). Her area of interests are image processing, artificial intelligence, neural networks, automation and networks intelligence systems. She has published many papers in National and international conference and journals. She was the part of MOU with the oracle. She has guided many projects.

Algubelly Yashwanth Reddy is working as a Head of the Department of Computer Science and Engineering at Sree Dattha Group of Institutions. He graduated in Computer Science and Engineering at Anurag Engineering College – JNTU Hyderabad, Telangana, India. He secured Master of Engineering in Computer Science at Sree Dattha Institute of Engineering and Science – JNTU Hyderabad, Telangana, India. He completed Ph.D. in the field of Cloud Computing at Sri Satya Sai Institute of Technology and Medical Sciences, Sehore, Bhopal, India. He is in the teaching profession for more than 10 years. He has presented 30 plus papers in National and International Journals, conferences, and Symposiums. He has 10 plus patents and copyrights. His main area of interest includes Cloud Computing, Wireless Sensor Networks, and the Internet of Things. Author of Books titled "Internet of Things: Technologies & Applications", "Programming in C++", "Cloud Computing". Awarded Certificate of Reviewing by ELSEVIER in recognition of the review contributed to the Heliyon Journal. "Best Research Scholar Award" by DKIRF. Distinguished Young Researcher Award 2020 by Green Thinkerz. International Young Researcher Award 2020-21 by I2OR. International Emerging Young Scientist Award 2021 by the Green ThinkerZ, India. Global Young Researcher Award 2021, instituted by the International Institute of organized Research (I2OR). Member in IEEE with a membership ID: 95061464. Member in CSI-Computer Society of India with a membership ID: CMSC018551. Life Time membership by International Society for Research and Development (ISRD) with a membership ID: M4150905243. Member in International Association of Engineers (IAE), International Computer Science and Engineering Society (ICSES), Member in Institute for Engineering Research and Publications (IFERP), Member in Institute for Engineering Research and Publications (IFERP), . Life time Professional Member in Institute of Scholars (InSc), Life Member in International Institute of Organized Research (I2OR).

K. Hemant Kumar Reddy was born in Berhampur, India. He received his M.Tech and PhD. and from Berhampur University, Bhanja Vihar, India. He is currently with the VIT AP University Andhra Pradesh, India, as an Associate

Professor Grade-II. Currently serving as an Associate Editor Journal of Intelligent Fuzzy System (JIFS). He has published several papers in High-Quality SCI/SCIE/ESCI Impact Factor Journals cum Scopus/ESCI indexed Journals, 25+ Papers in International Conferences indexed with ACM, Springer, IEEE Xplore. His current research interests include distributed and cloud computing, fog computing, IoT, Edge, Fog computing, and service-oriented architectures.

Preeti Sahu is currently a research scholar at Poornima University, Rajasthan, India.

Anil Sharma is working as Assistant Professor with College of Computing Sciences & IT, Teerthanker Mahaveer University, Moradabad, India. He holds a PhD and M.Tech degree in Computer Science & Engineering from Guru Gobind Singh Indraprastha University, Delhi, India. He has an experience of 3 years in IT industry as ERP consultant and 12 years as academician in India and abroad. He has published and presented 31 research papers in reputed international journals and conferences. He has authored one book on Information System Management and holds three Indian patents. He has delivered 4 one-week workshops and 5 FDPs in universities in India and Ethiopia. He has organized three international conferences as Convener. He is a reviewer for reputed Scopus, Web of Science and SCI journals International Journal of System Assurance Engineering and Management (Springer), International Journal of Intelligent Computing and Cybernetics (Emerald Publishing), IET Collaborative Intelligent Manufacturing (Wiley Publication), Journal of Sensors (Hindawi), Journal of Intelligent & Fuzzy Systems (IOS Press), International Journal of Services, Economics and Management (Inderscience) and many more. He is a member and reviewer for professional body Internet Society Delhi Chapter. His research interests include natural language processing, artificial intelligence, big data, rough set theory, information retrieval and semantic web technologies.

Bilal Toklu took his PhD degree from Loughborough University of Technology, UK in 1993. He works as a lecturer at Gazi University, Department of Industrial Engineering. His research interests are assembly line balancing, data envelopment analysis, multi-level lot sizing and optimization.

Mohammed Mujeer Ulla is an Assistant Professor at the Computer Science Engineering Department, School of Engineering at Presidency University, Bangalore.

Index

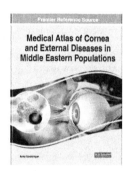
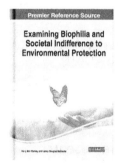
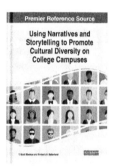

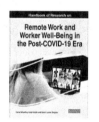

Printed in the United States
by Baker & Taylor Publisher Services